GEOLOGICAL

METHODS FOR

ARCHAEOLOGY

GEOLOGICAL METHODS FOR ARCHAEOLOGY

Norman Herz

Ervan G. Garrison

New York   Oxford • Oxford University Press   1998

Oxford University Press

Oxford   New York
Athens   Auckland   Bangkok   Bogota   Bombay   Buenos Aires
Calcutta   Cape Town   Dar Es Salaam   Delhi   Florence   Hong Kong
Istanbul   Karachi   Kuala Lumpur   Madras   Madrid   Melbourne
Mexico City   Nairobi   Paris   Singapore   Taipei   Tokyo   Toronto   Warsaw
and associated companies in
Berlin   Ibadan

Published by Oxford University Press, Inc.,
198 Madison Avenue, New York, New York 10016

Oxford is a registered trademark of Oxford University Press

Library of Congress Cataloging-in-Publication Data
Herz, Norman, 1923–
Geological methods for archaeology / Norman Herz,
Ervan G. Garrison.
p.   cm.
Includes bibliographical references and index.
ISBN 0-19-509024-1
1. Archaeological geology.   I. Garrison, Ervan G.   II. Title.
CC77.5.H47   1997
930.1'028—dc20   96-25472

1 3 5 7 9 8 6 4 2
Printed in the United States of America
on acid-free paper

Contents

Foreword

Many years ago at lunch with a group of distinguished classical archaeologists, an animated discussion started on the value of different strategies in archaeology. A few in the group strongly argued that the best approach to any archaeological problem was intuition, or as some art historians expressed it, "a strong gut feeling." When the British archaeologist Colin Renfrew and the geologist J. Springer Peacey suggested that scientific tests could be used to tell the provenance of Greek Aegean marble,[1] a dean of British archaeologists, Bernard Ashmole pooh-poohed the entire effort. He stated in 1970 that "the chances of any scientific method, present or future, being able to determine *with certainty* (his italics) the source of any given specimen is nil. Meanwhile . . . we shall have to rely on a method which is far more than eighty years old, and one which was employed by sculptors, quarry-masters, and marble-merchants for centuries before . . . namely that of using the naked eye and common sense."[2]

At that lunch, when several analytical tests were mentioned that could be carried out on different kinds of artifacts to tell provenance as well as time of manufacture, one enraged archaeologist let loose—"You scientist fellows are taking all the fun out of archaeology!"

Today, few archaeologists would argue against scientific methodologies to help solve their problems. Still, there are voices raised now and again that the pendulum has swung too far the other way and that too much science may not be a good idea. Spier, an art historian, argues[3] that "the proliferation of scientific tests has brought a vast number of analyses of all types of materials—metals, terracotta, organic material, and so on—and the conclusions from these examinations can be highly significant. Many technical and scientific studies, however, are not conclusive, especially in determining authenticity, and often appear to be invoked by archaeologists as a desperate appeal to the unattainable, 'objective', result rather than as a proper study."

Static also comes from some scientists. A prestigious Penrose Conference of the Geological Society of America was held in 1982 on Archaeological Geology. The conference, the first on the subject, was convened by two geologists working in the Mediterranean on problems of sourcing archaeological materials: Charles Vitaliano on tephra and Norman Herz on marble.[4] Because they were working with archaeologists trying to find solutions for important archaeological problems, they considered themselves "archaeological geologists." Similarly, chemists working on geological problems call themselves "geochemists," physicists "geophysicists," and so on ad naus. However, one of the attendees complained, admittedly good-naturedly, that neither organizer was a real "geoarchaeologist." We assumed by that he meant someone cast strictly in the mold provided by Michael Waters's recent book on geoarchaeology,[5] that is, with interests only in geomorphology or sedimentology.

The authors have blissfully ignored all debate concerning what is the correct level of science to use in archaeology, or the question of what is archaeological geology, and have gone ahead teaching a course since 1986 with the title "Archaeological Geology." We hope that our students represent the wave of the future, for they are truly interested in scientific applications to archaeology. Students coming from anthropology and classical archaeology and forced to pursue a B.A. track come into the course with great enthusiasm. To meet their curriculum requirements, which are heavy in humanities courses, they are often deprived of any kind of science. Sadly, these students seem to feel this lack greater than their home department mentors. There is a great need to enlighten, to proselytize, and to make sure that practicing archaeologists know that science, used properly, can be of great help and not a hindrance in their research. Geologists must also accept the fact that their methods can help solve many important archaeological problems, problems as interesting and as important as any in their own "pure" geological fields. Our approach to the subject has been one of inclusion, agreeing with Brian Fagan that "geoarchaeology is a far wider enterprise than geology."[6]

We hope that this book then will show archaeologists the many ways that geological sciences can help solve their problems. After all, much of archaeology is only the geology of that small part of the earth's surface that has been occupied by humans. We also hope that whatever snobbism has been associated by some geologists with the various definitions of geoarchaeologist, archaeogeologist, and archaeometrist, all will accept the fact that any applications of geology to archaeology can be a useful enterprise.

GEOLOGICAL

METHODS FOR

ARCHAEOLOGY

1

The Scope of Archaeological Geology

Physical scientists—geologists, chemists, physicists—have tried to solve archaeological problems longer than archaeology has existed as a recognized discipline. Few would argue today that scientists should stay away from archaeology and leave it, as it once was, reserved to the humanities. Nevertheless, the union of science and archaeology, although more and more a happy one, has on past occasions resulted in both minor and not-so-minor disasters. To avert disaster, the scientist must appreciate the problems and materials of the archaeologist, and the archaeologist should be knowledgeable about the virtues and pitfalls of scientific methodologies.

One of the earliest not-so-minor disasters that resulted from a scientist playing with unfamiliar historical materials occurred in 1818. In response to a plea from the king of Naples, the king of England sent Sir Humphrey Davy to help with the unrolling of papyrus scrolls.[1] The scrolls had been discovered more than 60 years previously in a villa near Herculaneum. A local friar devised a complicated contraption with which he could separate the papyrus sheets, but extremely slowly. At the rate the scrolls were being unrolled, the king of Naples felt it would take several centuries to get the job done. Europe was clamoring to know the contents of the first classical library ever discovered and had also become impatient with the slow progress of the local Neapolitan talent.

Sir Humphrey arrived, complete with his portable lab and chemicals. He started work immediately on the scrolls, armed with all the knowledge of early 19th-century science (to which he had contributed a significant portion). Eleven scrolls were selected, and Davy's rapid chemical method was applied. All eleven were destroyed in the process before any attempt could be made to decipher the inscriptions. Whether he was deliberately sabotaged (an English interpretation) or was unlucky in the choice of scrolls submitted for the experiment will never be known. However, the incident serves as a great object lesson for teaching scientists proper humility in approaching archaeological problems.

Archaeologists are largely concerned with the esthetics and function of objects and with the cultural and demographic implications of settlements. Scientists are more interested in the nature of raw materials and their related technologies and in the siting and paleoenvironments of early settlements. The primary tool of the field archaeologist has been a shovel; the scientist could not function without a laboratory.

Definitions: A Rose by Any Other Name . . .

The terms *archaeogeology, archaeometry, geoarchaeology,* and *archaeological geology* (as well as their *eo,* instead of *aeo,* varieties) have all been used to describe collaborative research efforts between the physical sciences and archaeology. The journal *Archaeometry* has been published at Oxford University since 1958, but the field is much older than the journal. The *Journal of Archaeological Science* (started 1973), *Geoarchaeology* (1986), and *Archaeological Prospection* (1995) are among several in the field (table 1.1). The differences in meaning among these terms are trivial, and there is no apparent agreement even among the practitioners on what the differences (if any) are.

Archaeogeology was first used by Colin Renfrew in 1976[2] to describe the contribution of the geological sciences to archaeology. The term became popular in North America but less so in Europe, where *archaeometry* was already favored. *Archaeometry* signifies the measurement of the physical and chemical properties of archaeological materials. In recent years in Great Britain and North America, it has become limited to two principal fields: (1) dating methods applicable to the late Pleistocene and Recent and (2) analytical methods for determining the technologies and the provenance of artifacts and their raw materials. Studies in the latter field are aimed at delineating trade routes, discerning esthetic tastes, and detecting forgeries. *Geoarchaeology,* at least as represented in the journal of that name, is primarily concerned with sedimentological and geomorphic processes. The Geological Society of America (GSA) established the Archaeological Geology Division in 1977, whose purpose was "to provide a suitable forum for presentation of papers on archaeological geology, . . . to stimulate research and teaching in archaeological geology." The GSA Division is all-encompassing, attempting to avoid the nit-picking distinctions described here. This

Table 1.1 Periodicals in the Archaeological Sciences and Year Founded

Advances in Archaeological Method and Theory, annual, 1977
Archaeological Prospection, 1995
Archaeomaterials, 1985
Archaeometry, 1958
Expedition, 1958
Geoarchaeology, 1986
Journal of Archaeological Science, 1973
Journal of Field Archaeology, 1973
MASCA Journal (Museum Applied Science Center for Archaeology, University of Pennsylvania), 1978
Quaternary Research, 1963
Revue d'Archéometrie, annual, GMPA (Groupe des Méthodes Physiques et Chimiques de l'Archéometrie), 1976
SAS Bulletin (Society for Archaeological Sciences), quarterly, 1977

book follows the broader definition of *Archaeological Geology,* avoiding semantics and attempting to be all-encompassing in subject matter.

The History of Science in Archaeology

M. H. Klaproth, the great German chemist and discoverer of the element titanium, made one of the earliest contributions to archaeometry.[3] In 1796 he published the results of chemical analyses of Greek and Roman coins, as well as of glass. In 1815 Humphrey Davy analyzed paint pigments that dated to the time of the Roman emperors. Close cooperation between the field archaeologist, Henry Layard and the scientist T. T. Philipps finally occurred during the excavations at Nineveh and Babylon. An appendix to the 1853 excavation reports included analyses of artifacts uncovered on the dig.

By the end of the 19th century, archaeology had developed its formal systematics, and excavations started to be carried out in an orderly fashion. Artifact hunting was abandoned, and many strategies from geology were adopted. Careful attention was paid to soil, sediments, and stratigraphic position; cooperation with scientists became more routine. Chemical and physical analyses of artifacts, pigments, and alloys began to appear, either in excavation reports or in scientific journals.

New discoveries made in physics at the end of the 19th century were applied to archaeological problems. In 1896 Roentgen, discoverer of x rays, noted their absorption by lead pigments in a painting by Dürer and thought that x rays might be used to analyze pigments and help to detect fakes. Archaeological geophysics was born the same year when an Italian, Folgheraiter, measured magnetic moments in Etruscan pottery. Aerial photography from a tethered balloon was used successfully for archaeological prospecting in England by Beazely in 1919 to locate many Roman and pre-Roman structures.

Both archaeology and geology were handicapped through the first half of the 19th century by the universal acceptance of biblical stories to explain natural phenomena. For example, unconsolidated sediments lying over bedrock were classified as either diluvium (that is, deposited by the Noachian flood) or alluvium (post-flood). Human skeletal remains in sedimentary deposits were considered to be part of an earlier population that had been wiped out in the Noachian flood, named *Homo diluvii testis* (man who witnessed the flood) by the Swiss geologist Scheuchzer in 1726, or they were thought to be accidental intrusions or burials.

In the early 19th century, J.-C. Boucher de Perthes discovered worked flints and stone tools in diluvial gravels of the Glacial period in the Somme River Valley at Abbeville in France.[4] The worked flints and extinct megafauna were found in situ in undisturbed gravel beds 5 m below the original surface. Clearly the human remains were the same age as the sedimentary deposits within which they were found and could be dated within the context of their enclosing sedimentary deposits. Archaeological remains could be analyzed by geological methodology; geological evidence was applicable to archaeological problems. The link between the two sciences was firmly forged with the establishment of the antiquity of humans.

Charles Lyell, one of the most important scientists of the last two centuries, visited the Abbeville site. He had already helped establish geology as a discipline with the publication of his *Principles of Geology* (1830–33). In 1863, after a visit to Abbeville, he published *Geological Evidence of the Antiquity of Man,* in which he used a geological context to document the known remains and artifacts of early humans. This book united geology and archaeology as kindred disciplines and established the important role that geology could play in archaeology.

Starting in the early 19th century, studies by geologists of ancient Greek and Roman marble quarries began to appear. In 1837 L. Ross described the quarries of Mount Pentelikon, near Athens.[5] In the 1880s G. R. Lepsius studied many classical marble quarries and listed criteria for distinguishing each. Then, using these criteria, he named the marble sources for over 400 well-known statues in European museums. In 1898 the American petrologist H. S. Washington warned archaeologists that Lepsius's criteria were not infallible and that a great range of textures and minerals were to be found in each quarry.[6] Again, these were warnings that scientific criteria should not be applied indiscriminately to solve archaeological problems. However, the caveats were not heeded and Lepsius's criteria were followed uncritically for the next 60 or so years. We know today, based on more-detailed, isotopic analysis, that many of Lepsius's identifications were incorrect.[7]

In 1905 Raphael Pumpelly, then president of the Geological Society of America, led an expedition to the Turkestan region of Siberia. His study of prehistoric sites was a pioneer effort to re-create paleoenvironments and served as a forecast of studies yet to come. Ellsworth Huntington, the eminent geographer, followed with similar studies at early sites in North and Central America and showed how geomorphic and archaeological evidence could be used to discern environmental and climatic change.

In the United States, arguments over artifacts and extinct fauna continued on well into the 20th century. In the 1920s and 1930s, E. H. Sellards, considered to be the first to practice geoarchaeology on North American sites, established the geological context of early human habitation near Vero, Florida, and near Midland, Texas. The so-called Midland Man (actually a young woman) is now known to date to 11,000 BC*.

The paramount role of geology in archaeological studies today can no longer be debated. Prospecting to locate ancient buried sites uses geophysics and geochemistry. Geomorphology can point out the most likely places for ancient habitation and help to determine paleoenvironmental settings. After digging starts, stratigraphy and sedimentological methods determine the sequence unique to the excavation site. Geophysics helps again to locate and interpret the nature of buried features and to point out the most promising places for careful excavation for walls, buildings, and artifacts. Palynology (the study of ancient pollen) and phytology (the study of siliceous plant remains) can help decipher ancient subsistence and agricultural strategies and the paleoclimate. Artifacts and human remains are analyzed chemically and isotopically; analysis of bone helps determine ancient diets and diseases. Geochemical, geophysical, and petrographic techniques provide the signatures needed to recognize the provenance of raw materials and to re-create the ancient technologies used to fashion ornamental and useful objects. These same techniques can help establish the authenticity of ancient artifacts.

Scope of This Book

Clearly no one book can hope to cover all the possible applications of geology to archaeology in any detail. To be a practicing geoarchaeologist, one must first be an accomplished

---

*Prehistoric dates in this book will be expressed as either BC (before the common era) or BP (before the present): BC calculated from the Julian calendar year 0, and BP calculated from AD 1950 (i.e., Julian calendar 0 + 1950 years).

practitioner of geology and then be versed in archaeology. And surely every archaeologist should at least understand enough about the physical sciences to know which methods can best be used to resolve a particular problem. The happy result of knowledgeable archaeologist-scientist pairs is that the former can question the techniques and challenge the results of analyses that fly in the face of sound archaeological data. Archaeologists skeptical of magical "black-box" geophysical methods that see under the ground and of absolute age determinations reported to the nearest year will go far to keep the scientists both "honest" and "on their toes."

The aim of this book is not to turn out a professional archaeogeologist but to serve as an introduction to the many fields where geology has been successfully applied to archaeology. Previous books with the word *geoarchaeology* in their titles have been devoted almost entirely to sediments, soils, and geomorphology.[8] Having mastered the details presented here, the archaeologist will not know how to carry out sophisticated stable isotope analysis but will know the available analytical techniques, including their shortcomings and virtues, that can be used to solve specific problems. Geologists will not learn any new principles of geology but will learn about fascinating applications of their profession that they perhaps never imagined.

The book is divided into four major parts: the archaeological site and environment (chapters 2–3), dating techniques (chapters 4–7), site exploration (chapters 8–9), and artifact analysis (chapters 10–14).

The Archaeological Site and Its Environment

In geological and archaeological studies, landscapes and environments of the past can be re-created through studies of geomorphology and sedimentology. Since a supply of water is essential for survival, as well as for many human activities, habitation is commonly near water: along floodplains or near rivers, around lakes, and coastal estuaries. In such environments, landscape changes can be extreme because of floods, drought, sediment deposition, erosion, tectonics, or changes in sea level. Landscape is subject to dynamic changes, so today's is not identical to landscapes of the past. Geomorphic and sedimentologic studies offer the tools to determine the appearance of ancient landscapes.

A favored site for urban civilizations, and indeed the birthplace of urban civilization, is along and in the floodplains of rivers. The great river systems that witnessed the development of the first advanced civilizations—the Nile, Tigris-Euphrates, and Indus—all show dynamic changes in their courses marked by channel migration, floodplain development, and changing deltas as they flow seaward and landlock coastal sites (figure 1.1). The changing course and delta of the Tigris-Euphrates resulted in political boundary changes and were directly responsible for the recent war between Iraq and Iran. Ur, which was an important deltaic port in the 3rd millennium BC, is now more than 200 km from the Persian Gulf. The present Nile has two main branches in its delta, but Herodotus reported five in the 5th century BC, and Ptolemy noted eight in the 2nd century AD. Sediment from the Danube delta carried southward by currents cut off the Graeco-Roman ports of Histria and Argamum from access to the Black Sea.[9]

Shorelines have always been prime real estate property for both play and settlement. However, the littoral is a dynamic environment, with changes in landforms brought about by erosion and deposition by the sea itself or more radical changes caused by storms and

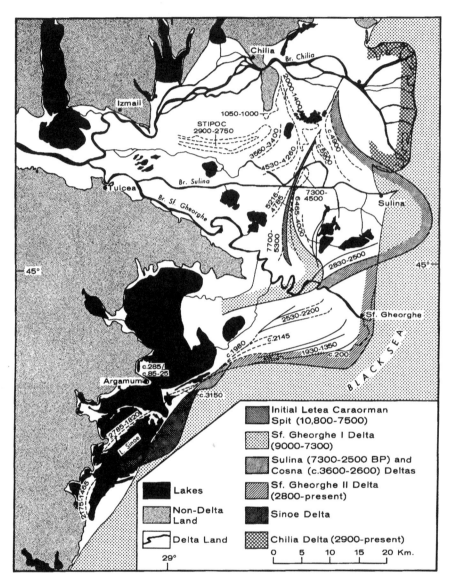

Figure 1.1 Radiocarbon dating of sediment accumulation from the Danube River delta (modified from Panin et al., 1983).

changing sea level. Methodologies that have been successfully used to reconstruct ancient littoral settings include

1. geomorphic surveying to detail the evolution of landforms and land-sea relations;
2. determining the original coastline relationship of coastal archaeological structures that are now submerged or elevated;

3. core drilling in coastal areas and radiocarbon dating of the sedimentological sequence to establish the changes in geological environments.[10]

Detailed studies of sediments in fluvial and deltaic settings have recorded the first environmental misuse of the land by humans. In the eastern Mediterranean, beginning about the time of the Early Bronze Age (ca. 5000 BP), rapid sediment infilling along river valleys and in deltaic regions took place. This was a direct result of deforestation, intensive agriculture, and grazing of newly domesticated sheep or goats—the first clear evidence that humans had arrived, settled down in the area, and changed their lifestyle from hunting and gathering to husbandry and agriculture.

Dating Techniques

Both relative and absolute dates are important for artifacts and remains and to establish the chronological history of a site. Relative dates can be obtained by *(a)* geological, *(b)* archaeological, and *(c)* chemical techniques. Absolute ages are obtained largely by radiometric and biochemical techniques but also, in some places, by archaeomagnetic measurements.

Relative Age Dating

Relative age can be determined by stratigraphic position, style, and absorption of transient elements.

Objects found in the same undisturbed stratigraphic horizon should have the same age, those in a higher horizon should be younger, and those lower should be older.

Style, including method of fabrication and ornamentation, can be used to correlate objects from a known stratigraphic column to other objects found in neighboring sites; that is, similar styles suggest similar dates of manufacture. A more precarious application of this technique occurs when archaeologists attempt to date far-removed objects, such as pottery, on the basis of ornamental or technological style.

More positive identification is possible by knowing that objects buried at the same time in a site should absorb equivalent amounts of transient elements, such as fluorine, uranium, calcium, nitrogen, and water. Differences in the fluorine content of bones allegedly found in the Piltdown site were decisive in exposing the hoax. All should have had about the same amount of fluorine if they were buried at the same time. Some bones, however, had a high fluorine content and were presumed to be more ancient than others which had a low fluorine content and thus were clearly modern and completely unrelated to the ancient bones.[11]

Absolute Age Dating

Absolute age dating by radioactivity is based on two general principles: damage by radiation and radioactive decay.

Radiation damage is produced in crystalline or other materials by *(a)* the emission of radioactive particles from disintegrating atomic nuclei and *(b)* the flight of heavier nuclei from fissioning atoms. The flight of these particles, especially of the heavy nuclei, will produce measurable damage in the material. By measuring the extent of the radiation damage in an artifact, an indirect measure of the age of the material or the time of manufacture can be obtained. The older the artifact, such as a ceramic or glass, and the higher the original radioactive element content, the greater will be the measurable radiation damage. Methods based on radiation damage include thermoluminescence, fission tracks, and electron spin resonance spectroscopy.

Radioactive elements and their daughter products produced through radioactive decay have a measurable half-life. If the elements produced in the decay series of the radioactive element and their half-lives are known, then a direct determination of the age of the material is obtainable. For example, radioactive potassium, ^{40}K, decays at a known rate to argon, ^{40}Ar. By measuring the amount of the radioactive parent, potassium, that is present in an artifact and the amount of its daughter product, argon, that resulted from the decayed potassium, the age of the artifact can be calculated.

Radiocarbon (carbon 14, ^{14}C) dating is the most widely used direct radioactive age determination method in archaeology. Traditional laboratory techniques of radiocarbon dating have been successfully used on materials younger than about 40,000 years, and now tandem accelerator mass spectrometer methods can be used for up to an additional 10,000–20,000 years. Uranium-series dating of bone and cave deposits for materials older than about 50,000 years and K-Ar dating of volcanic ash older than 100,000 years have been widely used.

Biochemical Dating

Amino acids in organic materials undergo a chemical change known as racemization that is time and environment dependent. Unfortunately, precise knowledge of the temperature history and other environmental conditions that prevailed during burial are required for a reliable age determination. Greater precision can be obtained only if the dates determined for some artifacts can be calibrated against other methods, such as radiocarbon. Then a calibration curve can be set up for use on artifacts and remains uncovered at that particular site.

Archaeomagnetic Dating

Archaeomagnetic dating of objects up to a few thousand years old has been carried out with great success. The method is based on the fact that the earth's magnetic field is constantly changing, both in direction (declination and inclination) and in intensity. In order to use this method, a paleomagnetic databank must first be established for the area in question. The paleomagnetic history is worked out by measuring the magnetic properties of accurately dated and spatially fixed samples such as pottery kilns. Then objects can be dated by comparing their magnetic properties to those of the dated samples.

Site Exploration

Archaeological Prospecting

The first decision that must be made in any new archaeological program is where to dig. In many cases, the decision can be based on historical records, by the visible remains of ancient construction, or by the discovery of abundant artifacts. Literary sources and records of taxation have also been fruitful. Schliemann thought Homer to be an accurate source for locating the site of ancient Troy and calculated the circumference of the city walls by the time needed for Achilles' pursuit of Hector. Details of the production records of each major silver-lead mine of Laurion found in ancient Greek inscriptions were used both for archaeological excavations and for 20th-century mineral exploration. To cover a larger area, remote-sensing imagery from earth-orbiting satellites is commercially available from American, French, and Russian agencies. The French satellite *SPOT, Haute Resolution Visible,* provides 10-m ground resolution (pixel level), a useful scale for archaeological decision making. Radar imagery acquired in the late 1980s by the space shuttle revealed an extensive stream system in Egypt that is now completely buried by the sands of the Sahara Desert. The pattern suggested prehistoric sites of habitation which were not apparent from surficial evidence.

However, to see deep enough into the subsurface so that decisions can be made on specific target areas and depths for excavation, and also to obtain an idea of what artifacts and constructions to expect, techniques involving geophysical "black-box" and geochemical methods are increasingly used.

Geophysical Prospecting

After an area has been selected as an archaeological site, but before excavation begins, information on the location and nature of the buried material can be obtained through geophysical techniques. The most common systems employed in archaeological prospecting include the following:

1. *Magnetic surveying:* measuring the intensity of the magnetic field on the site at selected points over a grid. Variations between the magnitude of the earth's magnetic field and that measured at the grid point will be due to the magnetic properties of subsurface materials, either natural, such as iron-rich sediments, or artificial, such as hearths, buried walls, and artifacts (figure 1.2).
2. *Electrical resistivity surveying:* measuring the current flow in the subsurface between two electrodes at different potentials. This is actually a measure of the electrical resistance and is highly dependent on water content; generally, higher water content in a formation or construction will yield lower resistivity. Stone structures have a higher resistivity than do unconsolidated sediments and soils.
3. *Electromagnetic surveying:* measures both electrical conductivity and magnetic susceptibility. This technique is excellent for locating large earth features such as mounds and refilled pits.
4. *Ground probing radar:* locates buried structures and artifacts directly. The sandier the soil, the more translucent to radar. Clay-rich layers tend to be opaque to radar. The exact nature of the object seen by radar must be interpreted archaeologically, however.

Geochemical Prospecting

Human occupation of a site leaves distinctive chemical signatures. Today, we are familiar with human intervention in the atmosphere, in surface and ground waters, and in the soil, all measured by the increasing presence of undesirable and even toxic chemical compounds. Detectable chemical changes in the soil of an archaeological site caused by habitation of ancient humans are principally the result of the decomposition of organic matter. The prin-

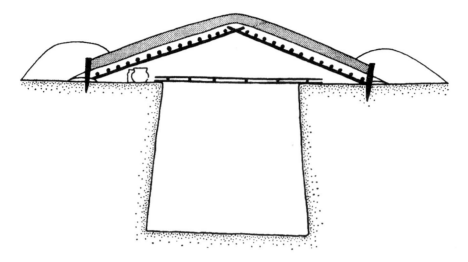

Figure 1.2 *Top:* Iron Age house (redrawn from Harding, 1970); *right:* Magnetic survey results, Oxfordshire, Great Britain, of Bronze Age circular ditch-and-barrow features and Saxon sunken houses.

cipal sources of this material are human and animal excreta, refuse, and decayed human and animal bodies. Phosphorus, carbon, and nitrogen are all released over time in large amounts from these sources and remain in the soil. Other elements are removed from the site by bacterial and chemical reactions. Today, the presence of excess phosphorus in the soil commonly indicates an ancient site of habitation.

Artifact Analysis

Determining the sources of raw material used in the manufacture of artifacts yields valuable information on ancient cultures that is not obtainable by any other means. The beginning of trade around the Aegean by about 5000 BC is clearly marked by the appearance of worked obsidian flakes; shortly thereafter, marble appeared in sites where there is no local obsidian or marble source. Much of this Neolithic obsidian has been traced to the island of Melos. Obsidian from Melos was identified by detailed chemical analyses of many samples that were used to produce a distinctive trace-element signature. This signature was different from the chemical signatures determined in obsidian from other sources in the eastern Mediterranean.

The development and spread of technology are clearly evident in the production of pottery at different sites in the Tigris-Euphrates Valley beginning about 5500 BC. The designs were universally used and emanated from single sites, so that archaeologically it appeared that the pots were being exported from those sites. However, since the materials used at each site were determined to be local, technology was exported and not the pottery.[12] Each site must have had its own production center, which took advantage of local raw materials; it also must have had a representative of the "head office" to ensure quality and artistic control.

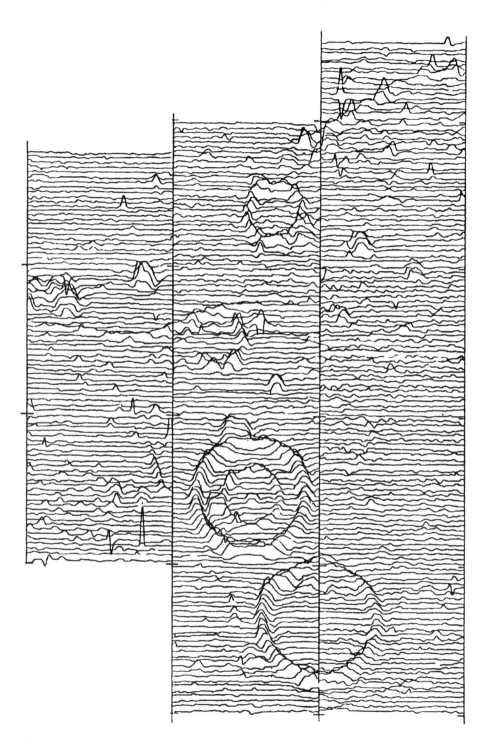

Figure 1.2 *(continued)*

To determine the provenance of artifacts, each potential source area must be "finger-printed" and a database built up of sources. The distinctive fingerprints can be based on pet-rographic, magnetic, geochemical, or any other distinguishing physical characteristics of the material. Artifacts can then be analyzed for the same properties and compared to the database of potential sources. Some techniques that have been successfully used to determine provenance include the following:

1. *Mineralogy:* This method was first used on hand axes by Dumour in 1865.
2. *Petrography:* In the late 19th century, Lepsius determined the characteristics of many classical marble quarries and used this information to source Greek and Roman statuary in European museums.
3. *Trace elements:* Lithic materials, ceramic clays, native copper, and metals have been analyzed by many methods, including optical emission spectroscopy, x-ray fluorescence, neutron activation, wet chemistry, and atomic absorption.
4. *Stable isotopes:* Isotopic ratios of oxygen, carbon, and strontium have been used for classical marbles[13] and strontium for alabaster and gypsum. Lead has been used for metal and glass artifacts to determine the ore-producing mine (figures 14.3–14.6).

A relatively new application of isotope geochemistry has been the determination of the diets of ancient humans. Human and animal bone and food, either refuse or foodstuffs burned onto pots, have been analyzed. Stable isotopes of carbon were first used to show when corn was introduced into the North American diet.[14] Nitrogen, carbon, and strontium are also used to distinguish hunting-gathering diets, which include large amounts of wild berries, from settled-agriculture diets, which include large amounts of grains.[15] Some chemical ratios, such as Ca/Sr, measured in human and animal bone have also been useful in diet determination.

PART I

THE ARCHAEOLOGICAL SITE
AND ITS ENVIRONMENT

2

Geomorphology
in Archaeology

Geomorphology is the study of the evolution of landforms. Analysis of surficial deposits provides much of the evidence for changes in landforms over time. These deposits may be residual materials, formed in place by weathering of underlying formations, or may have been formed elsewhere and then transported by wind, water, or humans to their present site of deposition. They include both sediments and soils, which are commonly confused in the field although each originates by different processes and each yields different kinds of information. Both geomorphology and surficial deposits are the principal subjects of several other publications[1] and will not be covered in great detail here. This book aims to cover in more detail fields that are universally acknowledged to be important for archaeology but are generally ignored in the "geoarchaeology" literature. Those seeking more information on geomorphology and surficial deposits should refer to other publications.

The kind and amount of surficial materials change with the changing land surface and climatic conditions and so offer the best evidence regarding the evolution of the landscape. An understanding of these changes on a site will allow a re-creation of the paleoenvironment at the time of occupation and a modeling of the prehistoric land-use patterns. Archaeological exploration in an area is facilitated by first pinpointing desirable habitation sites of the time and then targeting these sites for geophysical prospecting.

After a site has been discovered, geophysical and geomorphic-sedimentologic information can help develop excavation strategies. Such information commonly allows a better idea of the distribution and nature of buried artifacts and may explain anomalous surficial redistribution of artifacts, for example, by downslope wash or sediment burial.

The first study in a new area proposed for any detailed archaeological work should be geomorphic-surficial geology. It can be carried out in three distinct phases:[2]

1. Geomorphic mapping affords meaningful descriptions of the landforms, drainage patterns, surficial deposits, tectonic features, and any active geomorphological processes.

2. The erosional processes that carved the landforms—including soil formation, sediment removal or deposition, and tectonic uplift—are documented. Typical important indications of landform changes are raised stream terraces, old landslide scars, and changes in sedimentary deposits. Exposures of surficial deposits are studied and cores are collected to gather detailed information on the nature and physical characteristics of the deposits.

3. The landforms, climate, and surficial deposits of today are extrapolated back to reconstruct paleoenvironment and paleoclimate at the time of occupation of the site. The nature and rate of deposition of the sediments will yield valuable data for interpreting these physical changes over time. Information on probable habitation sites and also on the mechanisms of movement of artifacts, including their removal and reburial, will be obtained.

Features such as stream terrace remnants above a broad river floodplain or a deep V-shaped incision cut by the modern stream into older terraces are evidence for a different paleolandscape and perhaps also paleoclimate (figure 2.1). The first case suggests ample sediment-carrying ability of the stream through time, implying wet conditions and an ample sediment source. In the second case, a diminished sediment or water source or tectonic uplift could be responsible. Tectonic uplift in the southern Appalachians led to steepening stream gradients and increased the erosion ability of the streams. With the arrival of the Europeans, human-induced processes, specifically cultivation and plowing, accelerated erosion and changed the clear, sediment-free pre-Columbian streams to the muddy, sediment-charged ones of today.

In 1968, in a study of the Nile Valley of ancient Egypt, K. W. Butzer and C. L. Hansen listed the prime issues that should be addressed in a detailed geomorphic study.[3]

1. Initially, geologic-geomorphic relations of the area planned for the archaeological study are mapped in detail.

2. Then an analysis of surficial deposits, especially around valley margins, is made. Certain features, such as a broad alluvial flat, make certain sites better for settlement than others, such as those near landslide zones or on low, active floodplains.

3. Generalizations can be made from this study about the likelihood of site occurrence within the region studied. Whether or not the known preserved sites represent the original density of settlement can then be inferred.

4. Finally, and most important, the paleoenvironmental conditions present at the time of settlement are determined.

From figure 2.1 it can be seen that alluvial flats can be eroded away in a later stream cycle or they can be buried by younger alluvium. If ancient settlements existed on these flats, they, too, may either be eroded and their remains carried away and buried downstream or still be in situ but deeply buried.

Geomorphic Mapping

To produce a geomorphic map, special attention must be paid to the landforms, stream drainage patterns, and the nature of the surficial deposits. This mapping should be an important part of a general geological study that includes a description of the tectonic features and the nature of the bedrock outcroppings. A study of outcrops will give some information

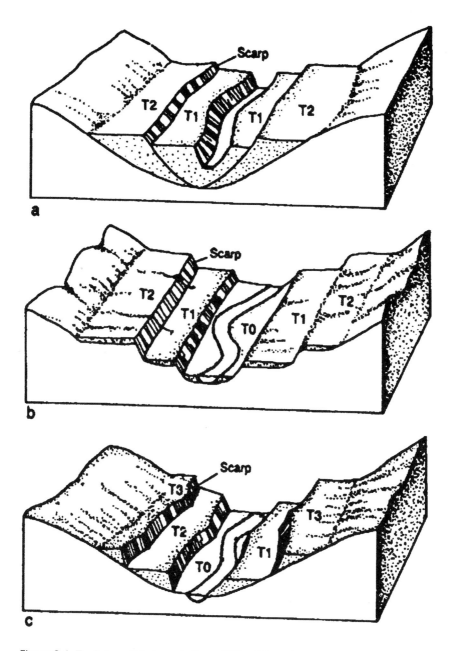

Figure 2.1 Evolution of drainage in Loess Hills of Iowa for the past 8,000 years (modified from Waters, 1992).

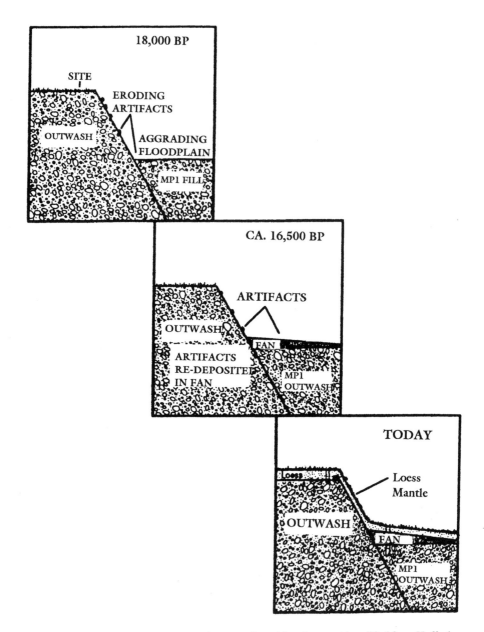

Figure 2.2 Migration of artifacts downslope to redepositional context (modified from Hoffecker, 1988).

on the thickness and nature of the surficial deposits, but, generally, quantitative information can only be obtained from drill cores. George "Rip" Rapp has recently stated that core drilling is the most underutilized technique in geoarchaeology. Geophysical techniques can also tell much about the nature and thickness of surficial deposits, as well as the distribution and nature of buried archaeological remains.

With this information, the spatial distribution of the landforms can be shown and evidence on how they have changed since the time of occupation can be made more obvious. Depending on the goals of the study, geomorphic mapping can be carried out either in a reconnaissance fashion for a region—that is, covering a large area at a small scale—or in great detail at a large scale to cover a site.

The kind of detail needed for a geomorphic map depends on the setting. In central Alaska, for example, in studies of the late Glacial age (14,000–10,000 BP) careful attention was paid to periglacial features, including glaciofluvial outwash, valley alluvium, and loess.[4] Potential source areas for artifacts that had been redeposited in outwash or alluvium were identified. Testing was then restricted to the identified areas, where the most success was expected in finding evidence of early humans (figure 2.2).

A good example of the application of a geomorphic-surficial geology study to a large region with known and prospective archaeological sites is that carried out on the Plain of Drama in northern Greece.[5] Figure 2.3 is a map showing the general geomorphic features and surficial deposits. The Plain is floored by limestone and is located in a fault-bounded graben structure that has sunk relative to the surrounding uplands. D. A. Davidson distinguished five general landform units: foothills, alluvial fans, other types of alluvium, lowland limestone, and peat and marshland. When the known Neolithic and Bronze Age tells were plotted, it became clear that most were concentrated in or near areas with an older alluvial cover. In this region, the alluvium is derived from weathering of the upland, whose varied rock units would yield a soil rich in elements such as potassium and phosphorus that are necessary for cultivation. In addition, the alluvial areas are better drained and possess better water sources than the lowland limestone, which weathers into a relatively poorly drained, impermeable clay.

Davidson's geomorphic map also provides useful information on the relative ages of alluvial deposits. Over time some sites were eroded away, others were covered by alluvium, and a third group has been exposed at or near the surface. The first group of sites was located on an older alluvium which suffered erosion over time. The second group is probably largely intact but has been buried by a younger alluvium. The third group was fortuitously preserved and lies on an alluvium that has not been eroded since the occupation of the sites. Since the peat and lake deposits are also younger, possibly deposited after Neolithic occupation of the area, they, too, might cover some sites.

Favored Landforms for Habitation

Four general environments have been favored for human habitation since the earliest of times:

1. alluvial environments, associated with streams and their floodplains, stream terraces, and depositional fans;

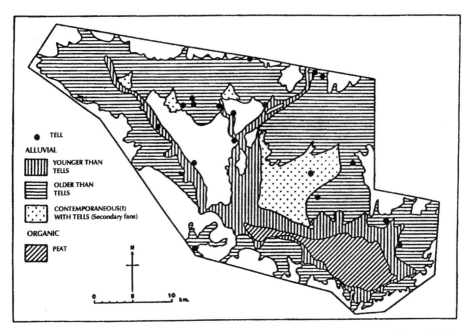

Figure 2.3 Distribution of tells in the Plain of Drama, Greece (modified from Davidson, 1985).

2. coastal environments, modified by erosional or depositional coastlines and islands, as well as by Quaternary sea-level changes;
3. eolian environments, such as deserts, sand dunes, and loess deposits, with habitation sites controlled by available water sources such as streams and springs; and
4. other terrestrial environments, including rock-shelters and lake littoral zones.

Important considerations for site selection for habitation were the following:

1. the availability of food resources, whether by hunting, fishing, husbandry, gathering, cultivation, or, usually, a combination of these methods;
2. availability of materials for artifact manufacture and construction, including the building of defensive walls and dwellings;
3. avoidance of sites prone to natural disasters such as floods, landslides, and, to a much lesser extent, earthquakes and volcanic eruptions;
4. good drainage and, if possible, a relatively agreeable climate;
5. access to important trade routes, an increasingly important consideration as more sophisticated societies evolved.

For most early cultures, the fluvial environment was by far the type of site favored for habitation.

Fluvial Landforms

Streams attract human settlement. Their floodplains are made up of sediment derived from an upland which is often rich in major and trace elements essential for cultivation.

Traditionally, "bottomland" has always promoted agricultural development; bloody feuds have been fought in the valleys of America's Appalachia over their possession and riparian rights. In addition, the pollution-free streams of ancient times provided drinking water, irrigation, water power, raw material for industry, and an important food source. Because the riverine environment has been used for hunting, gathering, agriculture, and settlements, it is a prime locus for archaeological remains incorporated into alluvial deposits.

Navigable streams were important as communication networks; wide streams provided natural defensive barriers. Stream banks and sediments are an important source of raw materials for pottery and bricks; many stream gravels are rich in colored and precious or semiprecious stones and detrital gold and platinum, all of which were used for personal ornamentation and jewelry making.

In Great Britain, an estimated 95% of Lower Paleolithic artifacts have been found in stream gravels. The figure is equally high for pre-Columbian artifacts in the southeastern United States, where archaeological inventory on the Oconee River in north-central Georgia has uncovered an enormous number of Cherokee and Creek artifacts in some sites that were later flooded.[6]

The principal controls of erosion by a stream are water velocity and grain size of the unconsolidated sediments. The critical erosional velocity to entrain material is obviously lowest for lighter, sand-sized and smaller particles and higher for heavier, coarse-grained sand and gravel. High stream flow velocity is also needed to uproot fine-grained silt and clay, which are highly cohesive. Particles will continue to move downstream as long as the current velocity remains higher than their settling velocity. Streams rarely maintain high velocities over a very long distance, so most sediments transported and deposited downstream include only sand and finer materials. Silt and clay are transported in suspension and so will move greater distances before settling out, commonly by flocculation. Gravels and other coarse material can move only during flood periods, when stream velocity is at a high. Eventually, all material will reach a point where its settling velocity exceeds the stream flow velocity and will settle out as alluvium following Stoke's Law. The critical velocity for movement of materials of different diameters on a streambed is typically shown on a Hjulstrom diagram (figure 2.4).

The fluvial environment is dynamic and constantly changing. In times of increased rainfall, more abundant and coarser-grained alluvial material will be transported and deposited, and the floodplain may grow. At other times, a depleted sediment source and low rainfall will result in incised stream channels and erosion of the floodplain. Since erosion and deposition are constantly taking place, the physical appearance of floodplains changes. A river can develop meanders, then later cut them off; it can downcut into its floodplain and create a terrace. Changes in the balance of the system from primarily erosional to depositional are commonly attributed to (1) changing climate over time, (2) increased tectonic activity that raises or lowers the stream profile, (3) rising or falling sea level, which will also change the stream profile, and (4) human land use and accelerated erosion due to deforestation or agricultural practices.

Alluvium deposited at different times by a stream furnishes a good record of the paleoenvironment. However, because of periods of nondeposition during drought, of erosion of previously deposited sediments, and of stability, the record of human occupation or use of the fluvial environment may not be complete. Absences or gaps of evidence of occupation in the archaeological record thus may be a result of natural stream processes and not actual abandonment of sites.

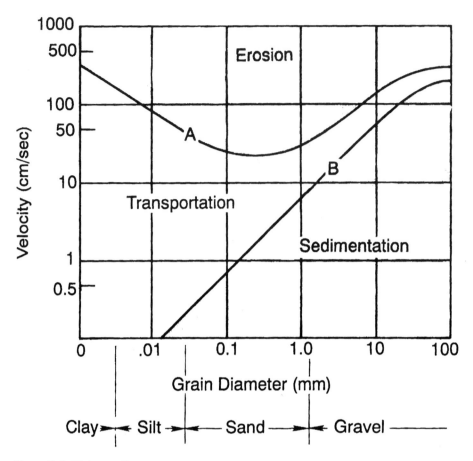

Figure 2.4 Hjulstrom diagram.

An understanding of the geomorphology in the Khabur Valley of northeastern Syria was the key to resolving the Early Bronze Age (3000–2500 BC) history of the site of Tell 'Atij.[7] The region today is semiarid with only 200–300 mm of rainfall annually, but this was not the case at the time of occupation. Although the Khabur River is not now navigable, abundant archaeological material was uncovered to reveal that Tell 'Atij had been an important trading post in the first half of the third millennium BC.

The Khabur is one of the main tributaries of the Euphrates, 50–55 m wide at the site and nearly 2 m deep, with a relatively slow discharge of 50 m³/s. The rocks of the area are Upper Miocene sandstone, clay, siltstone, and marl. The central part of the Khabur Valley is mostly covered by Pleistocene and Holocene age alluvium, which forms three terraces. The lowest terrace is Holocene with abundant gypsiferous sands and loams and extends to the bed of the Khabur River. It constitutes the alluvial mantle of the modern floodplain. The second is composed of alluvial and floodplain deposits, which stretch as an uninterrupted cover along the Khabur and Jagh-Jagh Rivers. The deposits consist largely of pebbly gravel and sand, cemented in places by carbonates. The pebbles are mostly allogenic limestone.

Numerous tells can be seen on the surface of this terrace. The third and highest terrace, which is 20–30 m above the floodplain, contains alluvial material including cross-bedded sandstone, pebble, and gravel beds. Coarse pebbles are abundant and range from 1–3 cm up to 10 cm. The pebbles are well rounded and made up of flint, quartzite, igneous rocks, sandstone, and limestone.

Now located on the eastern bank of the Khabur River, Tell 'Atij consists of two hillocks. The larger one is the principal tell and is 150 m long at its base, 40 m wide, and rising to 10 m above the valley floor. The secondary tell is 30 m to the east and is 200 m long, 40 m wide, and only 2 m high. A stratigraphic study of the site was facilitated by the exposure of the west side of the principal tell due to the erosive action of the Khabur River, which exposed archaeological and geological deposits. A cross section 25 m long and 6 m high was observed and allowed a study of the stratigraphy and the mode of deposition of the material. The bottom of the section was composed of alluvium consisting of gypsiferous gravel or pebbles and sand that formed cross-bedding in places. These are alluvial sediments of the ancient Khabur River which accumulated when the river flow was sufficient to transport such coarse material. To deposit this material, the river level had to be at least 4–5 m higher than it is today, reaching the 290-m contour. This level agreed with the deposits of the highest terrace, which showed that the area of Tell 'Atij was part of the Khabur River bed and was under water during the Pleistocene, 1.8 million years to 10,000 BP (figure 2.5, top).

During the Holocene (10,000 BP to the present), the climate has become progressively drier, resulting in a reduction in the river width. The course of the Khabur River became entrenched and meanders formed about 4000 BC. The summit of the tell became exposed and was first occupied by humans. The oldest occupied level, as seen in the studied exposure, is at about 289 m; the riverbed, about 286.5 m; and the water level, 288 m. Tracing these levels out shows that the tells at this time were islands in the river. Although the Khabur is presently unnavigable, navigation was possible at the time of occupation. Artifact evidence attesting to this includes stone anchors found at the site, as well as a silhouette of a sailing boat incised on a miniature terra-cotta.

Coastal Landforms

As every good land developer knows, almost everyone would like to live at least part of the year near a large body of water, preferably the ocean. This is not a new idea; the coastal environment was one of the earliest choices for human habitation. For hunting and gathering societies, in a presurfboard culture, the shore environment was attractive for its abundant resources of food and raw materials that were accessible within short distances. In addition, water transport via the sea, rivers, or lakes provided a means of contact with others. The earliest hominid remains, those of the East Africa Rift system, are found on or near the shores of ancient lakes. The search for hominid remains in East Africa has been helped by the knowledge that remains and artifacts of ancient habitats should be most abundant in deltaic, lacustrine, and floodplain deposits.

The marine coastal zone consists of the continental and island margins at the edge of the sea. Geomorphically, it is perhaps the most dynamic of environments, whose form changes with each storm. Changes are also brought about by the strong interplay between marine and terrestrial processes; most important are tectonic uplift or depression of the land surface and epeirogenic rise of sea level. In northern latitudes where the Pleistocene ice sheets were thickest, land that was once depressed by the weight of the ice is now rebounding iso-

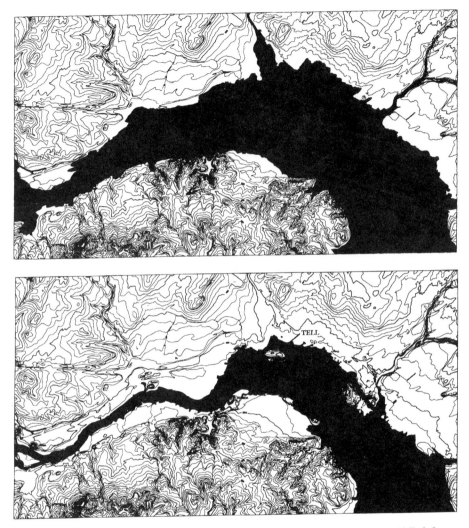

Figure 2.5 Khabur Valley, Syria: *Top left:* extent of the river in the Pleistocene; *middle left:* extent of the river in the Early Bronze Age; the tell is an island; *right:* present extent of the river (modified from Blackburn and Fortin, 1994).

statically. Melting and removal of glaciers have caused a rapid rise of the land, and in Canada and Scandinavia, settlements that were coastal only several thousand years ago are now far inland.

The intrinsic value of the coastal environment for human habitation has long been recognized. Some apparent reasons for the attractiveness of the coast include the following:[8]

1. The ocean and its embayments moderate extreme temperatures: coasts are warmer in the winter and cooler in the summer than inland continental areas at similar latitudes.
2. Major food sources of estuarine and marine fauna and flora are readily available.

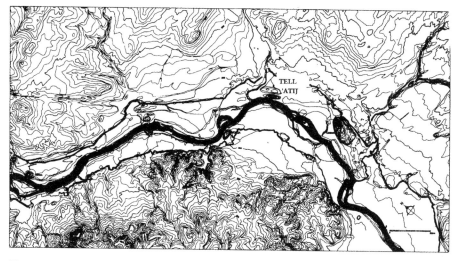

Figure 2.5 (*continued*)

3. Tidal fluctuations provide a natural flushing mechanism to disperse waste products to the open ocean.
4. Water transport routes follow the coast, the easiest and cheapest mechanism for commerce between societies.
5. Coastal sites are centers of commerce where cross-pollination of different cultures and ideas can occur.

An important factor affecting the coastal environment is the certainty that base levels will change over time. Landforms will evolve by erosion and deposition, and sea level will rise and fall. Waves, currents, and tides are the agents of change of the shoreline. Once sediment is entrained by water circulating beneath the waves, it will be transported away from the beach or roughly parallel to the beach. Certain parts of the beach may be depositional at times, accumulating sediment, while others can be erosional, losing sediment. Each major storm accelerates the process, so over long periods of time the outline of the shore will change substantially. Land and sea level go up or down, and erosion and deposition are constantly occurring (figure 2.6), so many coastal sites have been displaced in either a seaward or a landward direction or, over time, in both directions.

The forces responsible for change, including marine transgression or regression, in a coastal zone are changes in sea level and tectonic uplift or downwarp.[9] Eustatic changes are changes due to a worldwide rise or fall in sea level (figure 2.6). This can be brought about in two principal ways. (1) The amount of glacial ice can change. More ice means less seawater, as during Glacial times. (2) The shape of the ocean floor can change. A deeper floor and a sharper mid-oceanic ridge result in a greater volume during periods of slow ocean plate spreading. When the oceanic basins are deeper or more water is tied up in glaciers, the sea will withdraw from the coasts, a regression. When the ocean floor rises because of increased ocean plate spreading or melting glaciers releasing water to the sea, the excess seawater will flow over the land, a transgression. For the Pleistocene epoch, and archaeology in particular, the changing amount of glacial ice is much more important than the changing shape of the ocean floor.

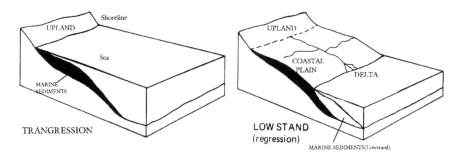

Figure 2.6 Marine transgression and regression in a coastal zone.

Tectonic uplift or downwarp will change the land height relative to sea level. After Glacial times, the removal of the great weight of glacial ice over the continents caused an isostatic rebound that was most noticeable in Scandinavia, Canada, and Alaska's Aleutian Island chain. Ancient coastal settlements of only several hundred years ago are now far inland. In Hudson Bay, the shore of the island on which Henry Hudson and his son were abandoned is now a high terrace above sea level. In Finland and Sweden, coastal areas around the Gulf of Bothnia have risen over 200 m since 6800 BC. The present rate of uplift around the northern shores of the gulf is 1 cm/yr.[10]

All studies of sea level for the Holocene have revealed a marked rise over the past 12,000 years. The most rapid rise took place during the first 6,000 years when sea level rose from about 40 m below its present mean and then either stabilized, according to some researchers, or rose at a much slower rate, reaching its present position by 4,000 BP. Sea level might have been at its lowest at the height of continental glaciation, around 18,000 BP, when it was between 90 and 130 m below its present position. As the ice sheets melted, sea level started to rise. The wide continental shelves, which had been exposed when so much water had been tied up in the ice, then was covered once again by seawater. Each sea-level curve has been worked out based on local conditions (figure 2.7) and should not be extended in fine detail to other regions where tectonics and Postglacial adjustments produce much different curves.[11]

Archaeological sites along the present Atlantic shoreline of the eastern United States where epeirogenic rise of sea level was important cannot be older than about 6000 BP. The oldest sites on the barrier islands of Georgia and South Carolina have a maximum date of 4500 BP, but most are much younger.[12] The oldest coastal sites of the Gulf of Mexico have similar dates. Sites that are older than about 4500 BP that show evidence of having once been coastal are now buried in submerged sediments offshore of the modern coastline.[13]

The offshore locations of sites that flourished during low stands of the sea and are now buried in sediment depend primarily on the slope and width of the submerged continental shelf. Along the Atlantic and Gulf coasts the shelf is wide and has a gentle slope. Some 12,000 years ago, the shoreline could have been displaced between 100 and 150 km seaward of its present position. As a result, most prehistoric sites of this coast are far offshore today. Still, the Pacific coast of California is narrow and steeply inclined, so that the ancient shoreline was only 0.5–10 km seaward (figure 2.8). Some pre-4500 BP sites that were adapted to the coastal environment are found situated on highlands or bluffs very near the modern coastlines of the mainland and nearby islands.

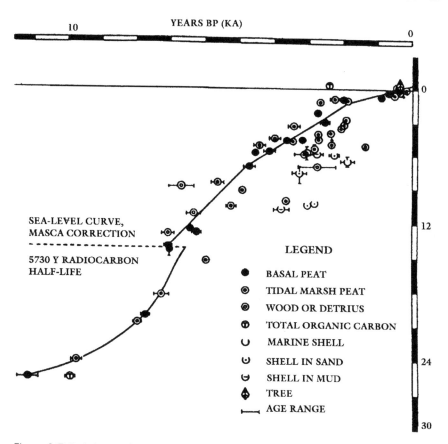

Figure 2.7 Relative sea-level curve, southern Mid-Atlantic Bight, North America (modified from Kraft, 1976).

For the Northern Channel Islands, off north-central California (figure 2.8), evidence of past sea level is direct. Theories for Holocene sea level in the basin postulate at least three periods of stable sea level significantly below that of today. The shallowest two ancient shorelines are now believed to be at 5- and 20-m depths. The deepest is roughly 40 m below present sea level (figure 2.9). The shallowest shoreline, or terrace, is dated to roughly 6000 BP, and the 20-m terrace dates to ca. 8000 BP. The deepest terrace is believed to coincide with the end of the Pleistocene era in North America at ca. 11,000 BP. Sea level in the Santa Barbara basin, at the glacial maximum, ca. 18,000 BP, was at least 100 m below present levels.[14]

The placement, spatially and temporally, of past sea level has implications for understanding the early human occupation of the Northern Channel Islands, of which Santa Cruz is the largest. As shown in figure 2.9, the impact of sea level on the islands' exposed landmass has been significant, with today's being only 26% of that at the glacial maximum. Some researchers believe that the earliest occupation of the islands was in the late

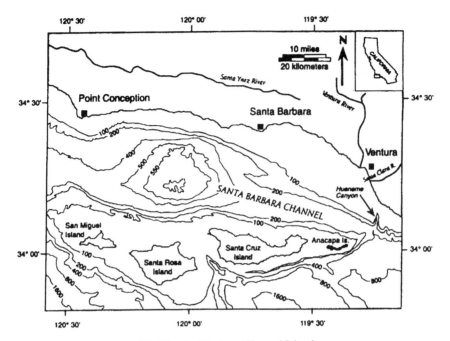

Figure 2.8 Pacific coast of California, Northern Channel Islands.

Pleistocene, perhaps as early as 29,000 BP, but such dates and the associated megafauna (dwarf mammoth) are controversial and have been challenged on the basis of geoarchaeology and the lack of definite human association. More conservative estimates have the first occupations in the early Holocene, ca. 8000 BP.[15] Many of the earliest sites have been exposed along present shorelines, suggesting that evidence for even earlier occupation lies on the drowned shelf.

Land also goes up and down through tectonic processes that are not well understood. The southeastern United States has been rising at a relatively fast rate this century, resulting in a shoreline that, except for periods of storms, is depositional. The New England states are going down, and erosion of the shoreline is more prominent.

Sea-level estimates are important in the preservation of archaeological sites. The speed of transgression, together with sedimentation rates before, during, and after transgression, determines in large part whether a site will survive on the submerged seafloor. The eastern Gulf of Mexico is considered to be a zero- to low-energy coastline created by small tidal variation (small waves) over a broad sloping plain. Rising Holocene seas flooded a sediment-starved, low-gradient (1:6,000), exposed karstified limestone surface of the then Gulf of Mexico coastal plain.[16] The transgression of the sea was probably not steady but moved in stops and starts during the Holocene. The present sea level was reached around 5000 BP.[17]

Climate changes less dramatic than glaciations and sea-level variations are responsible for other processes which result in the evolution of landforms. Sediment carried to the coast is a principal source of beach nourishment. Sediment load is controlled chiefly by (1) stream gradients and water volume and (2) human intervention. Deforestation and agricultural

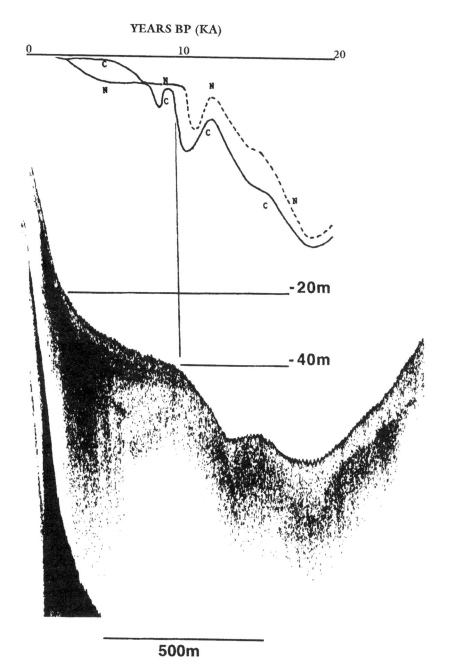

YEARS BP (KA)

Figure 2.9 Sea-level curve for the Santa Monica shelf, California (after Nardin), compared to subbottom profile of Santa Cruz Island shelf, Northern Channel Islands.

practices break up the topsoil, accelerating erosion and adding to the sediments washed into the streams.

Desert Landforms

Desert environments are characterized by eolian deposits and landforms. The wind is the primary agent of erosion, transport, and sedimentation of materials. Archaeological sites may be buried by eolian deposits and can also be later exhumed by the action of wind. Many important archaeological sites in which early cultures flourished are now and were desert areas at the time of occupation. The list includes the Nile, Tigris-Euphrates, and Indus Valleys, which were all highly dependent on the evolution of adjacent desert areas.

Wind is an agent of erosion and deposition wherever (1) it is strong enough to mobilize and transport sediment, (2) there is a source of unconsolidated sediment available, and (3) there are no obstacles to inhibit wind erosion such as vegetation protecting the ground surface. Wind erosion may take place anywhere, including in humid regions, but there the conditions are usually met only in restricted environments, such as stream and beach settings where the wind velocity is strong and unconsolidated material is readily available for transport. Eolian processes are most common and widespread in arid and semiarid deserts where all conditions for wind erosion and deposition are met.

Particles will move by wind action when a critical velocity is reached, which varies according to the particle size. Sand grain sizes between 0.1 and 0.84 mm are the most easily entrained. Higher velocities are needed for silt and clay because they are so cohesive; coarser sand and gravel are difficult to move because of their weight. Some of the factors that impede entrainment of particles include cementation of grains, vegetation, and moisture content. After particles are entrained by the wind, they will be transported by suspension, saltation, and surface creep, depending on the wind velocity and particle size. Saltation is the process by which sand grains move downwind by hopping or skipping along the ground. This process commonly occurs when falling airborne particles strike stationary grains on the surface and set them in motion. Particles larger than 0.84 mm can become entrained this way and continue to move downwind, striking and setting other particles in motion as they go. Saltation is the principal mechanism by which the wind transports sand-sized sediment. Coarse sand and gravel never become airborne but are moved by surface creep. This occurs when saltating grains strike the surface and shove or roll heavier particles along the surface. Surface creep is the most important mechanism for the transport of coarse-grained material.

Material finer grained than 0.1 mm, which includes silt and clay, is transported in suspension by the wind. They may be released into the wind by the impact of saltating grains and can then be carried to great heights by atmospheric turbulence. Once airborne, silt and clay will be transported by the wind in suspension for very great distances before finally settling out.

Wind-transported sediments form a variety of deposits, depending on their grain size. Finer-grained silt and clay produce loess deposits that cover the preexisting land surface. Under ideal conditions, which were commonly met in Peri- and Postglacial times, these sediments formed extensive deposits covering vast areas. Sand-sized grains can be concentrated as dune fields. Gravel, which travels only very short distances or not at all by wind action, will form a lag deposit at or near the source area. Sediments that are formed as the result of wind action are thus generally well sorted and will be accumulated according to grain size.

Studies of desert landforms can reveal evidence for important environmental changes:

1. Lakeshore sequences can display evidence of higher and lower water levels and saline evaporation deposits, which indicate former wetter or drier climates (for lakeshores and climate, see chapter 3).

2. Deflation basins are lag deposits of gravel and coarse-grained material from which much of the lighter sand and fine particles have been removed. Wind action may form lunettes of the finer-grained material.

3. Former river courses and fluvial terraces can give a good idea of the volume of water transported at different times in the past. Relatively dry periods will lead to narrow stream valleys and incision of older floodplains.

4. Relict sand dune fields and their distribution can be compared to modern active ones. Shifts of up to 900 km in sand dune fields have been recorded. The greatest development of dune field cover in the Near East was synchronous with the ice sheet maximum, that is, about 30,000–12,000 BP.

5. Studies around wadis and modern deserts using techniques of remote sensing have proven fruitful. The most probable sites of agriculture and settlement are near ancient wadis, where any available water was to be found. Remote-sensing techniques have been able to make out ancient stream courses that retain some moisture under a modern desert cover.

6. Study of the soil can reveal the formation of calcrete ($CaCO_3$-cemented sand) and saline soils. These can form by human activities such as *(a)* irrigation, which may carry salts from saline-rich waters or soils, and *(b)* clearing of vegetation in semiarid areas, which may result in a rise in the water table. Salinization in these cases is facilitated by capillary action carrying salts in solution upward into the soil, followed by excess surface evaporation.

Detailed studies in Mesopotamia have recognized three distinct periods of salinization: 2400–1700 BC in southern Iraq, 1300–900 BC in central Iraq, and after AD 1200 east of Baghdad.[18] Intensive irrigation and clearance of vegetation are considered the main reasons for accelerated processes of salinization in the region.

Rock Shelters and Caves

Both rock shelters and caves are important sites for at least seasonal or temporary occupation. Rock shelters are ledges protected by overhanging bedrock, whereas caves extend beyond their openings into networks of passageways and chambers. Both are controlled by local lithologies. Rock shelters are generally developed in sedimentary sequences where a more resistant cover rock, commonly a well-cemented sandstone, overlies another formation such as shale which is more easily eroded.[19] Caves are present largely in carbonate bedrock—limestone or marble—which is relatively soluble compared to most other sediments, and where the formations are relatively flat lying. Some small caves have formed in other than carbonate rocks. They can develop in places in sandstone; the famous Ajanta caves of India are actually large lava tubes that formed in basaltic flows of the Deccan Traps.

Human occupation was generally restricted to the mouth of the cave, within the reach of sunlight and influenced by outside weather conditions, whose effects are dampened to some degree. The interiors of most caves are dark and wet or at least very humid and are not conducive to year-round human habitation. These interior areas were generally restricted to artistic or ritualistic activities.

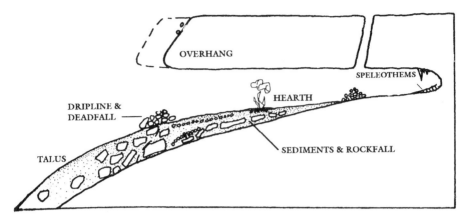

Figure 2.10 Generalized cross section of a rock shelter (after Farrand).

Terminology describing rock shelters is shown in figure 2.10. The brow overhangs the rock shelter. Directly below the edge of the brow is the dripline, an accumulation of debris where water running off the brow strikes the ledge. Talus and colluvium will accumulate downslope from the entrance.

The sediments that accumulate in a cave or rock shelter are largely produced internally. Disintegration and fall of material ranging from fine particles to large blocks from the walls and roof, as well as chemical deposition of speleothems, form the bulk of the deposits. Speleothems are (1) dripstone, stalagmites, and stalagtites, which are composed of calcium carbonate deposited when chemically saturated water drips from the ceiling, and (2) flowstone, which is similar material that precipitates on the walls and floor. Exogenous sediments, which enter from outside the shelter by the action of water, wind, or human and animal activity, may also be present.

Archaeological excavation of rock shelters and caves must take into consideration the dramatic changes that can take place over time. Because of the inevitable retreat of the brow of a shelter, materials recovered from the mouth might have been located well back toward the wall at the time of occupation. Because of rockfall in shelters and caves, cultural materials may be covered by debris or sediments and go undetected. Depending on the location of the shelter, it may become submerged by lake or marine water, which may fill the cave with its own sedimentary record. For a more detailed discussion of cave sediments, see chapter 3.

Geomorphic Processes and Archaeological Sites

Geomorphic analysis may make apparent a pattern of potential archaeological sites in a region. An understanding of the distribution of landforms and their evolution should be the first step in archaeological prospection of an area marked out for detailed study. Once a site has been located, it should be considered in its stratigraphic context: it cannot predate its underlying landform and is older than overlying sediments. Erosion or deposition are major problems and can remove or bury the remains of sites. A geomorphic study can help one to understand the changes in the landscape that have taken place since the time of occupation and to determine whether sites of different ages are expected to be buried or eroded away.

Many studies have found an absence of sites in relatively steep terrain. It is true that if sites were on steep slopes with high erosion rates, they would tend to be carried away and disappear. However, more compelling is the idea that steep slopes were avoided for habitation, given both the physical difficulties and the problems in construction. When the culture was sufficiently advanced, terraces were constructed on slopes to provide land for cultivation, as at Machu Picchu and throughout the Mediterranean area, but, typically, dwellings were not constructed on steep slopes or artificial terraces.

Thus, sites can be eroded away or buried, depending on local rates of erosion and deposition and their position relative to the terrane. Artifacts will be redistributed according to the movement of downslope wash, the weight and nature of the artifact, and the slope angles. Processes of erosion are important controls for artifact distribution. Depending on the climatic conditions and the terrane, artifacts can accumulate much as lag deposits on some surfaces. In studies of surface sherd concentrations in Iran and Mexico, sherds were found to increase in the first 50–100 years after site abandonment and then decrease exponentially. In these relatively dry regions, removal of lighter sedimentary material by water and wind left heavier sherds as a lag deposit. Eventually, however, sediment deposition covered the surface, and rare flood or sheet-wash episodes transported many sherds away. We must be skeptical of archaeological studies that make use of surface-sherd count ratios to extrapolate population numbers or settlement patterns but do not take geomorphic evolution into account.

Many sites develop their own geomorphic features and undergo their own evolution. Tells in the Middle East are low hills containing the remnants of successive settlements. Much of the sediment forming the tell consists of the decomposed remains of mud-brick houses that collapsed and over which new mud-brick houses were built. Over time, the accumulated debris and sediment of each succeeding occupation formed the tell. Thracian and Amerindian mounds are other good examples of artificial structures that form geomorphic features. In Thrace, mounds were constructed as burial grounds for kings; Amerindian mounds were constructed for their religious significance or as burial sites.

Some workers have applied the techniques of geomorphology in their studies, considering the mounds as artificial landforms. Over time, the mound profile will be modified, with the rate of change dependent on local factors such as climate and mound slope. At the time of occupation, the mound should have a noticeable hill profile. With abandonment, erosion will lower the profile. Efforts have been made to calculate the age of a mound by estimating the original mound height and profile versus its present height and profile; this is done by factoring in the local rate of erosion.

Archaeological Site Conservation

Geomorphological analysis is also used to protect archaeological sites threatened by erosion. Any campaign of protection or rescue of a threatened site should begin with a geomorphic study.

In the United States and elsewhere, coastal erosion, especially in sand dune and barrier island areas of the Atlantic and Gulf coasts, has been a serious problem for the preservation of seashores, forts, and lighthouses. The rising sea level has exacerbated the problem of preservation in the dynamic coastal environment.

Pollution is a serious problem for some sites due to its deleterious effects on both the historical monuments themselves, such as the buildings of the Acropolis and the Taj Mahal, and their sites. The accelerated chemical weathering processes have contributed to the instability of many sites.

Some important historical monuments were constructed on hills, which are subject to higher rates of erosion than the depositional plains below. Slope instability is threatening historic monuments of every culture and period. Monuments as far apart as the classical temples in Greece at Delphi and the colonial period baroque churches in Olinda, the first capital of Brazil, have serious problems of preservation.[20] To develop a program of mitigation to protect the site, a detailed study is first made of the climatic factors, the geomorphology of the site, and the mechanisms of weathering. Delphi and Olinda, as well as the Acropolis of Athens, are underlain by limestone, in places cavernous, and are also subject to accelerated chemical weathering in today's polluted environment.

In dry or desert areas, the salt that rises with capillary water from the soil into the foundations of a monument creates serious problems. This process is affecting the monuments of Mohenjo-Daro in the Indus Valley and the Egyptian Sphinx, which has already lost its nose (see chapter 10). The problem is how to prevent the salt from rising by osmosis and evaporating in pore spaces in the rock monument itself. The salt crystals that form expand into the pore spaces of the rock, which then cracks and spalls off large pieces of the monument. Since the problems of rising salt in these environments is almost insoluble, attention has been paid to ways to preserve the stone itself. Many remediation measures have been tried, such as filling pore spaces with a nonreactive medium and divergence of drainage patterns. Clearly, for archaeological preservation, there is a strong need to evaluate the physical landscape with reference to the nature of the resident human society.

3

Sediments
and Soils

Sediments are *not* soils. Sediments are *layered, unconsolidated* materials of lithic and/or organic origin.[1] Soils are mixtures of organic and lithic materials *capable of supporting plant growth*.[2] Sediments that overlie bedrock and cannot support plant growth are termed *regolithic*. Soils begin as rocks. If they are eroded or altered by diagenetic processes, they can become paleosols. The study of sediments and soils has great importance for the archaeologist because these materials can tell us about the conditions that led to their formation, thus giving much information on paleoenvironments and climate.[3]

Not too long ago, the soils of Holocene archaeological sites and the paleosols of more ancient Pleistocene ones were no more than the annoying material concealing the objects of prime interest. More recently, sediments and soils achieved their own importance as the stuff of archaeological geology.[4] In more narrow definitions of the field, they are generally the only subjects of study. The emphasis of archaeological geology is decidedly not the rock. In this text we define archaeological geology in its broadest context such that the linkages between the geological, sedimentological, and cultural (i.e., archaeological) can be explored in all their depth and texture. Ultimately, the linkage of these aspects of earth science and humanity can be categorized under the term *ecology*, as first coined by Ernst Haeckel in 1866.[5] Here the interrelationships of all areas of the earth system are sought in order to conceptualize stability and change wherever they occur.

Sediments and soils typically occur as *strata,* which are described by principles of superposition, horizontality, continuity, and succession:

Superposition: In a series of layers and interfacial features as originally created, the upper units of stratification are younger and the lower are older, for each must have been deposited on, or created by the removal of, a preexisting layer.

Original horizontality: Any layer deposited in an unconsolidated form will tend toward a horizontal position. Strata that are found with tilted surfaces were originally deposited horizontally or lie in conformity with the contours of a preexisting basin of deposition.

Original continuity: Any deposit as originally laid down and any interfacial feature as originally created will be bounded by a basin of deposition or may thin down to a feather edge. Therefore, if a deposit or interfacial feature is exposed in a vertical view, a part of its original extent must have been removed by excavation or erosion, and its continuity must be sought or its absence explained.

The principle of stratigraphical succession: A unit of stratification takes its place in the stratigraphic sequence of a site from its position being older than the units that lie above it and younger than the units that lie below it and with which the unit has physical contacts.

Stratigraphic units can be defined in a variety of ways. The most familiar, from a geological standpoint, is the rock stratigraphic unit. This concept is independent of time and is a boundary of lithological change. Its basic unit is the *formation,* and parts of it are termed *members.* Within the member, the *bed* is the smallest subunit. A *group* consists of two or more formations. It is important to point out that a group is not a series. The term *series* is not applied to rock-stratigraphic units, but rather used only as time stratigraphic nomenclature.

Soil stratigraphic units are soils with physical features such that stratigraphic relations permit their consistent recognition and mapping as stratigraphic units. Here again, these are independent of time. They are distinct from pedologic units and may comprise one or more pedologic units or parts of units.

Biostratigraphic units are characterized by the content of fossils (plant and animal remains), which are, in turn, indicative of environment. The fossils that identify these units are not redeposited (allochthonous). The largest biostratigraphic unit is called a *zone* and is defined by a fossil taxon or taxa such as phylum, class, or order. Examples are the Cenozoic (the zone of mammals) and the Mesozoic (the zone of reptiles). Within zones are *assemblages,* or strata. Each assemblage is characterized by a certain group of fossils. This is clearly the origin of the term *archaeological assemblage,* which is defined on the basis of contemporaneous artifacts and features.

Time stratigraphic units (e.g., series) are subdivisions of rock sequences that are records of a specific interval of geologic time defined by succession, paleontology, radiometry, and lithology. In terms of soil stratigraphic units, they can be either time parallel or time transgressive, and calibration is done with radiometric or paleontological techniques. Correlation of units can be done using fossils, artifacts, direct dating, pollen, or lithology.

Principles of Sediment Formation

J. K. Stein, following others, has defined four stages in the "life history" of any sediment: (1) source, (2) transport history, (3) deposition, and (4) postdepositional alteration.[6] Sediment sources can usually be pinpointed by analyzing texture and composition. Sediments tell us much of the history of an area or site, as well as its geology. Once the parent rock is disintegrated and decomposed, it is transported by wind, water, ice, or gravity to its depositional site.[7]

Terrigenous sediments are land-derived, composed of detrital rock fragments and clays; chemical precipitates including carbonates and sulfates; iron and manganese oxides; organic sediments including lignite; and phosphates and nitrates. Of special interest to paleontologists and archaeologists are the remains of organisms. Where these organisms have completely deteriorated, they often leave impressions, or "ghosts" in the encapsulating sediment.

Description of Sediments

Sediments point backward to climate and environmental processes. The proper study of sediments and their successor—soils and detritus—can tell us much about the past. The impact of climate conditions is *weathering.* Dry, cold glacial climates promote physical disintegration of rocks through freeze-thaw cycles, resulting in fracturing due to pore water expansion. Dry, hot climates produce exfoliation but little fracturing and restricted chemical disintegration. Where rainfall is sporadic and sparse, water in the soil rises through capillary action, and helps form *caliche,* a hardened calcic horizon, by accumulation of secondary calcium and magnesium carbonates.[8] Moderate climates—cool, dry to moist, temperate regimes—produce physical and chemical disintegration rates between the extremes of harsh climates (figure 3.1).

Granite, one of the most durable of building stones, disintegrates readily by the processes of exfoliation and granular disintegration. Granite boulders of the Wichita Range, in western Oklahoma, are examples of the former. Exposed to cool nights and long, hot days, the

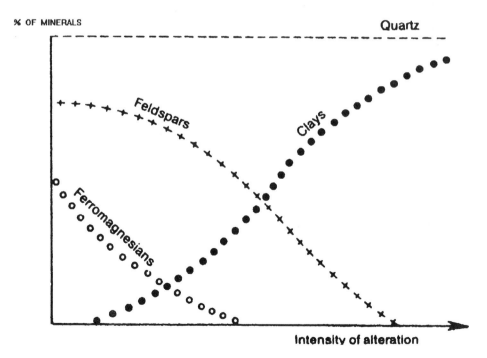

Figure 3.1 Alteration of a granitic sand over time in a temperate environment (modified from Weller 1960).

boulders commonly shed both thin and thick laminae parallel to the rock surface. Wind and rain take over from there. Granular disintegration of granite at times allows the transport of the particles by wind and water downslope and across the landscape.

Where a granite is weathered in place, clays form from the silicate minerals. In situ weathering of this type is responsible for the deep regolith formed in the Georgia piedmont. Elsewhere, limestones and shales can weather to even greater depths: 60–120 m. Limestones weather in humid regions, where dissolution by hydrolysis forms karst features such as caves and sinkholes. In dryer, montane areas like the Italian Dolomites, this stone forms resistant ridges and peaks. Even in water-rich regions, the limestone can resist solution unless organically derived acids (carbonic, silicic, or humic) increase the solvent power of the water.

Because of the great resistance of quartz to weathering, most quartz-rich sandstones will weather to almost pure sands. Feldspar-rich sediments such as arkose will also, over a long period of time, weather in a similar fashion. Each rock type has its own weathering characteristics, depending on its mineralogy, petrography, and environment of weathering. Most quartz-poor rocks, such as basalt, lose their primary minerals and weather to a soil rich in ferromagnesian components.

Soil

Soils form from unconsolidated sediments, including quartz-rich sand. Soils derived from the regolith are the result of chemical and biological activity. Soils form in situ or are redeposited by eolian and fluvial processes. Soils originate and change in response to climates and organisms, and hence they are good indicators of environmental change. The formation of soil is termed *pedogenesis*. Little soil is older than the Tertiary; most is no older than the Pleistocene.

A *pedon* is the smallest volume of an individual soil bounded by other soil bodies. Pedons are aggregates of individual soil particles (peds) with planes of weakness between them. A *soil profile* is composed of all the layers that influenced pedogenesis. While geologic horizons are static, soil horizons are dynamic—their structural and chemical properties change over time. Soils are divided into O, A, E, B, and C horizons (figure 3.2). The *solum* is composed of the A and B horizons and the E where formed. A particular soil is a function of five factors: climate, topography, biology, parent material, and time. In the case of anthrosols, humans are a sixth factor. Soils allow some degree of reconstruction of the processes that formed them. Ancient and buried soils are termed paleosols; they are more typically found in the middle latitudes and are rarely seen in the Arctic or the Tropics. An interesting aspect of paleosols, in the United States at least, is the general lack of well-defined A horizons. By 1938 the 10 major soil groups were defined.

In temperate and tropical forest soils, the acidic nature of the forest floor with its rich organic litter of leaves and microorganisms produces acids that leach the A horizon (figure 3.2). Prairie soils are rich in organic content due to the annual grasses, which can deposit as much as 0.3 m of dark A horizon. E horizons are zones of leaching, with transitional coloration between the A and B horizons. The B horizon will typically be heavily clay enriched. Spodsols form rapidly in cool, organic-rich environments where acidic solutions are efficient in transporting minerals to lower horizons, the mineralogic fractions broken down by bacteria. Bacteria can exist in spodsols in concentrations of 10^6 organisms/g. Mesofauna

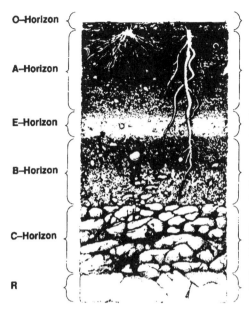

O–Horizon

A–Horizon

E–Horizon

B–Horizon

C–Horizon

R

Surface litter:
Freshly fallen leaves and organic debris
and partially decomposed organic matter

Topsoil:
Partially decomposed organic matter (humus),
living organisms, and some inorganic minerals

Zone of leaching:
Area through which dissolved or suspended
materials move downward

Subsoil:
Accumulation of iron, aluminum, and
humic compounds, and clay leached
down from the A– and E– horizons

Parent material:
Partially broken-down
inorganic minerals

Bedrock:
Impenetrable layer

Figure 3.2 Generalized soil profile.

(nematodes, snails, worms, and insects) are important in mixing soils. Indeed, some soils are created by worm action. The mesofaunal activity is called *bioturbation*. Ants are particularly good soil mixers in temperate and tropical climates, whereas large animals are relatively unimportant to soil formation.

Whatever the source of sediment, soil is a product of sediment and organic matter, climate, topography, and time. Climate affects weathering rates and the biota, both flora and fauna. The fauna of importance are not the ecosystem's consumers—predators, herbivores, and the like—but the multitude of humbler organisms such as worms, larvae, and bacteria. These organisms break down plant and larger-animal remains, which release products that are the stuff of chemical exchanges and which control organic weathering and humus buildup. A feedback loop begins with the colonization of sediments by small plants and organisms. Then soil fertility increases, and larger organisms arrive to contribute more biomass and hence more soil.

Analytical Parameters of Sediments and Soils

Texture/Particle Size

The lithological fraction of sediments and soil can be characterized by individual particle size. The phi (φ) scale, commonly used in grain-size studies, is logarithmic and ranges from −8 (gravel) to +8 (clay). The scale is inclusive from clay-sized particles (less than 0.0039 mm) to pebbles, cobbles, and boulders (greater than 2 mm). In between lie silt (0.0039–

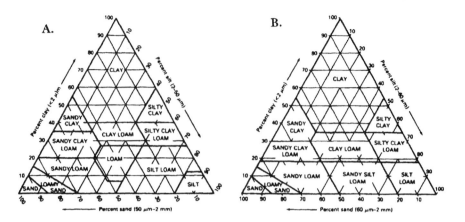

Figure 3.3 Particle-size ternary plots for (A) United States and (B) Great Britain.

0.0625 mm) and sand (0.0625–2 mm) (figure 3.3). In Britain size fractions (in μm) are as follows: clay, less than 2 μm; silt, 2–60 μm; and sand, 60–2000 μm (figure 3.3b).

Grain size in sediments is a function of sorting, which reflects the depositional history. Where there is poor sorting and the mixture is a jumble of large and small, course and fine, we suspect torrential or flood deposition. Outwash of glacial meltoff reflects this type of sorting. In many instances the intensity and duration of a flood event can be inferred from the sorting of sediments from coarse to fine. The velocity gradient of the current can be deduced as well: coarse fractions indicate higher current velocity, and finer fractions indicate lower current velocity. Multiple or repetitive events are shown by the intergradation of course and fine fractions.

Most beaches tend to be highly sorted, with abundant fine materials. Still, there are dramatic cobble beaches in many parts of the world, such as Great Britain, Hawaii, and Canada, where the shorelines are rocky and there are few sediment sources—rivers, bays, and estuaries—to provide a finer-grained sediment supply. Lacustrine beaches show the same sediment trends, which provide insights into past winds, currents, and erosional and depositional regimes. One such area will be examined in our discussion of Lake Neuchâtel, Switzerland.

In particle-size studies the emphasis is generally on those size fractions readily extracted by use of multiple screens below 2 mm in size. Above this size the fractions are classified as pebbles, gravel, cobbles, and boulders.

Since the mid-1970s, in the course of large-scale rescue excavations in western Switzerland, detailed study of the larger size fractions has led to a better understanding of their origin and role in lakeshore Neolithic and Bronze Age settlements. B. Arnold and C. Monney carried out granulometric and petrographic analyses of piles of stones found during the excavation of Auvernier-Nord, a Late Bronze Age settlement (ca. 1000 BC).[9] Similar stone heaps were seen by one of the authors (Garrison) during 1988 excavations of the Eneolithic-Chalcolithic village of Saint-Blaise (ca. 2500 BC) which helped form general stone pavement under dwellings. Arnold and Monney, following J.-P. Portmann's analysis of Würmian moraines,[10] concluded that the stones forming the piles had been taken from an ancient beach situated at a higher altitude than the villages at both Auvernier and Saint-

Blaise. The types of stone found at the beach were limestone, sandstone, quartzitic limestone, quartzite, granite, gneiss, serpentine, and schist. Arnold and Monney concluded that the morainic deposit had been reduced and leveled into a cobble beach by wave action, creating a source of lithic material for later human settlers. The stone heaps were thought to be of two varieties: refuse piles containing a considerable amount of organic and cultural debris and raw-material heaps. At the earlier village, Saint-Blaise, the stone piles were generally of the former type and situated under the dwellings.[11]

Shape and Surface

The degree of angularity or roundness can tell the investigator much about the history of a grain. Quartz grains when freshly weathered are typically very angular, whereas those that have experienced eolian or fluvial transport are rounded. The degree of roundness and the texture of the grain surface (matte to glossy) are clues to the distance a particle has traveled and the environments through which it has passed (figure 3.4). Three categories of grains can be identified by shape and surface appearance.[12] Unworn grains indicate little or no transport. Dull, translucent surfaces indicate water transport, and rounded, matte surfaces are generally produced by wind transport. The percentage of grains with dull surfaces can distinguish between marine and freshwater transport (more than 30% of total sand fraction—marine; lower than 20%—probably riverine or lacustrine).

Eolian transport produces *windkanter,* which are particles with planar surfaces divided by crests or ridges. Glacial transport causes tiny striations due to abrasion during mass transport by the ice. Based on particle alteration, R. L. Folk defines four stages in the life of a sediment: (1) *immature* sediments are rich in clays and micas, and the particles are angular; (2) *submature* sediments are still composed of angular grains, but fine-grained particles are mostly gone; (3) *mature* sediments have subangular or subrounded particles and no clay; (4) *supermature* sediments have well-rounded grains and no clay.[13]

Contact Relations

Taken en masse, the pedological or sedimentological stratum's relation to adjoining horizons can be described. A *sharp* contact is less than 0.5 cm in thickness. The contact is *abrupt* if the juncture is between 0.5 and 2.5 cm thick, *clear* when between 2.5 and 6 cm, *gradual* at 6–13 cm, and *diffuse* if more than 13 cm thick.

Color

Color is determined by use of Munsell Soil Color Charts. The charts are typically used to describe moist samples, but dry values can be taken. Each chart is divided along three dimensions: hue intensity, chroma, and value. A typical Munsell description is 10YR3/4, which is read, "10 Yellow Red Hue, 3 Chroma / 4 Value." Much of the Munsell reading is in the eye of the beholder and thus somewhat subjective. Imperfect, it remains the field standard for color descriptions and provides more definition than just a description like "dark brown earth." Color is often utilized to differentiate A and B horizons, as in assigning B to the color that is one Munsell unit redder (YR) and one unit brighter than the A.

Figure 3.4 Four varieties of quartzitic sand: *upper left:* multicyclic (Wisconsin); *upper right:* immature, subrounded (Great Sand Dunes, Colorado); *lower left:* poorly sorted beach sand (California); *lower right:* well-sorted, mature beach sand (Virginia Beach, Virginia). (Samples courtesy Professor Bruce Railsback.)

Structure

Structure refers to the macroscale—that is, to the physical organization of particles in a horizon or stratum. In soils the size, shape, and aggregation of soil particles *(peds)* define this dimension. Structureless soil has no aggregation; weak structure has barely observable peds; moderate to strong structure refers to the degree of ped formation and adhesion upon disturbance. Coresistance (resistance to breaking) and strength are related aspects.

Coats (Cutans)

Coats are studied at the micromorphological level and are one of the commonest pedological features. Some researchers consider them to be direct evidence of paleosols.[14] Clay coats are termed *argillans;* sand or silt coats are *skeletons;* sesquioxidic coats are *sesquars, mangans,* or *ferrous.* Organic coats are *organans;* carbonate coats are *calcans.*

Occupation Surfaces

T. Gé et al. have used the micromorphology of sediment facies in archaeological contexts to define occupation and nonoccupation surfaces.[15] For occupation surfaces they attempted to link structural variation in sedimentary microfacies with human activity. They examined the role of natural processes, especially pedological ones, in altering anthropogenic contexts. This microanalysis echoes, in microsedimentological terms, Michael B. Schiffer's call for the determination of the role of cultural and natural formation processes in archaeological sites.[16]

At the macromorphological level, other investigators, such as those studying the French Paleolithic, have attempted to identify occupational surfaces in the very complex stratigraphies of caves and rock-shelters.[17] Occupation units consist of the following:

1. *The topmost surface:* a layer of a few centimeters to a few millimeters thick, consisting of microartifacts and particles derived from the underlying occupation floor
2. *The artifactural surface:* the original occupation surface modified by human activities, for example, trampling, refuse, and artifacts
3. *The undersurface:* consisting of disturbances relating to the occupation, such as incorporation of artifacts from the artifactural surface by trampling and bioturbation[18]

Gé et al. described three microstratigraphic units as follows:

1. *The passive zone* is characterized by a homogeneity of material mass, randomness in the coarse fraction, and either large tubular voids lined with phytoliths or elongated cavities and vesicles. When less than a centimeter thick, the voids are generally subhorizontal. Thick voids are subhorizontal nearer the top of the void.
2. *The reactive zone* is characterized by a greater abundance of subrounded anthropogenic constituents than in the passive zone. The microstructure is fissural, with voids subhorizontal. Thickness varies with the intensity of human activity.
3. *The active zone* is characterized by densely packed subrounded-to-rounded, well-sorted microaggregates mixed with abundant anthropogenic remains (bones, charcoal, shell, coprolites, phytoliths, etc.). The two varieties of microfabric are (a) randomly distributed aggregates and (b) laminated, both with well-developed subhorizontal fissural porosity.

Four sediment microfacies compose these zones: *structured, residual, relict,* and *natural.* The structured variety relates to humanly induced modifications in the structure. Where the underlying deposits are cemented or well structured (dense loam, silty clay), the anthropogenic modification is less obvious than in the case of ground materials with less cohesion. Here, compaction and the development of subhorizontality are more pronounced. Sharp contacts are observed between zones, or at least the three zones are clearly present.

Residual microfacies differ from the structured microfacies in the lack of microscale lateral continuity. The microfabric has been altered by postdepositional processes; well-defined units are juxtaposed with less organized units in which the three zones (passive, reactive, active) cannot be identified. Bioturbation is one process that accounts for the clouding of microstructure.

Relict microfacies are seen where occupation surfaces are more altered than in the foregoing two types. Heterogeneity in the microstructure can be attributed to pedological, alluvial, or colluvial processes, as well as to human agency, such as dumping or secondary deposition.[19] The zonal nature is less clear, and the occupation surfaces can vary in reworking.

The natural microfacies do not contain anthropogenic constituents and are deposited and transformed by natural agencies. These facies divide the occupational surfaces in sediment stratigraphies. An excellent discussion of these natural processes is presented by W. R. Farrand in Stein and Farrand's *Archaeological Sediments in Context: Peopling the Americas.*[20] The overall role of soil micromorphology is discussed in a 1991 joint symposium of the Royal Society and the British Academy. Farrand noted that sediments are the result of all kinds of deposition at a given site: natural geological processes, animal activity, and anthropogenic activity. They are, thus, the prime evidence of site stratigraphy, past environments, and the rate of environmental change with or without human agency.

Mineralogy

Since rocks are aggregates of minerals, so are sediments and soils (table 3.1). Sources of these can be determined by close evaluation of their mineral content. The most abundant minerals are feldspars (orthoclase, microcline, and plagioclase), quartz, pyroxenes, amphiboles, olivine, and the micas (biotite and muscovite). All can be readily identified optically under a polarized-light microscope. Heavy minerals include garnet, tourmaline, zircon, kyanite, and staurolite, which occur as accessory, or less abundant, minerals. They can often be identified by their color, for example garnet (red); epidote (green); amphibole (gray); rutile (red or black). Many heavy minerals, including rutile, are called resistate minerals because they are highly resistant to weathering. Since minerals weather at different rates, given the same or similar source, younger sediments tend to have more accessory minerals than older sediments. In fact, original sources are the principal control of the mineral content of sediments and soil. In Lake Constance on the Rhine River smectite decreases from the upper to lower ends of the lake reflecting the local nature of the clays as opposed to the illite/mica dominated fractions from the upper Rhine basin (figure 3.5). Thus the clay fractions in this freshwater lake behave in a similar fashion as detrital minerals and reflect the mineral composition of rocks and soils in the surrounding drainage basins. At Lake Constance smectite is derived from local volcanic rock while illite/mica and chloritic minerals are extra-local metamorphic and igneous rock products.

Minerals are composed largely of the same twelve elements that comprise 99.23% of the

Table 3.1 Classification of Varieties of Sedimentary Rocks

Clastic varieties		Chemical-organic varieties	
Texture (grain size)	Varieties	Chemical or biochemical composition	Varieties
Granule-, pebble-, cobble-, or boulder-size grains (2mm or larger)	Rounded fragments are called *conglomerates*: quartz pebble conglomerate; Angular fragments are called *breccias*: quartz breccia	Carbonate minerals, e.g., calcite ($CaCo_3$) Sulfate minerals	Limestone Gypsum
Sand-size grains (0.0625–2 mm)	All varieties are called *sandstones*: quartz sanstones	Iron-rich minerals	Hematite
Silt-size grains (0.0039–0.0625 mm)	All varieties are called *siltstones*: quartz siltstone	Siliceous minerals	Chert
Clay-size grains (<0.0039 mm)	All varieties are called shales or claystones: red shale or claystone	Organic products	Coal

earth's crust. These are, in decreasing order of abundance, oxygen, silicon, aluminum, iron, calcium, magnesium, sodium, potassium, titanium, hydrogen, manganese, and phosphorus. The remaining 94 naturally occurring elements exist in much smaller amounts down to trace levels (typically at the μg/g level). Assessment of these trace elements can characterize sources, sediments, or soils to a high degree of reliability using common instrumental analytical techniques such as x-ray fluorescence, atomic absorption, and so on (see chapter 11).

Chemistry

In soil and sediment studies the focus of geochemical analyses is on relatively easy, routine procedures for a few key elements. Factors such as pH, cation exchange, and conductivity are measured as well. The most common analyses are for total organic carbon and calcium or magnesium carbonates. Nitrogen is measured in concert with carbon and expressed as the ratio C/N, which is a useful indicator of the degree of decomposition—a C/N of 10 is well decomposed; a C/N of 50 is poorly decomposed. Iron, aluminum, and sulfur indicate leaching and acid potential. Phosphate analysis, particularly of the bound fraction P_2O_5, is a good indicator of anthropogenic soils (see chapter 9).[21]

A multitude of wet chemistry and instrumental analytical techniques are available today for the geochemical characterization of soils and sediments. Instrumental methods read like an explosion of acronyms—AAS, XRF, XRD, ESR, IR, FTIR, NAA, ICP, PIXE—to name some of the most common. The techniques can either be for elemental or mineralogical analysis. In the former, individual elements such as Zn, Fe, Cu, and P are routinely measured; in the latter, molecules such as TiO, K_2O, CuO, FeO, and P_2O_5 are determined from mineral compositions. Both are needed to fully characterize soils and sediments. Samples can be studied in dry or wet states—that is, the sample may be ground or powdered and then

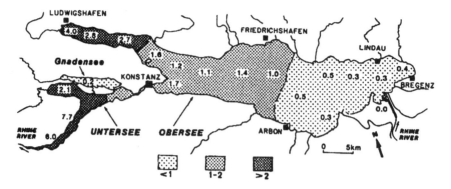

Figure 3.5 Smectite-illite ratio in Lake Constance, Switzerland/Germany (modified from Muller and Quakernaat, 1969).

examined, or it may be chemically digested by an appropriate reagent (HNO_3, HCl, HF) and then analyzed. The information obtained by any combination of procedures allows one to link a soil or sediment sample to its parent lithology, its geographic origin, and its geomorphology, as well as to characterize an anthropic modification to which it may have been exposed. In chapter 11 we shall examine the more commonly used instrumental techniques in greater detail.

Paleoclimatology

Contained within sediments and soils are indicators of environment and climate, principally by their minerals and morphology. In addition, organic inclusions in the matrix further point to climatic factors. In this section we shall discuss the remains of plants: pollen grains and plant macrofossils and microfossils.

Pollen

T. Webb III and R. A. Bryson made the primary assumptions that (1) climate is the ultimate cause of changes in pollen, (2) a connection exists in the climatic response of plants and, thus by inference, in the resulting pollen types, and (3) there is a statistical correlation between pollen types and climate.[22] Following these assumptions quantitatively, they discovered that the most dramatic climatic shift during the late Quaternary in North America's Midwest occurred at the Pleistocene-Holocene boundary. A second important finding was that the peak of the mid-Holocene warming period was more a shift in moisture than a change in temperature.

Pollen is extracted from cores and excavated soils and sediments using techniques that have been standardized over the century. These procedures include measurements of (1) relative pollen frequency and (2) pollen influx. The analysis of pollen is of European origin, having been first attempted on peat deposits in Sweden by Gustav Lagerheim in the early 1900s. His student E. J. Lennart von Post, a Norwegian geologist, presumed that pollen directly reflected vegetation history.[23] Von Post set forth five basic principles, all of which have been validated since that time:[24]

1. Many plants produce pollen or spores in great quantities, which are dispersed by wind currents.
2. Pollen and spores have very durable outer walls (exine) that can survive for long periods of time.
3. The morphological features of pollen and spores remain consistent within a species; different species produce their own unique forms.
4. Each pollen- or spore-producing plant is restricted in its distribution by ecological conditions such as moisture, temperature, and soil type.
5. Most wind-blown pollen falls to the earth's surface within a small radius, roughly 50–100 km (30–60 mi) from where it is dispersed.

While von Post was correct in assuming that most plants rely on wind dispersal (anemophilous), other classes of pollen-producing plants, such as hydrophilous, cleistogamous, and zoophilous exist. While the anemophiles produce vast quantities of pollen (oak, elm, and birch produce 100,000 grains per anther), some zoophilous plants such as clover generate only 220 grains per anther. Autogamous plants, or self-pollinators, are low pollen producers also. Hydrophilous plants are water pollinated, and cleistogamous plants are self pollinating. These plants contribute little pollen to the pollen record: hydrophilous pollen has no durable exine, and cleistogamous plants produce their pollen before they flower. Cleistogamous or zoophilous pollen appear only by human or animal transport. Even airborne varieties appear differentially, because the grains have different weights and sizes, which affects their dispersion. Pine and birch pollen can be transported many kilometers, whereas fir pollen travels only a few hundred meters.

In 1941 Johannes Iversen, a Danish geologist, awakened European archaeologists to the value of pollen studies when he cooperated with Peter Glob at the Barkaer site.[25] Barkaer is a site in northern Denmark that Glob, well known for his work on "bog-people," suspected to be a locus of early Neolithic settlement in Scandinavia.[26] Iversen examined core samples from the bog surrounding the site for fossil pollen and found clear evidence for local transformation from a hunting-gathering economy to one of agriculture (figure 3.6.) The pollen spectrum shifts dramatically from arboreal pollen (AP) types (tree species such as oak, elm, linden, ash) to herbaceous varieties and grasses (fig. 3.7). This nonarboreal (NAP) scenario is repeated in pollen diagrams for Neolithic sites throughout the Old World.[27]

Preservation, Concentration, and Analysis

Soils and sediments directly influence the amount and kind of pollen that is recovered for analysis. V. M. Bryant lists three principal factors active in the preservation of pollen grains: (1) mechanical, (2) chemical, and (3) biological.[28] Degradation of pollen grains by the first factor occurs during transport and deposition. Abrasion and crushing can damage or destroy the exine. Table 3.2 lists the relative susceptibility of some tree pollens to destruction of the exine.

Damage due to corrosion of the exine is closely related to pH, whereby in sediments of 6.0 or greater fossil pollen rarely survives. R. H. Tshudy studied the effect of oxidation potential on pollen preservation and found that preservation is enhanced in reducing environ-

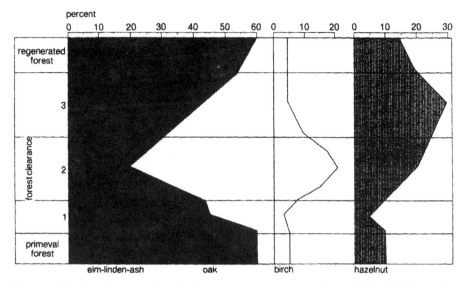

Figure 3.6 Pollen diagrams of Danish bog cores showing changes from 2500 to 2300 BC at the Barkaer site (modified from Bryant, 1991).

ments.[29] The composition of the exine determines chemical reactivity as well. A. J. Havinga noted the close relationship between the cellulose and sporopollenin in the exine.[30] With higher sporopollenin, a relatively inert polymer, leading to better preservation. Low sporopollenin combined with high oxidation potential leads to poor pollen preservation.

Biological degradation is carried out by fungi and bacteria. Damage to the exine occurs from the microorganisms' attachment to the grain and infiltration to the cytoplasmic portion. This is supported by Havinga's study of preservation in specific sediments where deposits containing leaf mold (highly organic) showed dramatic destruction of pollens of alder, ash, poplar, willow, and elm.[31]

Combined, the three factors can significantly alter the amount and kind of pollen found in samples of soils and sediments. Because of differential preservation, the pollen spectrum cannot faithfully reproduce a picture of the life assemblage, or biocenose. The pollen grains are members of the death assemblage, those organic survivals that compose the thanatocenose. Therefore, paleoenvironmental reconstructions based completely on pollen may be misleading. Overpeck has demonstrated the conceptual strength of paleoenvironmental modeling in his paper on forest changes in the American East during the last eighteen thousand years using multivariate statistics to analyze the dynamics of plant-pollen correspondence.[32]

Specific arboreal species will control the percentages of pollen detected. In the typical forest where pine and oak are equally represented (50:50), the pollen spectrum will yield a ratio of 85:15, with pine dominating. Table 3.2 reveals why this is so: the pollen corrosion resistance of pine is 19.6, whereas for oak it is 5.9, a more than 3-fold difference. The additional disparity can be chalked up to relative productivity. Pine is a voluminous pollen producer, much more so than oak. Other species may also be underrepresented in preserved pollen versus actual community percentages. For example, in a 50:50 beech and oak forest

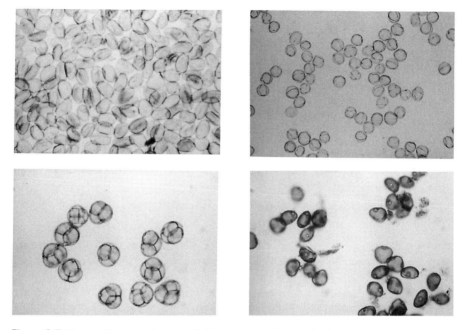

Figure 3.7 Four pollen varieties: *(top left)* grass, *(top right)* oak, *(bottom left)* rhododendron, and *(bottom right)* beech. (Courtesy George Brook.)

Table 3.2 Susceptibility to Corrosion in Relation to Sporopollenin Content

	Genus and species	Sporopollenin content (%)
Low	*Lycopodium clavatum*	23.4
	Polypodium vulgare	—
	Pinus sylvestrics	19.6
	Tilia cordata	14.9
	Alnus glutinosa	8.8
	Alopecurus pratensis	—
	Corylus avellana	8.5
	Betula verrucosa	8.2
	Calluna vulgaris	—
	Carpinus betulus	8.2
	Ulmus carpinifolia	7.5
	Populus sp.	5.1
	Quercus robur	5.9
	Taxus baccata	6.3
	Fagus sylvatica	—
	Fraxinus excelsior	—
	Acer pseudoplatanus	7.4
High	*Salix* sp.	—

Table 3.3 Correction Factors for Pollen Spectra of
Northern European Trees within Forests

Genus	Correction factor
Quercus, Betula, Alnus, Pinus	1:4
Carpinus	1:3
Ulmus, Picea	1:2
Fagus, Abies	1:1
Tilia, Fraxinus, Acer	1×2
Corylus as un understory shrub	1:1
Corylus as a canopy tree	1:4

Source: S. T. Andersen, 1970, The relative pollen productivity and
pollen representation of north European trees, and correction factors
for tree pollen strata, *Damarks geologiske undersogelse, 2 raekke*, p. 96.

we see a pollen percentage of 40:60, and in a 50:50 maple and oak forest the numbers are
even more disparate, 0:100, with maple nonexistent. Table 3.3 shows some correction factors for northern European forest pollen spectra.

It is beyond the scope of our text to extensively review sampling and precise analytical
approaches in pollen studies. However, before extensive pollen processing begins, pretesting procedures spare wasted time and money. The pretest procedure relies on the observed
correlation between organic matter content (O.M.) and pollen preservation; for example,
where there is less than 1% organic matter, the recovery of pollen is highly problematical.[33]

The size of sediment samples depends upon the deposit being examined, as well as the
amount of recoverable pollen per gram or milliliter. Some workers, suggest a "pinch
method," whereby several small (1–2 ml) samples are collected over a surface rather than
a single bulk sample. A minimum of 200 fossil grains per sample is generally needed to
guarantee a statistically valid analysis.[34]

A somewhat debated aspect of pollen study concerns the detection of "seasonality" in deposits. The phenomenological assumption is straightforward: specific plants flower or produce their pollen at specific times of the year, and a fine enough analysis of the recovered
pollen should be able to detect such indicators. The problem in such analyses is the difference in the behavior of trees, herbs, and grasses. Arboreal pollen (AP) dominates in nonanthropogenically altered environments where edaphic and climatic factors allow forests to be
the climax vegetation. It is only where the forest has been cleared or otherwise altered that
we see an increase in the pollen of herbs and grasses (NAP). The shift in AP/NAP ratios led
Iversen and others to infer the presence of Neolithic farming activities on the basis of pollen.

We must also be cognizant of the taxonomic level to which pollen grains can be identified. This is particularly obvious when discussing grass pollen, which often cannot be identified below the level of family.[35] Trees, such as oak, can be identified to genus level but
species-level determination is not reliable.

The relationship of pollen concentration and sedimentation is termed *pollen influx*, the
concentration divided by the sedimentation rate. Where the sedimentation rate is well
known through calibration with independent chronometry, the interpretation of the pollen
record is greatly enriched. A fine illustration of the use of pollen influx is the work by
H. Richard on pollen deposits at the two French Jura lakes Chalain and Clairvaux.[36]
Dendrochronology, augmented by cross-dated ceramics, tools, and adornments (bracelets,

Table 3.4 Pollen Zonation, Station 2AC, Lake Chalain

Zones	Observations	Interpretations
6	AP	Break in anthropic deposits
Samples 100–110	Low anthropic signs	Desertion of dwelling
		Presence of nearby cultivations
5	AP	Local deposits
Samples 61–99	NAP	Anthropic deposits
	Increase of Poaceae	Fire archaeological
4	NAP important	Anthropic deposits
Samples 49–60	Filipendula and Umbelliferae	Important riverside "megaphorbiaie"
3	NAP important	Typical results for analysis
Samples 30–48	and regular	made in the center of a
	AP low and stable	dwelling
2	AP	
Samples 12–29	Anthropic deposits	Agricultural occupation
	Poorness (bad	Frequent exundations
	conservation)	
1	Important domination	Beginning of exploitation
Samples 1–11	of AP	
	Modest anthropic signs	Other(s) nearby dwelling(s)

Source: H. Richard, 1993, Palynological micro-science in Neolithic lake dwellings, *Journal of Archaeological Science* 20:257, Figure 10.

spirals, etc.), fixed the duration of cultural occupations of the lakeshore villages. For example, one level at Clairvaux was occupied for 10 years, with a buildup of 36 cm of sediments. Another level lasted 55 years, with a sediment thickness of 22 cm. By sampling at 2-mm intervals, Richard achieved a microlevel of pollen analysis. His data approach most nearly the type needed to deduce seasonality, but he shys away from inferring annual layering. Table 3.4 summarizes the pollen microanalysis of 110 samples taken from 22 cm of sediment of level C, Lake Chalain, Station (site) AC, ranging from colonization in zone 1 to abandonment in zone 6.

Plant Macro- and Microfossils in Sediments and Soils

Where preserved, plant remains have always been studied in modern archaeological excavations. Common macrofossils are the seeds of plants. In this context we shall consider *macrofossils* to mean those elements of plants that can be seen with the naked eye or with little magnification. *Microfossils* are plant remains, other than pollen, that cannot be seen without the aid of high-power magnification (100× or more). We shall first examine the archaeological study of macrofossils and then the realm of microfossils, or more specifically *phytoliths*.

Plant Macrofossils and Paleoethnobotany

Paleoethnobotany (also termed archaeobotany) is the study of the human use of plants in the past. The botanical remains studied by specialists of this relatively new archaeological science are preserved portions of both domesticated and wild plants. In most archaeological deposits, plant remains are rarely recovered unless they have been carbonized, sealed in anoxic, waterlogged deposits, or found in desiccated environments such as dry caves or deserts, either hot or cold. Information on paleoenvironment, diet, subsistence, and disease can be obtained from ancient plant remains.

The most commonly found macrofossil is the seed; other plant elements, such as the flowers, stems, leaves, and roots, are less commonly found, if at all. In general, the more woody or fibrous the plant element, the better its chance of survival in sediments and soils. This "woodiness" is directly related to the cellulose content and the siliceous content of the plant's cells (intracell silica is the stuff of microfossils).

As with much of the history of archaeological science, we can trace the recovery and analysis of plant remains to Egyptology's early days. Charles Kunth (1826) analyzed dried fruit, grains, and seeds found in tombs.[37] O. Heer (1865) reported on the preserved plant materials found in Swiss lake villages.[38] Both these areas meet our expectations with regard to preservation environments. Egypt is arid, and the Swiss lakes provide a waterlogged sediment matrix. J. M. Renfrew summarized other early studies of plant remains in the archaeology of the 1800s.[39] In North America interest in paleoethnobotanical studies languished until the 1930s and the establishment of the Ethnobotanical Laboratory of the University of Michigan by Melvin Gilmore.[40]

The recovery of plant remains from sediments is generally accomplished by flotation using specific gravity and buoyancy to separate plant remains in water. Generally it is done in a pump-fed tank that uses various screen filters to trap the different-size fractions in the water stream while the coarsest, heaviest elements in the sediments (rocks, minerals, etc.) collect on the tank bottom. The various size fractions are taken from the flotation traps and dried. Then they are weighed and the individual fractions are analyzed in the laboratory.

An important source of dietary information comes from plant elements that have survived the digestive process and are preserved in *coprolites*. As inelegant as the subject strikes most people, the fossilized remains of feces, human and animal, preserve direct evidence of what individuals were eating in a given environment.[41] Coprolites are composed of components that can provide useful information for archaeology. Pathogenic organisms such as bacteria, parasites, and viruses have been found in coprolites.[42] Pollen is also generally well preserved in coprolites, at least in desiccated environments such as the American Southwest. One is cautioned in inferring too much from coprolitic pollen about paleoenvironment or climate. Unlike the typical pollen spectrum, that seen in a coprolite reflects the eccentricity of the host's dietary habits, as well as his or her mobility across a landscape. Plant and faunal remains are major elements in coprolites. Seeds, in particular, survive the digestive process well due to the hard seed coat. Seeds of large fruits—peaches, pears, dates—are less frequent in coprolites due mainly to their size. Leaves, roots, stems, and bark are also less common in coprolites, being affected by both mastication and digestion. The smallest plant residues—the phytoliths—due to their siliceous nature, join seeds and pollen as the most reliable survivor in the coprolite. Vertebrate and insect remains such as hair, bones, feathers, shell, scales, and exoskeletons are also preserved in coprolites.

Coprolites of great antiquity have survived in such well-known archaeological sites as Lazaret in France, Olduvai Gorge, and other African hominid find sites. In those sites with hominid coprolites, the paucity of inclusions, together with the effects of diagenesis, renders the data difficult to interpret. Bryant asks: (1) What should a human coprolite contain after thousands (even millions) of years of weathering? (2) Do the apparent differences in data reflect different digestive patterns between earlier and later species (or subspecies) of humans?[43]

Plant Microfossils (Phytoliths)

The study of plant phytoliths is a relatively young but growing field of paleoethnobotany. Phytoliths are the microscopic bodies formed in plants from silica that is absorbed as monosilicic acid and forms an opaline silicate ($SiO_2 \cdot nH_2O$).[44] At first glance, the silica bodies have a morphology as diverse as that of the plant kingdom itself. Still, there are shapes within the welter of phytolithic forms that are regular enough to allow the recognition of plants at different taxonomic levels. Like pollen and plant macrofossils, the survival of these elements allows the application of botanical systematics in the analysis of phytoliths. As with pollen's exine, the opaline silica phytoliths generally survive diagenesis and even thermal events in sediments.

In 1908 H. C. Schellenberg first noted the presence of phytoliths in soils of archaeological sites in the North Kurgan.[45] He was preceded by F. Netolitzky (1900), who identified wheat and barley phytoliths in soils of European sites.[46] Other workers did not follow these pioneering demonstrations of the utility of phytolith research in archaeology. Phytolith studies had no Iversen to do what he and others did for pollen. Nor did phytolith research have the body of detailed work that pollen research had enjoyed for over two centuries since the development of the microscope.[47] In the late 1950s H. Helbaek's work in the Middle East and N. Watanabe's studies of cereal cultigens in Japan demonstrated, anew, the value of phytolith identification in archaeology.[48]

In the New World the publications of I. Rovner, D. M. Pearsall, and D. R. Piperno provided important impetus to phytolith studies.[49] The field now has a large literature, focusing principally on paleoecology and cultural plant usage. In the main, phytolith studies follow protocols very similar to those in pollen work. First, the phytoliths are separated using particle size and specific gravity (sp.gr.). Phytoliths are small—on the order of a few microns—and are present in fine silt and clay fractions (5–20 μm). Settling typically removes these fractions. Chemical pretreatment is done before sieving, using hydrogen peroxide for organics and sodium hexametaphosphate for clays. After centrifuge treatment or sieving, the phytoliths are floated in a liquid with a specific gravity of 2.3–2.4. The sp.gr. 1.5 liquid then sediments the materials. Flotation and sedimentation are repeated to maximize recovery of the phytoliths.

The phytoliths are then mounted in slides and identified as to "morphotype." Quantities of each are determined by counting under optical or scanning electron microscopes. The morphotypes are diverse: in grasses one can expect 36 phytolith categories, ranging from triangular to dumbbell shapes (figure 3.8).[50] Phytolith patterns for the grasses originate in the leaf blade material; for rice and maize in the motor cells of the fleshy part of the leaves. The morphotypes are quite dissimilar. Wheat phytoliths come in both multicell (husks) and single-cell (leaves) forms.

As a rule of thumb, bilobate and cross-shaped phytoliths are characteristic of tall grasses of the Panicoid subfamily (maize, a grass, has a cross-shaped phytolith), saddle-shaped forms occur in short grasses of the Chloroid subfamily, and trapezoidal phytoliths are characteristic of the Pooid subfamily (e.g., *Festuca,* or fescue grass). The cereals, along with some trees, also produce multicell phytoliths. A number of families and genera of dicotyledons (deciduous trees) produce unique types of phytoliths,[51] as do pine, fir, larch, and other coniferous varieties.

Appropriate methodology in measurement is a recent problem to arise in phytolith studies. In essence, it is the specialist hunched over his or her microscope versus impersonal,

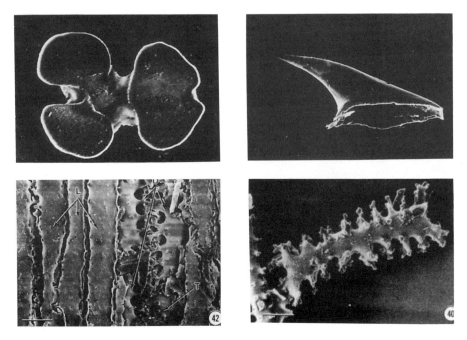

Figure 3.8 Scanning micrographs of selected phytoliths: *(top left)* chloridoid "cross-body," *(top right)* tricom grass "spikelet" cell, *(bottom left)* festucoid (wheat) cell, *(bottom right)* festucoid cell. (Courtesy Irvin Rovner.)

computer-assisted image analysis. On the "traditional" side are Pearsall and Piperno, and on the "modernist" side are Russ and Rovner. Speed and replicability will adjudicate the issue. The complexity and variability of phytoliths necessitate great skill gained from considerable experience in identification. Expertise matched against diversity of form results in subjectivity. Perhaps the surest way to ameliorate the identification problem is to use digital analysis, as Russ and Rovner suggest. One only has to recall the problems that fission-track counting presented to the nuclear scientist-turned-microscopist to realize that phytolith analysis is a quantum leap beyond in terms of difficulty. Nuclear tracks, which we will discuss in a later chapter, are almost always two-dimensional but are difficult to quantify accurately without automated assistance. Quantification and identification of phytoliths by automata seems the natural solution.

Interpretation of Sediments in Archaeological Geology

Seismic and Sedimentological Studies in Inland Waters:
Lake Neuchâtel, Switzerland

An extensive application of subbottom profiling to the study of inundated archaeological sites was conducted at Lake Neuchâtel, Switzerland.[52] Extensive shallow-water (2–20 m) surveys have been conducted over large stretches of lake bottom that contained several archaeological sites. The principal reasons for this undertaking include scientific interest and a recognition that the combined deleterious effects of erosion caused by engineered modifications of the level of Lake Neuchâtel and shoreline development would soon result in a loss of these sites forever. Many important geological and geomorphological details relating to the lake and its associated archaeology were recorded, resulting in a new understanding of post-Pleistocene environmental change in Alpine Europe. Archaeological, hydrological, and sedimentological studies of Lake Neuchâtel provide a window into the complexity of climatic change in the Quaternary of Europe. Situated on the Swiss Plateau between the geologically young Alpine massif and the older Juras, the lake basin was a corridor for the expansion of the Rhône glacier, as well as for smaller ice sheets of the Juras. The rise and fall of the lakes of the plateau reflected the glacial cycle of the Pleistocene. Neuchâtel is the largest of these lakes, 43 km in length and over 150 m in depth today.

In the heights of the canton of Neuchâtel, at the cave site of Cotencher, archaeologists have found traces of humanity extending back to the Middle Paleolithic (ca. 35,000 BP) and the Würm glaciation. High up in the Areuse River valley, the site is within easy access (~5 km) of the lake. Clearly the lake has affected and been affected by the human component of the Quaternary ecosystem. Twenty millennia after Neanderthals sheltered at Cotencher, other sapiens returned to the shores of a lake basin fresh from the grip of Würm glaciation. Upper Paleolithic hunters of the Magdalenian culture colonized the reborn lake, to be closely followed in time by Epipaleolithic Aziliens and much later by Neolithic and Bronze Age fisher-farmers. The lake has remained a repository of archaeogeological information in the form of geomorphological features—streams, terraces, and inundated shorelines, together with the sediments and soils in and around it. The dynamism of the lake was recognized early and has been studied for over 100 years.

Correlations in sediment, pollen, hydrology, and culture are difficult given the size and overall complexity of the lake's environmental regime.[53] For instance, only now do we have data from the southern end (Yverdon), the central area (Cortaillod and Auvernier), and the northern end (Champréveyres and Saint-Blaise). Each region of the lake is governed by unique local climatic and hydrological variations. Yverdon, at the southern extremity, is not influenced by the prevailing winds (south-southwest) and has, relative to the northern and central lake, only nominal amounts of fluvial and colluvial deposition. As one proceeds northward along the lake, both erosional and depositional effects increase, reaching their highest levels at the inflow of the Broye River, a small river, co-opted in the past by the Alpine-draining Aare River. The Aare's meanders between Aarberg and Kerzers (figure 3.9) strongly altered lake levels and their associated sediments between the Preboreal (10,000–9500 BP) and the Atlantic (8200–5900 BP), raising inflow by a factor of 5 and lake levels by as much as 3 m.

Combining our newfound knowledge allows us to address regional and extraregional questions of climate: "crises," fluctuations of centennial to millennial scale, and "anomalies," decadal scale. From overlapping sediment facies, lake levels, cultural events, and chronology (AMS-radiocarbon, varves, and dendrochronology; see chapters 6–7), a new appreciation of the paleoclimatology of the Jura lakes and western Switzerland can be formed extending from today's Lake Neuchâtel to its ancient predecessors.

A larger lake, comprising the Jura lakes of Neuchâtel, Bienne, and Morat, termed "Lake Solothurn," had shorelines as high as 510 m (80 m higher than today's lakes), with a radiocarbon date of 27,550 ± 1130 BP obtained for the 445-m level.[54] Streams draining the great ice sheet's last melt after 10,000 BP carved now-drowned channels across the morainic

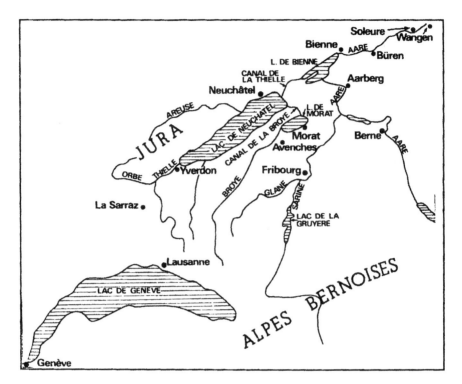

Figure 3.9 Map of the Jura lakes, western Switzerland.

deposits. In and along these streams is evidence of the oscillations in the early Holocene climatic regime, as well as the remains of camps of Upper Paleolithic and Epipaleolithic hunters, evidence of a dramatic cultural transition.

Cortaillod-Est and Les Esserts: The Sedimentary
History of a Lake, Part 1

The twin sites of Cortaillod-Est and Les Esserts represent different villages of the Late Bronze Age along the shores of Lake Neuchâtel. From 1981 to 1983, Cortaillod-Est was excavated in toto by the Archaeological Service of the Canton of Neuchâtel under the direction of Béat Arnold.[55] In 1984 27 sediment cores were taken and in addition, a sediment profile was made on the slopes above the lake at 445 m, which yielded the radiocarbon date of 27,550 ± 1130 BP.[56]

The sedimentary sequence began at the end of the Würm glaciation on the Swiss Plateau. In the wake of the retreating Rhône glacier, meltwaters filled the basins of the Jura lakes—Neuchâtel, Bienne, and Morat (figure 3.9). The sequence shown in figure 3.10 has 10 major phases deduced from cores 8 and 14; major human occupation occurs at phases VI and VIII. These phases correspond to the Subboreal period (5900–2250 BP), with a Neolithic occupation after 5280 BP and a Bronze Age village in the 3rd millennium BP. Ironically, the cores did not find definitive habitation layers for either the Neolithic or the Bronze Age. The

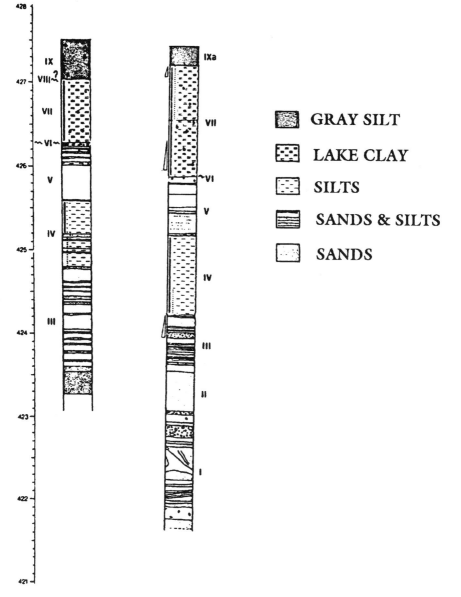

Figure 3.10 Sediment cores from Lake Neuchâtel showing major sediment phases at Cortaillod (modified from Brochier, 1991).

early phases, I to V, comprise alternating amounts of alluvia and sands. Phase VI is a lacuna of several millennia (14,790–5280 BP) interpreted as a time of the lake's regression to a low-stand of 426.5 m. Neolithic farmers of the Cortaillod culture had settled on exposed shore-line by the mid–7th millennium BP. By the early part of the 6th millennium (5790 BP), the lake had returned in a transgression that lasted until the late Neolithic (5203 BP). After a short time (ca. 100 years) another, more extensive transgression (phase VII) lasted into the 3rd millennium BP, represented by the eroded level of phase VIII at Cortaillod, and the time of settlement of two villages. After the mid–3rd millennium BP, the lake again rose to cover the fields and villages. In the last phase, IX, a gray loam covers the site as a polyphased secondary lacustrine deposit created, in large part, by erosional processes begun with the artificial lowering of the lake in the late 19th century after rerouting the Jura water system (Aare, Thielle, and Broye Rivers). Characterizing the sediments in cores, as Brochier has done, is a first step; correlating the cores, which were taken systematically, is next. Next at Cortaillod the extent and thickness of the sediments was mapped using geophysical methods—specifically, seismic subbottom profiling.[57] The detailed characterization of the multiple cores allowed a precise calibration of the seismic records and a ready identification of the acoustic reflectors.

Sediment phase V, composed of sands and silts, represents a significant change in the acoustic image and contrasts with phase IV, which is dominated by fine chalk and silt fractions more massive and acoustically dense than the overlying sediments. These "lake clays" mark early transgression(s) of the lake in the late Glacial (Bølling?). Phase V extends from the edge of the deltaic front of the Areuse across Cortaillod-Est and is less pronounced or absent at Les Esserts, which is slightly higher in elevation (427.5 m). The only radiocarbon date of sediments at Cortaillod (14,790 ± 550 BP) is from the end of phase V.[58] Phase VII, succeeding the ephemeral phase VI, marks another transgression in the early Subboreal. This sedimentary layer is ubiquitous along the lakeshore and is truncated by the eroded phase VIII and the "modern" secondary deposits of phase IX.[59]

Hauterive–Saint-Blaise and Champréveyres:
The Sedimentary History of a Lake, Part 2

Subsequent to the coring at Cortaillod (1984), simultaneous extensive archaeogeophysical and sediment studies were undertaken at the sites of Hauterive, Champréveyres further up the north shore of Lake Neuchâtel (1985). In 1987, offshore of the archaeological site of Champréveyres, geophysical data were taken as part of the same overall study which included Cortaillod. As at Cortaillod, Moulin described the late Glacial–early Holocene sedimentary sequence for this northern end of Lake Neuchâtel.[60] These data have given us a comprehensive view of the fluctuations in lake levels during the early to mid-Holocene (figure 3.11), and a description of the early Postglacial lake to which the Magdalenian hunters of Champréveyres came. The 1987 seismic study discovered buried paleochannels paralleling the shoreline off Hauterive, Saint-Blaise and Champréveyres-erosional features of the Older Dryas. The paleochannels and fill facies allow an environment reconstruction that is less conjectural, particularly for the early Holocene. Our knowledge, based only on archaeological deposits, fails to explain the later Holocene lacunae identified by Brochier at Cortaillod and Auvernier and confirmed by Moulin. These gaps are in those Neolithic, Bronze Age, and post–Bronze Age sites where the lake has eroded away some indices and mixed others (e.g., phase IX at Cortaillod).

The geometry, lithology, and chemistry of sediments suggest the following history of the lake:

> Early Older Dryas: Lake level is below 420 m, with strong erosion of a pure Glacial landscape forming the channels (1–3) at Champréveyres and originating many of the offshore paleochannels at Hauterive–Saint-Blaise ca. 15,000 BP.

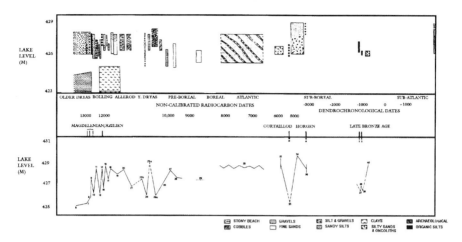

Figure 3.11   Lake Neuchâtel Holocene sediments and lake levels (modified from Moulin, 1995).

Late Older Dryas: Lake level is below 425.5 m. Warming continues, with advancement of vegetal cover reducing erosion, particularly that of colluviation. Appearance of the first encampments of Magdalenians at Champréveyres (13,050 ± 155 BP). In the succeeding 500 years, Magdalenians reoccupy the site twice more until 12,510 ± 130 BP. The channels are partially filled, providing living surfaces for the human groups. The lake is relatively calm and oligotrophic.[61]

End of the Older Dryas: The living areas of the Magdalenians are inundated by the first transgression of the lake in the Postglacial. The lake reaches a level of 427.5 m.

Transition between Older Dryas and the Bølling: Deltaic deposits are seen at Champréveyres, with torrential activity promoting erosion and activity in the channel system. The lake varies between 426 and 427 m.

Early Bølling: A high point of 428.5 m after 12,670 ± 40 BP is reached. Regression occurs between 426 and 426.5 m, followed by transgression before 12,330 ± 100 BP. The Azilien culture succeeds the Magdalenian culture. The Azilien encampments are dated to 12,550 ± 130 BP and 12,120 ± 170 BP. One encampment is at an altitude of 427.5 m during a regressive phase. The Azilien occupation is followed by transgression.

End of the Bølling: Sediments are dominated by chalky silts typical of transgression. Organic materials yield a date of 12,090 ± 70 BP.

Allerød: In the sediments are found wood and pine cones. The sediments are sand passing laterally into gravel bars. The lake is below 427 m. Radiocarbon dating yields 11,820 ± 40 BP.

Transition from Allerød to Younger Dryas: With the onset of the colder Younger Dryas at 11,000 BP, the level of Lake Neuchâtel falls to near 426 m. An erosional beach of large cobbles covers much of Champréveyres. This regression corresponds to return of Glacial cold (0°C isotherm at 200 m altitude) and dry conditions.[62]

Preboreal, Boreal, Atlantic, and Subboreal: The lake returns, in transgression, with warmer, wetter conditions in the Preboreal, radiocarbon dated at 10,090 ± 80 BP. The lake remains above 427.5 for millennia. In the Atlantic period the Neolithic farmers of the Cortaillod culture settle at Champréveyres (3810–3793 BC). A transgression occurs between the end of Cortaillod settlement and the beginning of the late Neolithic Horgen occupation (3200 BC). The Subboreal represents a lacuna in cultural activity at Champréveyres and elsewhere along the lake. The late Subboreal sees the establishment of Bronze Age villages at Champréveyres, Cortaillod, Auvernier, and elsewhere on the shoreline. Soils formed in time for the Neolithic peoples to settle, but the height (425.5 m) of Postglacial deposits places them at risk to transgressive/regressive cycles, of the littoral and as seen at Cortaillod and Champréveyres, by the absence of stratified deposits and the formation of a coastal plain. The beginning of the Subatlantic, a period of wetter, cooler conditions, sees the lake rise to levels in excess of 430 m until the present.

The archaeogeological study of lacustrine sediments in Lake Neuchâtel provides a rich and varied picture of climatic variation from the Postglacial to the mid-Holocene. Absences in the data point to where we must do more work to complete the rich cloth the many workers have woven for us.

The Zübo 80 Project: Quaternary Geology of Lake Zürich

In 1980 geologists from ETH-Zürich and the University of Geneva began the deep-lake drilling of Lake Zürich at a site named Zübo.[63] The interdisciplinary project was an ambitious attempt to investigate the Quaternary geology of the Zürich and Alpine areas. Ken Hsu (ETH) was Principal Investigator and Frederico Giovanoli (Geneva) was Project Manager. Unlike the Lake Neuchâtel studies, the Zübo Project concentrated its efforts at a 201m-deep borehole in the mid-lake of Lake Zürich. Before drilling, the site was seismically surveyed using air-gun and subbottom profilers. The combination of seismic survey and sediment boring is the same methodology used at Neuchâtel, but the greater depth achieved at Lake Zürich (201m versus and average of 4m) promised data on the climatic and sedimentary history in the Quaternary rather than just the late Pleistocene and Holocene.

The results of the project were mixed. The sediments of the Zürich basin turned out to be no older than the Riss-Würm (ca. 120ka) as the lake itself first came into existence as a subglacial lake in the Würm glaciation.[64] It only became an ice-free lake toward the end of the Pleistocene so the upper 30m of drilled sediments represent the Alleröd-Bölling. The Zübo core established a sedimentary history during advances and retreats of the Würm glaciation. The attempt to obtain a complete climatic history of that last Ice Age failed due to the young age of the lake and the allochtonous nature of the fossil pollen in the Zübo sediments below 30m. The pollen was not indigenous to the Zürich basin until the Alleröd or early Holocene when steppe vegetation began to give way to forests (cf. Lake Neuchâtel).[65]

The sedimentological response at Zübo is clearly climate driven—the gravels and sands of the lowest lithostratigraphic unit (LZ-alpha, 154–137m) are glacio-alluvial of Interglacial origin; the sediments of the middle lithostratigraphic unit (LZ-beta, 137–30m) are glacial and subglacial deposits beginning around 75 ka and ending ca. 15ka at the lower boundary of the upper-lithostratigraphic unit (LZ-gamma, 30–0m) composed of Holocene muds and marls representative of fluctuation in the modern climate cycle.

The fluctuations in the lakes, Neuchâtel and Zürich, recorded in their sediments, are indicators of climatic fluctuations. The transgressions correspond to oceanic-driven influences, and the regressions are the products of continental influences.[66] This is the "large" picture but within this frame operates regional to local variation. At Lake Neuchâtel, at least until the Atlantic period, the shifting of the Aare River's course in response to alluvial buildup near its junction with the Rhine and erosion caused by variation in discharge strongly influenced the lake and sediment deposition rates. The Aare linked the lake to its principal water source after the Postglacial—the Alps. Changes in snow cover, meltoff, and reflectivity (recall 1991 and the discovery of the "Iceman" due to increased glacial melt produced by higher absorbance of the sun's warmth caused by windblown sediments from Africa) are key factors of the larger climate system well removed from simple regional perturbations. Evapo-transpiration rates are modulated by forests as well as by atmospheric factors. Different tree species have differing influence on these rates; pines are more demanding of groundwater than are deciduous species. The sediments of the Postglacial developed after herbs, shrubs, and hardy grasses had taken root on the eroded slopes of the Juras. Once the soil was prepared, the tree species of *Juniperus, Salix, Hippophae,* and *Betula* established young tundra woods (Middle Dryas, ca. 12,000–11,800 BP) giving way to the supremacy of pines in the Alleröd (12,000–11,000 BP). The climatic shock of the Younger Dryas—seen floristically and sedimentologically—reversed this trend but did not bring the return of the glaciers. Grasses remained as wardens of the young soils until the return of favorable conditions in the Preboreal (10,000 BP).[67]

Anthropic Soils: "Dark Earths"

Dark A horizons that are greater than 40 cm thick and are "man-made" are termed "cultisols" in England, and "plaggens" (sod) in the United States.[68] In these soils, the A horizon is the direct result of man-made or induced amendments such as manure, organic debris, and industrial refuse. Most anthrosols can be characterized by color, texture, carbon content and the carbon/nitrogen ratio. The depths of these deposits are enhanced by earthworm activity.[69]

In urban settlements, particularly in Roman Britain, "dark earths" were developed as observed in recent excavations in central London by P. Rowsome.[70] They begin in the 3rd century and are postdated by 4th-century construction. J. Schofield found late Roman dark earth around abandoned buildings of the 2nd century, some of which stood into the Saxon period on dark earth areas.[71] Constituents of the deposits are brick, mortar, pottery, cinders, bone, iron slag, and large amounts of water-soluble phosphate from bone and animal residue.

Relative Age Determination of Alluvial Deposits Using Soil Color, Mottle Patterns, and Organic Content

E. A. Bettis has suggested that a robust relative age determination scheme for alluvial valley deposits can be developed using sediment characteristics including soil color, mottle patterns, and organic content.[72] By these criteria a deposit can be characterized as being early to middle Holocene (EMH), late Holocene (LH), or Historic (H). The application of his model "permits division of landscape elements comprising modern valleys into three groups: (1) those older than about 3000 to 4000 years (EMH), (2) those younger than (1) but prehistoric (LH), and (3) Historic deposits." The data gathered by Bettis were backed up by extensive radiocarbon dates from detailed investigations of alluvial fills and apply only to Iowa and southern Wisconsin which were uniformly affected by the last glacial period. However the method does appear to have usefulness elsewhere.

To Bettis the soil color is a key determinant of the age of a deposit. Color changes during drying, and different rates of drying of the materials in a section can be very useful in showing up layers or horizons of different texture, structure, or organic content.[73]

Table 3.5 Outline of Criteria Used to Group Upper Midwestern Alluvial Deposits into the Age-Morphologic Groups

Age-morphologic group	Landform(s)	Bedding	Weather zone	Mottles	Surface soil (horizontal sequence; B horizon color)
Early to Middle Holocene	Terraces, alluvial fans; colluvial slopes	Restricted to lower part of section	O; MO; R; or U in part of some sections	Common; brown, reddish brown, and/or gray	A-E-Bt; A-Bt; brown B horizon
Late Holocene	lowest terrace; floodplain	usually restricted to lower part of section	color usually 10YR hue, values 4 or less, chroma 3 or less; disseminated organic carbon imparts dark colors; may be oxidized or unoxidized but matrix colors are dark because of organic carbon content	rare—usually not present	A-Bw; dark-colored B horizon
Historic	Floodplain; fence lines; foot slopes; buried older surfaces	present throughout section if > 50 cm in thickness	O; Mo; R; some sections dark color because of high organic carbon content	Can be present or absent; brown, reddish brown, or gray	A-C; no B horizon

Abbreviation Key: O: organic; Mo: mottled; R: red coloration; A: surface soil horizon; E: light colored subsurface horizon, characterized by removal of clay and iron oxides; Bt: subsurface horizon with significant clay accumulation also termed argillic; B: subsurface horizon, general designation for horizon between A-C horizons; Bw: subsurface horizon with low or no carbonates.

When a deposit is continuously waterlogged over a period of time, it can develop mottling due to change in the oxidation state of iron. In this environment of saturation, iron may be reduced and migrate out of the sediment and later be oxidized.[74] Identifying mottles is important to Bettis's model because it can establish if the deposit is early to middle Holocene or Historic.

Table 3.5 illustrates the classification developed by Bettis and its application to the determination of "age-morphologic groups." Using Table 3.5, soil can be classified by the color and the mottle pattern as an (O,R . . .) symbol. The symbol can then be matched to the column labeled weathering zone. The surface soil descriptions are used to establish the types of horizons that may exist. This is the final step in determining the relative age of a deposit. Once this is completed the investigator can map out the sequence of alluvium and its associated soils to create a landform sediment assemblage isopach map.

PART II

DATING
TECHNIQUES

4

Chemical Methods

"Time is nature's way of keeping everything from happening at once" (anonymous). Time is a continuum—we sense this continuum as a succession of events. In archaeological matters it is one of the most salient attributes. To determine time accurately the archaeologist must rely on modern dating techniques.

Age determination by chemical methods relies on the constancy or predictability of rates of chemical processes. For instance the oxidation of iron—rust—could be used for dating purposes if one could determine a chemical rate, in this case that of oxidation, that applied to more than the singular event. Unfortunately, the rate of the oxidation of iron is highly variable, being affected by temperature, available moisture, and the particular type of iron (mild, cast, stainless, etc.).

Another common chemical change is the patination of certain types of glass. Yet here, too, the process is highly variable, making dating impractical. Still, there have been attempts to use patination and rock "varnish" for archaeological dating, as we shall see.

In the main, chemical dating is used to determine relative ages since absolute ages require calibration for each sample and its find site using independent dating measures such as radiometric or dendrochronological techniques. We shall first discuss the relative techniques based on the uptake or decrease in fluorine, uranium, and nitrogen found in bone. This is most appropriate because these chemical techniques played a key role in unmasking one of the most famous frauds in the history of science: Piltdown Man. Next we shall examine the two most accepted chemical processes utilized in absolute age determination, which are based, respectively, on amino acid racemization and obsidian hydration. Finally, we shall examine a few techniques that show some promise for the dating of archaeological materials or deposits, such as those using patination ("varnish") and cation ratios.

Temporal Units

Our points of reference are those events we view as, in some sense, marking a change in the state of things. Stylistic or formal change in an archaeological facies can be a chronological landmark for the archaeologist and allows us to divide the continuum of time into discrete segments or phases.[1] Geological facies change as well, but at rates generally slower than those of archaeology.

The nature of events or things under study by archaeologists determines the specificity of time units used. Changes in political affairs—the rise of states, the fall of empires—may require finely drawn time units, decadal or less. Broad paleoclimatic change does not require so fine a scale; centennial or millennial may be adequate.

Accuracy and precision are elements of all temporal units, again largely determined by the specificity of the temporality needed. They in turn determine the type of dating technique that is appropriate. These terms are often misunderstood in their usage in dating. "Accuracy" refers to how close a date is to the actual age. "Precision" refers to the reproducibility of that date. An age determination procedure can be highly precise but be hopelessly off in the true age of something.

Relative Dating Techniques

Relative dating techniques attempt to ascertain the correct order of events. The magnitude of the events or the time units is *not* the primary question in this type of temporal study, only succession or antecedence. If our belief that "Smith's Principle" (Law of Superposition) is correct, then stratigraphic relationships imply relative age—that is, lower strata are older than upper strata. How much older is not the central issue in relative dating. What relative dating techniques adjudicate is contemporaneity. The first relative techniques were developed for archaeology to resolve the issue of contemporaneity for fossils of now extinct fauna found in association with the remains of early humans.

F-U-N Techniques

F-U-N stands for three chemical dating tests: the fluorine, uranium, and nitrogen tests. Typically three facets of age relative to fossil remains can be ascertained:

1. the association of human skeletal remains with an artifact assemblage,
2. the association of human skeletal remains with the skeletal remains of extinct fauna, and
3. the determination of whether bones presumed to belong to the same individual are, indeed, of comparable antiquity.

Fluorine Test

Skeletal remains undergo change after burial in a soil matrix. When water percolates about and through bone with a solution of minerals drawn from the soil, two things happen:

1. The principal component of bone, hydroxyapatite, undergoes alteration through elemental substitution.
2. Fossilization (the trapping of minerals in solution in the pores of the bone) occurs, often by addition of $CaCO_3$ or FeO, which increases the mass and density of the bone.

Fossilization occurs at varying rates. Hence, relative dating based on progressive fossilization of bone is not reliable. Elemental substitution, such as the alteration of hydroxyapatite, can be utilized because this change is slow and irreversible. Fluorine occurs in trace amounts in most groundwaters. Fluorine ions (F^-) displace hydroxyl ions (OH^-), creating fluorapatite, a less soluble compound. By measuring the hydroxo/fluorapatite ratio by x-ray methods (diffraction), contemporaneity can be examined in stratigraphically associated materials.[2] Age determination using fluorine has application for relative age dating where only the most general chronology is sought. As an age determination method, there are significant problems related to changing amounts of compounds due to absorption and loss of elements during diagenesis and burial history.

Hille et al. dated bones from Austrian caves.[3] Figure 4.1 shows the relative uncertainty in the ages so determined. X-ray diffraction by Bartsiokas and Middleton of bones of different ages indicate a predictable relationship between crystallinity and sharpness of the diffraction pattern.[4] Schmurr used fluoride-selective ion electrode analysis of solutions to measure fluoride concentration in his study of bones from the Angel site (Ohio). The site was a Middle Mississippian culture center with mounds, village, and stockade walls. The frequency plot of fluoride concentrations from 38 burials suggests internment within a relatively short period of unknown duration (figure 4.2).[5]

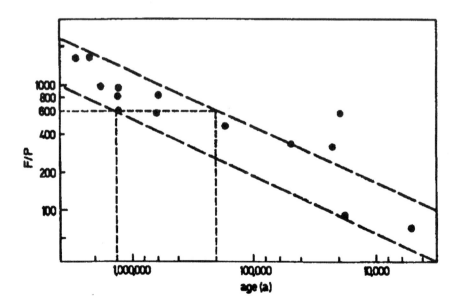

Figure 4.1 F/P ratio dating curve for bones found in Austrian caves (modified from Hille et al., 1981).

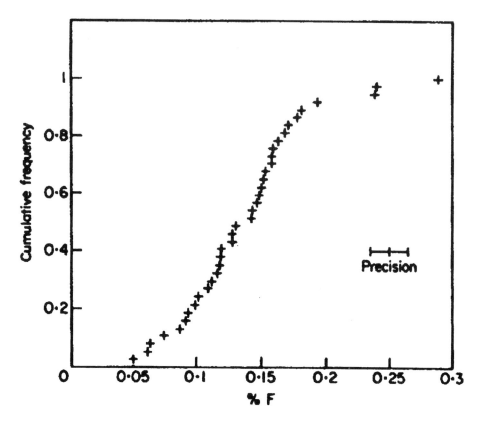

Figure 4.2 Fluorine content in 38 burials from the Angel site (Ohio) (modified from Schurr, 1989).

Uranium Test

Groundwater also contains uranium. Calcium ions in hydroxyapatite are displaced by hexavalent uranium ions. This is also a cumulative process, so the longer a bone has lain in the soil, the more uranium it will contain. Uranium is radioactive (see the discussion in chapter 5); that is, it decays by emission of a radioactive particle or particles. These particles can be either beta or alpha particles. Associated with beta decay is a gamma ray (2-particle decay). A beta particle is an electron; a gamma ray is a high-energy x ray. An alpha particle emission is a short-range particle with an atomic number of 2 and a mass of 4, or $Z = 2$, $A = 4$ (i.e., a helium nucleus). Counting the beta emissions or alpha particles by a proportional counter measures the concentration of uranium in parts per million. The greater the activity, the longer the sample has been buried.

Nitrogen Test

Bones lose organic constituents when buried: protein (collagen) and fat. The fatty portion is lost soon after burial; the collagen more slowly. Nitrogen, fixed in collagen, is an excellent measure of the residual organic matter remaining—that is, the less nitrogen, the older the bone.

The Piltdown Hoax

The Piltdown Hoax began with the "discovery" in 1911, by amateur archaeologist Charles Dawson, of what appeared to be flaked stone tools at Piltdown Common, in shingles used to gravel a road. Dawson showed them to A. Smith Woodward of the British Museum. Together, Dawson and Woodward found skull fragments and half of a damaged jaw as well as fossils of extinct fauna.

The find consisted of the right half of a lower jaw with two molars, a canine, and large fragments of the frontal, parietal, temporal, and occipital bones of a cranial vault. The occipital condyle was missing from the apelike jaw. The teeth were ground down apparently by chewing. The find did not look like any other fossil previously discovered. More important, it bore directly on the ongoing debate on how the human form evolved. Some held that the brain evolved before the rest of the body did, as was thought to be the case with Pithecanthropus, another "Pliocene" form. The Piltdown find was, indeed, significant—the kind that confers instant prestige on those who announce such a discovery. On 12 December 1912 they introduced *Eoanthropus dawsonii* to the Geological Society of London: "Dawson's Dawn Man."[6] The skull appeared modern, but the jaw was apelike. A concern about the differing maturity of the parts was raised: middle-aged skull, young jaw, old teeth!

The veracity of the fossil was debated until 1949. Exposure of the hoax had to wait until 1949, when the overwhelming number of fossils made the Piltdown find "stand out like a transistor radio in a collection of handaxes."[7] First to test the idea of fraud was J. S. Weiner, lecturer at Oxford's anatomy department. He mentioned his suspicions to Le Gros Clark, who doubted at first. Weiner then took a chimp jaw and filed the teeth to look like the Piltdown Man. Le Gros Clark then organized a thorough reexamination.

When Kenneth Oakley examined all the vertebrate specimens from the Piltdown gravels by fluorine dating, the bones of the nonhominid fauna contained 2–3% fluorine (3.8% is maximum). The Piltdown Man bones contained only 0.1% (skull) and 0.4% (jaw). While not disproving contemporaneity between skull and jaw, clearly the hominid remains were not contemporary with the other faunal remains. In 1953 further tests revealed that the teeth and jawbone contained no more fluorine than modern bone. The skull bones contained enough fluorine to be older. Nitrogen tests indicated that the jaw and teeth were modern and freshly dissected, confirming the hoax (table 4.1).

The relative age tests clinched the fraud case. An orangutan jaw and fragments from different modern skulls had been placed along with ancient bones in the gravel bed. Who did it? Dawson died in 1916, taking his story with him.

Amino Acid Dating Technique

Amino acids are molecules used to construct long-chain proteins which compose bone collagen in hair and feathers, as well as dentine in teeth. Some of these amino acid molecules, including aspartic acid, alanine, and isoleucine, are optically active, a property directly related to a molecule's ability to rotate polarized light waves.

Amino acid molecules, except glycine, have a central, tetrahedral carbon atom with four

Table 4.1 Chemical Comparison of Piltdown and Other Bone

	F (%)	N (%)	U (U_3O_3 ppm)
Fresh bone	0.03	4.0	0
Piltdown jawbone	0.03	3.9	0
Neolithic skull (Kent)	0.3	1.9	—
Piltdown skull	0.1	1.4	1
Piltdown "Elyphas planifions" molar	2.7	nil	610

Sources: J. S. Weiner, K. P. Oakley and W. E. LeGros Clark, 1955, Further contributions to the solution of the Piltdown problem, Bulletin of the British Museum of Natural History (Geol.) 2(6): 225–87, Table pp. 262–65; F. Spenser, Piltdown, A Scientific Forgery. (London: Oxford University Press), pp. 144–45.

different atoms or groups. Only chiral (four groups), or asymmetric, amino acids are optically active.

Enantiomers rotate polarized light either right (D or dextrorotary) or left (L, or levorotary) (figure 4.3). Today the convention is to use S and R (sinestra/recto) rather than L and D. Diastereomers, such as isoleucine, can rotate polarized light as well. Because all living organisms use S isomers, the ratio of R isomers to S isomers *(R/S)* is zero. Outside living organisms, the ratio is 1. At death the R/S ratio shifts from S to R diastereomeric forms by a process called *epimerization*. Another term commonly used in the literature of amino acid dating is *racemization*. When an enantiomeric mixture reaches 1, or equilibrium, it is termed *racemic*.

Rates for racemization are strongly dependent on the burial matrix and material. Shell, bone, teeth, and wood have different rates. Following P. E. Hare and P. H. Abelson, we have the following Alle/lle (alleisoleucine/isoleucene) *(R/S)* ratios for *Mercenaria* shells from Florida: 0.2 (ca. 4500 BP), 0.4 (ca. 34,000 BP), and 0.6 (ca. 110,000 BP).[8]

The most important control in this process is absolute temperature, which follows 1st-order kinetics. We have the integrated rate expression for a 1 chiral carbon atom, amino acid:

$$\ln \left\{ \frac{1 + R/S}{1 - R/S} \right\} = 2kt + c \tag{4.1}$$

where k = racemization rate constant, t = time or age, and c = constant of integration ($c = 0$ if $R/S = 0$ at $t = 0$). An Arrhenius plot of log k versus the reciprocal of the absolute temperature is done once for a species or specific material. k, the rate constant is most influenced by temperature: in isoleucine, a 25°C shift in temperature (0° vs. 25°C) can alter the rate, k, from 0.8×10^6 to 120×10^6 which will have a dramatic effect on the ages determined.

In the early application of amino acid racemization chemistry to archaeological age determination, this is precisely what happened.[9] With the first studies of fossil shells by Hare and Abelson in 1968, the correlation of amino acid racemization and age was clearly demonstrated.[10] This early success continued on other marine specimens such as foraminifera.[11] The first results were obtained using what is termed the "uncalibrated" technique, where the rate constant, k, is deduced analytically in the laboratory. In the deep-water marine sediments, it was realistic to use an average or fixed temperature, but it was misleading for other, less temperature-stable environments. As has been subsequently shown, the rate constant

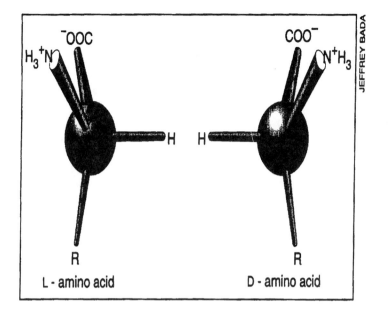

Figure 4.3 Enantiomeric rotation (chirality) in amino acids (modified from Bada, 1995).

doubles with a 4°C increase in temperature in some amino acids. Aspartic acid is the most volatile in this sense: a 1°C change equals a 25% shift in k. In terms of "half-life," where the R/S ratio has reached 1:3, we see a value of 430×10^3 years at 0°C and 3×10^3 years at 25°C. For isoleucine, the values are 6×10^6 years at 0°C and 5×10^4 years at 25°C. For isoleucine, the relationship is illustrated in equation 4.2 for the half-life:

$$t^{\frac{1}{2}} = 0.693/1.75k \qquad (4.2)$$

It is unfortunate that the early success, and hence promise, of amino acid dating was perhaps marred by the success of more well established, radiometrically based techniques such as radiocarbon. Central to radiometric techniques, as we shall see in the following chapter, is the fundamental invariance in the value of an isotope's half-life. In racemization reactions, the invariant half-life concept was unwisely extended to the much different chemistry of amino acids, that is long-chain molecules rather than single atoms as with radiochemistry. Where precise temperatures were incorrectly assumed or unknown, the uncalibrated technique foundered in a spectacular fashion, dooming a potentially useful age determination method to ridicule and early obscurity. Like Piltdown and "cold fusion," the amino acid dating story is a cautionary parable about unqualified and uncritical acceptance in science.

Calibrated Amino Acid Dating and Early Humans in the New World

The question of the antiquity of human presence in the New World has been, and still is, a contentious issue in archaeology.[12] As nonspecialists have learned, this is not a debate to enter into lightly, particularly with a new, and relatively untried, dating technique such as amino acid racemization. Nevertheless, in the 1970s this is exactly what happened. Jeffrey Bada, a young chemist at the Scripps Institution of Oceanography, was fresh from the first successful application of the amino acid dating technique to an animal bone (a goat bone from Muleta Cave, Mallorca, Spain).[13] Bada's use of the amino acid isoleucine was done using the uncalibrated procedure. Utilizing an alleleucine/isoleucine ratio of 0.083 at an average burial temperature of 19°C, he was able to obtain a date of 26,000 years, which was in good agreement with a radiocarbon age of 28,000 years.[14]

To avoid the temperature problem, Bada and his coworker, P. M. Helfman, chose to use an R/S ratio determined by use of independently dated samples, thus "calibrating" the rate.[15] The implicit assumption was that the effective chemical temperature was the same for the sample to be dated as it was for the calibrated specimens. On the face of it, this approach has merit: it avoids the problem of precisely determining k using laboratory means or assuming a burial environment that was temperature stable over millennia.

The specimens Bada and Helfman wanted to date were bones from two hominid skeletons found in California, at Sunnyvale and Del Mar. The amino acid dates reported for these finds were between 40,000 and 50,000 years. The ages were calibrated using another famous skeleton, Laguna Man, dated at 17,000 years by the conventional radiocarbon technique.

Without foreshadowing too much our discussion of the vagaries of the conventional radiocarbon dating technique and its own teething problems, it is important to point out that the controversy that followed the reporting of the amino acid dates turned on the calibration date—not the chemistry or the reliability of the investigators. Bada and Helfman uncritically accepted the 17,000-year age for the Laguna skeleton, which itself was debated among archaeologists. Being chemists, they relied, perhaps naively, on both an archaeological interpretation and the reliability of a radiocarbon bone date.

As pointed out in our Piltdown example, reputations have been badly singed when researchers claim more than they can deliver. What the amino acid dates for the Sunnyvale and Del Mar finds did was place humans in the New World far earlier than most archaeologists were ready to accept. It was this skepticism of the Bada and Helfman dates that kept the question open so that another new dating technique could adjudicate the dispute.[16] This newcomer was AMS—radiocarbon dating or more correctly—accelerator mass and charge spectroscopy (MACS, see chapter 6). While MACS is a more physically accurate acronym, AMS has come to be the preferred usage.

The AMS dates for the Laguna skeleton dismantled the argument for the greater antiquity of the California hominids by determining its age to be closer to 7,000 than 17,000 years. The Del Mar and Sunnyvale dates dropped a whole order of magnitude from 50,000 to 5,000. A comparison of racemization dates for bones and teeth from La Chaise-de-Vouthon, France, in 1983 showed a failure of these dates to agree with known ^{234}Th/^{234}U dates for these specimens.[17] In addition, it was shown that diagenetic alteration of the fossils had produced nonlinear racemization. With this second result, it became clear that racemization did not follow a kinetic order such that a given rate was constant over time. The decision was in: racemization will not date bones or teeth.[18] British archaeologist Paul Mellars summed up the situation succinctly in his appraisal of "wrong" dates in general: "To work with wrong dates is a luxury we cannot afford. A wrong date does not simply inhibit research. It could conceivably throw it in reverse."[19] The ghost of Piltdown surely was in his thoughts as it was in the controversy that has dogged amino acid dating.

Amino Acid Dating beyond Laguna Man

Fortunately, amino acid dating has moved beyond the early controversy. New methods of measurement such as high-performance liquid chromatography (HPLC) allow greater sensitivity as well more standardized and less complex chemical procedures. Sample size has decreased from several grams to less than 1 g, much akin to the progress seen in AMS dating.

A greater understanding and appreciation of the chemistry of racemization and a reduced variety of materials to be dated have bolstered confidence in the results of new dating studies.[20] One important aspect, now recognized, is the role of diagenesis and the hydrolysis of polypeptide molecules in burial contexts. Additionally, the contamination of specimens by amino acids from other sources—fungi or bacteria—is recognized. Degradation and contamination are factors much better understood today than in the 1970s.

Eggshell (ostrich) and teeth remain the best materials for use in racemization studies.[21] This is because these materials—particularly teeth enamel, not the more porous dentine— are more like "closed" systems; that is, diagenesis and contamination affect such samples less than they do more "open" bone. The important factor is the difference between the proteins: collagen versus, say, hydroxyproline, a high-molecular-weight protein less susceptible to alteration. Tooth enamel may turn out to be the most desirable and reliable organic material for amino acid dating, as it has become for other age determination techniques (e.g., U/Th, AMS, and electron spin resonance). Other materials, such as coral and wood, have been relatively dated (Klassies River), but the problems of temperature and linearity preclude absolute dating.

Amino Stratigraphy, Paleoclimate, and Sea Level

The earliest use of the racemization reaction was in relative age determination of marine shells, where a relative chronology of stratigraphic sequences was constructed: amino stratigraphy.[22] For paleoclimate and environmental reconstruction research, amino stratigraphy linked to raised beaches has been demonstrated.[23] Finally the dependence of k on temperature has a useful role in evaluating paleotemperature variation. For instance, it would be useful to examine R/S ratios of isoleucine/alleisoleucine for same-species marine molluscs either side of the Younger Dryas period (11,000–10,000 BP) for evidence of temperature shifts.

Free Amino Acid Content and Relative Dating of Shell

Recognizing the problems inherent in the "traditional" approaches to the use of amino acids in dating, Powell and others have proposed a relative dating procedure using free amino acid (FAA) accumulation and organic matrix (shell) decomposition rates.[24] The procedure utilizes the breakdown of the organic material of a shell due to hydrolysis, which releases free amino acids like serine and threonine. The increase of FAA is measured and compared with the amount of stable, or bound, amino acid (BAA) in the form of the ratio FAA/BAA. The more stable amino acids are also the more familiar ones (e.g., aspartic acid, glycine).

The procedure is straightforward. The shell is treated to remove contaminants and then dissolved in acid. The dissolution releases free amino acids to be measured in an amino acid analyzer. The sample is then hydrolyzed to free all amino acids and reanalyzed to determine the total fraction of BAA after subtraction of the FAA.[25] This technique has been applied to archaeological sites in the Chesapeake Bay area of Maryland that range from the historic (1690–1715) to the prehistoric (ca. AD 450). The results of Powell and coworkers suggest a relatively linear response in the ratio FAA/BAA over the time range examined.

As promising as these preliminary results appear, it is important to recognize, particularly with an eye toward the checkered history of amino acid dating, that much more research is needed on factors such as diagenesis, kinetics, and the behavior of specific amino acids. *Promise* and *caution* are watchwords in the tale of amino acid age determinations in archaeology and geology.

Obsidian Hydration Dating Technique

Basic Principles

The surfaces of glass and obsidian will absorb water to a depth of a few tenths of a micron to nearly 20 μm (a micron is a millionth of a meter). Present techniques are able to measure this surface thickness to within ±0.07 μm. The accuracy of the dates obtained must necessarily be limited by the thickness measurement. The uncertainty in age which this determination imposes depends on whether a linear or a quadratic formula is used to calculate the diffusion rate. When the technique was proposed, there was some dispute over which type of formula applied.

The obsidian technique was first investigated in 1960 by Friedman and Smith who found that a fresh obsidian surface will absorb water from its surroundings.[26] The absorption produces a well-defined surface layer which may hold as much as ten times more water than the interior of the sample. The water diffuses inward taking advantage of microcracks perpendicular to the surface of the obsidian. The water produces strains in the hydrated obsidian which can be clearly seen in polarized light as a luminescent band, an effect called *strain birefringence*. The hydrated layer can also be identified with phase-contrast microscopes, without difficulty (figure 4.4). The ease of identifying the thickness of surface layers is perhaps the one aspect of obsidian dating that enticed many early investigators.

A small sample is sawed from the material at right angles to the surface, ground smooth and then studied under a microscope. There is little evidence that the processing of the sample affects the hydrated surface layer.

Important questions to be answered in hydration dating include the following:

1. Is the process by which the water migrates into the surface of the obsidian, basaltic glass, or other material a true diffusion process? This question was by no means easily answered despite some rather enthusiastic early reviews of the technique.
2. If the process is a diffusion process, can the approximation which produces the equation $X = kt^{1/2}$ be used? Here, X is the depth of diffusion from the surface measured in some units of length; t is the time for the diffusion to take place; and k is a constant related to the diffusion constant. It was not clear whether this approximation or an approximation of the type $X = kt^{3/4}$ or of the type $X = kt$ could be used. All had been proposed. Table 4.2 lists those variations used by several early investigators.

Even if the process could be identified as a diffusion process and the correct equation resolved, questions remained as to whether any equation could be applied with confidence

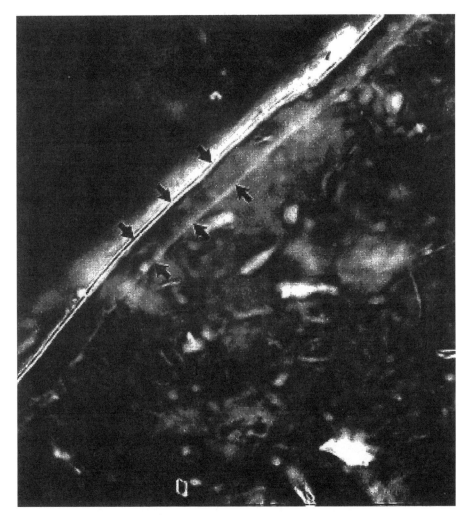

Figure 4.4 Hydration rind in a California obsidian; magnification is 490× (modified from Thomas, 1991).

under the physical conditions that have to be assumed to use the equations. In order to discuss hydration dating at all, a uniform diffusion process had to be assumed. However physicists were usually not very familiar with the solutions of diffusion equations (i.e., they had little intuition as to what approximations must be made to yield a tractable solution). Hydration experiments must measure the number of water molecules per cubic centimeter, (N), assuming that the number within the obsidian results from an average density of water molecules outside the obsidian (N_o).

It is more or less agreed that a definite boundary exists to which the depth of hydration from the surface of the material can be measured. As N builds up in the hydrated material, a certain critical value of N is reached where microfractures are produced and cause the ob-

Table 4.2 Some Early Published Hydration Depth vs. Time Formulas to Determine Age of Volcanic Glasses (Obsidian, Except as Noted) Grouped in the Order of Faster Penetration Reading Downward

Formula type	Investigators
depth $= kt^{\frac{1}{3}}$	Kimberlin 1976
depth $= kt^{\frac{1}{2}}$	Friedman and Smith 1960, Friedman and Long 1976
depth $= kt^{\frac{1}{3}} + K'$	Findlow et al. 1975
depth $= kt^{\frac{3}{4}}$	Clark 1961
depth $= kt$	Ericson 1975, Morgenstein and Riley 1975

Source: I. Friedman and F. Trembour, 1978, Volcanic glass dating, *American Scientist* 66:44; J. Kimberlin, 1976, Obsidian hydration rate determination on chemically characterized sampler in Advances in Obsidian Glass Studies, R. E. Taylor, ed. (Noyes Press); I. Friedman and W. Long. 1976, Hydration rate of obsidian, Science 1991: 347–52; F. J. Findlow, V. Bennett, J. Ericson and S. DeAtlay, 1975, A new obsidian hydration rate for certain obsidians in the American southwest, American Antiquity 40:344–48; D. Clark, 1961, The application of the obsidian hydration dating method to the archaeology of central California. PhD dissertation, Stanford University; J. Ericson, 1975, New results in obsidian hydration dating, World Archaeology 7(2):151–59; and M. Morgenstein and T. J. Riley, 1975, Hydration-rind dating of basaltic glass: a new method for archaeological chronologies, Asian Perspectives 17(2): 145–59.

served birefringence. The dating equation is that which best predicts the relation between X and t for the movement of a fixed boundary where the microfractures may start. In hydration, x is measured and t in the dating equation:

$$x = kt^{1/2} \qquad (4.3)$$

Theoretically, this dating equation conforms most nearly to diffusion predictions.

Sources of Variation, Error, and Precision

It has been pointed out that local temperatures and relative humidity may affect the results. Friedman, Smith, and Clark assert that obsidian of the same chemical composition buried in wet tropical Ecuador hydrates at the same rate as obsidian in a tomb in Egypt, point which implies that the monolayer of moisture on the surface must be uniform for any relative humidity.[27] Because so little is known about the thermodynamics of monolayers, it is difficult to say whether, from a physical point of view, a saturated monolayer could be independent of local relative humidity. Where artifacts are found in the ground or in water, chemicals in the ground or water can be expected to change both k and N/N_o. In some cases, this may be immediately apparent from the position of the hydrated material when the obsidian is discovered. In other cases, it may not be apparent whether alteration of the diffusion process due to the local chemical composition of the surroundings took place.

In a study of Mazer et al., obsidian from three sites was studied for the effects of relative humidity (RH).[28] The obsidian samples were taken from several surface flows and were unhydrated at the time of sampling. The samples were thin-sectioned and placed in specially designed containers and heated to simulate the effects of RH on hydration. Observations were made at 60%, 90%, and 100% RH. The thin sections were then cut to allow measuring of the hydration rind. It was found that at 60% and 90% RH, the rate did not increase greatly, but when an RH of 100% was reached, the rate of hydration increased rapidly. The humidities were reached by using elevated temperatures which did affect the rate of hydration and must be taken into account when considering the validity of this study. The effect of the temperatures was not conclusive but may have increased the rate of hydration at lower RH percentages.

In another study, Friedman et al. found similar evidence that the obsidian hydration rate is affected by RH.[29] Samples were taken from several sites to test the effects of composition on the rate of hydration. The samples were then ground to a fine powder, placed in special containers for hydration, and then weighed to determine the amount of hydration. Any increase in weight was determined to be caused by the addition of water. The powder method was used due to the difficulty of measuring hydration rinds in thin sections. It was found that there was a lower percentage of error by using this method to measure the water uptake. Friedman et al.'s conclusions supported those of Mazer et al. in their research on RH and obsidian hydration.

Friedman and Trembour also studied the effects of temperature on the rate of hydration.[30] Although it was known that temperature would affect the rate of hydration, how it would affect the rate was unknown. They found that the rate of hydration rises exponentially with increasing temperature and amounts to about 10% of each 1°C increase. The question was resolved whether the hydration rate was a square root, exponential, or a complex function. This rate would be easy to ascertain if the obsidian was allowed to hydrate at a constant temperature. The need to know the average temperature of the area in which the sample was taken and the burial history for a sample is critical. The chemical composition of the obsidian must also be taken into account when figuring the hydration rate for a sample. If these factors are known, the accuracy of the date will be increased substantially.

Thus, it has become apparent that the obsidian hydration rate constant, k, is source dependent. Values of k as given by Friedman, Smith, and Clark in units of microns^2/thousand years are as follows: Egypt, trachytic, 14; coastal Equador, 11; Egypt, rhyolitic, 8.1; Temperate (no. 1), 6.5; Temperate (no. 2), 4.5; Subarctic, 0.82; Arctic, 0.36.[31] Friedman and Long calculated a series of k values for obsidian in 1970.[32] From these figures, a minimum hydration age can be estimated. The error equation can be shown to be

$$\Delta t/t = 2\Delta x/kt^{1/2} \tag{4.4}$$

The quantity Δx is 0.07 μm as given by Michels,[33] although Lee et al.[34] suggested this may reach 0.02 μm.[35] This is the greatest accuracy with which the layer of hydrated material can be measured. It can be seen from the formula that the error in time is minimum with larger values of k. Using the trachytic obsidian figure of $k^2 = 14$, the ratio $\Delta t/t$ becomes approximately $0.04/t^{1/2}$ where $t^{1/2}$ is in thousands of years. Suppose that $t^{1/2}$ is 1 (i.e., 1,000 years,) then the error expected from obsidian dating under the best of circumstances is about 4%. If $t^{1/2}$ is 500 years, this error becomes about 6–7%. If $t^{1/2}$ is 4,000 years, the error expected from measurement decreases to about 2%.

Since the 1970s, obsidian hydration rate constants have been shown to be unique to each type of obsidian, allowing the establishment of a corpus of data for these sources.[36] Establishing the relationship of a specimen to its source chemical characterization is done. The final factor to be considered is the thickness that the material will allow, which determines the upper limit for a hydrated specimen's date. This appears to be roughly 50 μm in most obsidians. If the hydration rate is 1 μm / 1,000 years, then it is easy enough to determine the upper bound for such a sample. In deeper deposits, with lower temperatures, the rate will be less and will allow older ages to be determined.

"Varnish" and Cation-Ratio Dating Technique

Basic Principles

"Varnish" or patination is a glazelike coating on stone and glass that can be used to determine relative antiquity. It can form as follows:

1. silica leaching and redeposition on surface as a chertlike silcrete;
2. bleaching: leaching of silica and its replacement by lime salts in pits and cracks on the rock surface;
3. adsorption of limonite clays and salts on surface;
4. accretion of layers of clays, manganese, iron oxides, and trace elements, called "desert varnish";
5. removal of ion salts (arid environments), forming a "crust."

The process is slow; varnish requires hundreds to thousands of years to form.

The differences in cave painting and carvings should be obvious. Carvings or etchings are just physical alterations of preexisting rock faces. How is one to determine the age of the surface exposed by the carving process rather than the age of the rock itself? Previously, dating could only be done relatively by stylistic criteria and by radiocarbon age determination of associated artifacts. These methods were, for the most part, ineffective. The accumulation of rock varnish has long been hoped to provide some mode of absolute dating. It has been known that engravings in arid and semiarid regions will accrue a layer of varnish on their surfaces within 100 years of their production. This varnish is an accretion of clay minerals, manganese, iron oxides, and many trace elements caused by manganese-oxidizing bacteria.[37]

Rock varnish can be evaluated by chemical analysis. The trace elements potassium (K^+), calcium (Ca^{2+}), and titanium (Ti^{4+}) are leached out of the varnish and their ratios ($[K + Ca]/Ti$) decrease systematically over time. The cation ratios can be calibrated by using radiocarbon dating, in particular the AMS method.[38] Additionally, the uranium-series dating of varnish holds promise.[39] Cation-ratio dating, a procedure proposed by Ronald Dorn in 1983,[40] measures this decrease of the ratio of ($K^+ + Ca^{2+}$)/Ti^{4+}, assumed to decrease at a steady rate in a given region. The ratio is measured then compared to a AMS dating of the organic material (if present) in the varnish[41] to provide and age for an engraving.[42]

Francis, Loendorf, and Dorn have used cation ratios in concert with radiocarbon calibration to derive a cation-leaching curve (CLC) for rock varnish on petroglyphs in the Bighorn area (Wyoming and Montana). The curve is shown in figure 4.5. The dating equation to fit this curve is the least-squares regression:

$$\text{varnish cation ratio} = 15.42 - 2.26 \log_{10} (\text{age}) \qquad (4.5)$$

This equation was found reliable for rock varnish on petroglyphs older than 1,000 years. Younger specimens were found to be problematic.

Problems in Cation-Ratio Dating

In 1990 Julie Francis collected samples for both AMS and cation-ratio dating from petroglyphs at sites in Bighorn Basin in Wyoming and Montana. Details on the dating methods used can be found in Francis et al.[43] This was landmark research in that it gave new insight into the effectiveness of these two procedures. Cation-ratio dating is a very delicate proce-

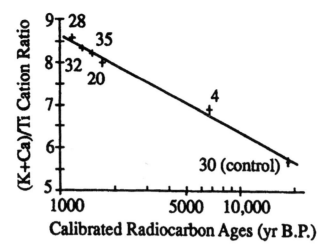

Figure 4.5 Cation-leaching data and curve for Bighorn area (Wyoming and Montana) petroglyphs (adapted from Francis et al., 1993).

dure subject to many environmental conditions. Francis et al. found that chalking was the
biggest problem, because it considerably damages the varnish. Further, the petroglyphs in
Bighorn Basin have been handled previously by other researchers. Latex molds and cloth
rubbings have been taken of many of the glyphs sampled by Francis et al. which rendered
them undatable.

Cation-ratio dating has been critiqued on many of its aspects, from the underlying theory to the procedure used to calibrate its results. S. L. Reneau and C. D. Harrington assert
that the decline in potassium/calcium cations is more a function of weathering of the underlying rock and that the rate is related to the thickness of the varnish and thus diminishes
over time.[44] In addition, certain environmental occurrences, like forest fires, can reset the

rock varnish.[45] Varnish formation differs over the area of the rock. The accumulation of a single layer can take anywhere from 100 to 1,000 years. While it is easy to establish a minimum age, this wide range puts limits on the accuracy of the chronology of the work.[46] Any differences in weathering of the rock face would make cation-ratio dates meaningless. For etchings older than 1,000 years Francis et al. have found a strong correlation between cation-ratio dates and AMS dates of organic matter on the surface of the rock. Their oldest date went back to 18,590 BP and the relation was still good.[47] However, for etchings less than 1,000 years the relation skews badly. Francis et al. point out that this may be a result of historical meddling or local environmental conditions.

Another assumption made by cation-ratio dating is that all the varnish accumulated on the rock surface before production of the etching was removed during the process. This is difficult to deal with, not knowing the methodologies used by ancient rock artists. These techniques are sure to vary from culture to culture and over time. The end result would be the dating of an older exposed surface and not the etching itself.[48]

Karen Harry presents data on rock varnish cation ratios for artifacts from an archaeological site in the Mohave Desert. In her study she raises questions concerning the use of the cation technique on individual lithic artifacts, questioning the rate and regularity of the varnishing process. In particular, it was observed that varnish formation is more or less responsive to the rock substrate. A microcrystalline flint forms less varnish than a coarser-grained material.[49] This was observed earlier by Dorn, Dorn, and Whitley, and Dorn and Krinsley.[50] Other factors affecting the formation of rock varnish on artifacts as opposed to natural outcrops are eolian abrasion and burial. Wind abrasion can reduce or retard varnish formation, and burial/exposure cycles can interrupt the rate at which varnish forms. Given these cautionary notes, it is not surprising that any accurate age based on cation ratios of varnish may well be only a minimum age estimate of the artifact or exposure surface.[51] Clearly, the different time of onset for varnish formation reduces our ability to determine the date of manufacture of a stone artifact or feature, as well as the exposure face of an outcrop.

5

Radioactive Methods

R adioactive decay can best be explained by accepted models of atomic or microscopic structure. By the mid 20th century most people understood that the atom was the smallest particle to which a homogeneous macroscopic sample could be subdivided and retain the physical characteristics of the original sample. The atom was called a microscopic particle because its dimensions were on the order of 10^{-8} cm. It is now known as *mesoscopic,* and the *microscopic* world is the subnuclear, or less than 10^{-13} cm, the distance across most nuclei. Physicists develop their intuitive feel for nature using such characteristic distances as 10^{-8} cm, called an *angstrom* (abbreviated Å). If you lived at the mesoscopic or microscopic level, you would choose this distance unit because it would be convenient. Today these levels are discussed as the *nanoworld* or *nanostructure* level.[1] We shall refer to the microscopic/nanoscopic under the general term *microworld.*

The energy required to separate two atoms coupled together is of the order of 0.1 *electron volts.* This is an energy unit characteristic of atoms, abbreviated as eV. In the macroscopic world, whose dimensions are most familiar to us, characteristic distances are of the order of 1 cm, which is 100 million times that of the microworld. The macroworld energy unit we are most familiar with is the food *calorie,* which is 1,000 heat calories, which is, in turn, about 4,180 joules. In the microworld, the electron volt is 1.6×10^{-19} joules.

In the microworld, atoms consist of nuclei, which are about 10^{-13} cm across and which contain almost all of the mass of the atom. The nuclei, in turn, consist of protons and neutrons. These are two of the four elementary particles with which we will be concerned. Nuclei can be thought of as built up of *nucleons* or baryons, members of the larger class of elementary particles, hadrons. *Baryons* are composed of even smaller particles known as *quarks.* Particles like electrons, muons, and neutrinos are known as *leptons.* Neutrons and protons have almost the same mass, about 1.6×10^{-24} g, but a neutron is without charge and every proton has a charge of $+1.6 \times 10^{-19}$ coulombs. This is called the fundamental charge, symbolized by e. The proton has a charge of e^+. The number of protons in a nucleus is Z, the atomic number. The number of nucleons (total number of neutrons and protons) is A, the mass number. A little arithmetic gives you the number of neutrons in the nucleus: $A - Z$ = number of neutrons. Note there is no special symbol for the number of neutrons in the nucleus. From this picture of nature we can obtain an idea of the number of atoms in a macroscopic sample. A sample of matter 1 cm $\times$ 1 cm $\times$ 1 cm in the form of a cube would have a volume of 1 cm^3. Now multiply together the number of atoms in a side. If the atom is about 10^{-8} cm "long," there are 10^8 atoms in a side, or $10^8 \times 10^8 \times 10^8 = 10^{24}$ atoms in the sample. Since no atom is exactly 10^{-8} cm across or square, a very exact calculation would yield between 10^{22} and 10^{24} atoms, so we can settle on an average of 10^{23} atoms and feel reasonably comfortable.

If each of these 10^{23} atoms had about 10^{-16} coulombs of charge (or even a small fraction of that amount), the charge of the macroscopic sample would be of the order of 107 coulombs. This is an enormous charge. It would be virtually impossible to keep two samples of this size in the same county because of electrostatic repulsion. From this we must conclude that the atom is uncharged. If this is the case, every atom must have Z charges of $-e$ in addition to the Z charges of $+e$. Likewise, surrounding an atom having Z protons in its nucleus are Z electrons, but they are not in one place like the protons and neutrons in the nucleus. They circle in orbits about the atom, with differing numbers of electrons in each orbit. The average diameter of any particular orbit is about an angstrom. Atoms on the earth have Z ranging from 1 for hydrogen to 94 for plutonium. As a rule of thumb, the number of neutrons is about equal to the number of protons, being less than half for low-Z atoms and slightly higher for high-Z atoms. This gives rise to descriptions, by nuclear scientists, of atoms either being "neutron deficient" or having a "neutron excess." The average distance across the atoms ranges about a factor of 10 between hydrogen and plutonium.

The volume of the nucleus of plutonium is about 94 times the volume of the nucleus of hydrogen. However, the size of an object goes up only as the cube root of the volume, so that the distance across the plutonium atom nucleus is no more than 4–5 times the distance across the hydrogen atom nucleus. The distance across a nucleus is only 1/10,000 of the distance across the atom. Electrons have masses about 1/1800 the mass of a nucleon (9.1 $\times$ 10^{-28} g). Therefore, the mass of the atom is concentrated in a very small nuclear space and is surrounded by a large volume of electrons occupying mostly empty space. When particles are emitted by a radioactive nucleus, they will travel mainly in this "vacuum," and since this space has the atomic electrons in it, interactions of the emitted particles will be with the electrons.

Let us briefly review the structure of matter. In solid, or "condensed," matter, atoms array themselves almost inevitably in orderly rows. In many materials, these rows extend such long distances that they affect the appearance of the material and form a crystal. Most natural materials comprise agglomerates of small crystals oriented in all directions. Atoms are held together by a balancing of cross-forces. The attractive force of the positive charge of the nucleus (Z) and the negative charge of the electrons of neighboring atoms is balanced by the electrons of the two atoms, which repel each other and keep the material from collapsing. Therefore, in solid materials, a delicate balance is struck between attractive and repulsive forces. The distribution of electrons in their orbits about the nucleus determines the optimum spot for the neighboring atoms to be, and so the type of array in which the atom finds itself depends on the electron structure. In crystalline materials, which includes almost every mineral, this atomic structure is rigidly regular. As we shall see in later chapters, this order allows us to instrumentally and chemically characterize and identify these structures

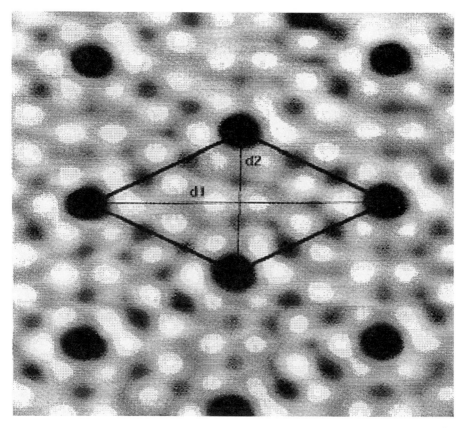

Figure 5.1 Scanning tunneling microscope image of silicon (111). Image size is 78 × 78 Å; depth of image is about 2 Å. The unit cell is outlined (modified from Giessibi, 1995).

in archaeologically interesting materials. When you look down on any of these structures, as with a scanning tunneling electron microscope, you are really peering at the empty space in the atoms defined by the electron "cloud" (figure 5.1). The small, by comparison, nucleus is somewhere within the cloud near its center. A particle leaving the nucleus of one atom will normally travel a considerable distance in the material before striking another atom's nucleus.

Nuclei and Isotopes

The number of protons, Z, in the atom, determines what the element is. This number also determines the number of electrons the atom has—just enough to be neutral. When two atoms are brought together to form a molecule, the electrons interact first and determine the chemistry of the atom or element. It is important to remember that there are no electrons in the nucleus. Each element has an accepted abbreviation—H for hydrogen, He for helium, Li for lithium, and so on—and Z is commonly shown as a subscript to the left of the abbrevation. Thus, hydrogen is $_1$H, helium is $_2$He, and lithium is $_3$Li. Very early in this game, it was realized that the rate of chemical reactions was not the same for all the hydrogen atoms in any naturally occurring sample. Since rates of chemical reactions depend on the masses of the species, it was apparent that the average collection of hydrogen atoms was a mixture of atoms that have different masses. More detailed investigation indicated that the mass of one of the species was almost exactly twice the mass of the other. This was an early indication, although there were others, that some hydrogen atoms contained neutrons in their nuclei. Another, and perhaps more convincing indication, was that although the mass of an atom should be proportional to its atomic number, so that masses of different atoms should be whole-number multiples of the mass of the most simple atom (hydrogen, with one proton), they were not. A hydrogen atom with a neutron has $A = 2$ nucleons (1 neutron, 1 proton); it is called deuterium and symbolized as $_1^2$H (mass number is written as a superscript to the left of the element abbrevation). This isotope, like most isotopes, is not radioactive—that is, it does not spontaneously change into another atom or into another isotope of hydrogen. There is also an isotope of hydrogen with two neutrons. It is called tritium, $_1^3$H, and does spontaneously change and is, thus, radioactive. Isotopes, then, have the same atomic number, Z, but differ in their mass number, A. Eighty-one elements have stable isotopes, some more than others. For example, in any representative sample of copper, one would find 69% $_{29}^{63}$Cu and 31% of $_{29}^{65}$Cu. Calcium has six stable isotopes. The distribution of isotopes tells us something about the genesis of the universe because it is now apparent that many isotopes were produced through the decay of radioactive elements constructed in stars—geocosmochemistry.

Radioactivity

In 1896 Becquerel discovered that salts of uranium left in contact with photographic plates in a dark room caused exposure of the plates. Roentgen had observed similar behavior with x rays. Most uranium is $_{92}^{238}$U. The Curies found that polonium ($_{84}^{210}$Po) and radium ($_{88}^{222}$Ra) exhibited the same behavior. These elements were called radioactive by someone in this period, although we are not sure who or why.

Gamma Rays

Among the particles emitted by Becquerel's uranium minerals were high-energy x rays. These x rays, emitted by a nucleus, are now called gamma rays. They have had other names, such as "nuclear rays" and, more colorful and sinister, "death rays," so called by early science fiction writers. These writers were not exaggerating: extended exposure to these radiations could and did kill. Madame Marie Curie was one of the first casualties.

Gamma rays have no mass and no charge and are in a sense raw electromagnetic energy which interacts relatively weakly with electrons in matter. Because of the weak interaction and the small probability of interacting with electrons, gamma and x rays will penetrate several centimeters of material. They have energies characteristic between 0.1 and 4 or 5 MeV (million electron volts); several electron volts are required to remove an electron from an atom. Gamma is simply a high energy x ray going through matter that will interact with electrons in one of three ways:

1. *Photoelectric effect:* Albert Einstein won his only Nobel prize (1905) for discovering and describing this type of interaction. The gamma ray, or high-powered x ray, trans-

fers all of its energy to an electron, lifting the electron from its atom. This requires very little energy, ca. 5 eV. The remaining several hundred thousand electron volts of energy from the gamma ray are converted to motion, or kinetic energy. Now this very fast electron goes through matter but loses energy by collision with other electrons.

2. *Compton effect:* Another Nobel Prize was awarded for the discovery of this effect— this time to Arthur Compton (1927). In this type of collision, the gamma ray loses only a fraction, possibly about one-fourth to one-half of its energy, to an electron and is reflected, now with lower energy, to react in a subsequent Compton event or in a photoelectric event.

3. *Pair production:* Very high energy gamma rays can simply disappear and produce a pair of particles simultaneously (this phenomenon clearly shows the equivalence of matter and energy as Einstein described in $E = mc^2$). One particle is a normal electron, and the other is a positive electron. To do this, the gamma ray produces two masses, one for each electron, each having an energy ($E = mc^2$) of 0.511 MeV. It also provides the kinetic energy with which the electron and the positron (the positive electron) leave the scene. Therefore, the parent gamma ray must have over 1 MeV in energy.

Beta Particles

In addition to gamma rays, the naturally occurring radioactive sources which the Curies and Becquerel had discovered also produced beta particles, electrons emitted by the nucleus. Identical radioactive nuclei will each emit a beta of different energy, ranging up to the maximum value of 1 MeV. The maximum value of the beta energy is characteristic of the particular isotope emitting the beta. Gamma rays are emitted monoenergetically (i.e., with discrete energies). Like betas, gammas are characteristic of particular isotopes and form the basis for neutron activation analysis. If one nucleus in a collection of identical radioactive nuclei, a particular isotope, emits three gamma rays having energies of, say, 0.1 MeV, 0.44 MeV, and 0.54 MeV, every nucleus in that collection will emit the same gammas. A drastic change takes place in the nucleus when it emits a beta particle.

Let us use, as an example, cosmogenic $^{14}_{6}C$, which is an isotope of carbon found in samples that have been incorporated into biological systems. This isotope is a beta emitter. The nucleus loses a negative charge when the beta particle emerges, although it does not lose any mass. The number of nucleons remains the same, but since the original nucleus, with a charge of +6e, lost a negative charge, the charge of the new nucleus, with an added $-e$, must equal +7e. The element with a charge of 7e and a Z of 7 is nitrogen. Note that the mass of the atom, which is essentially the mass of the nucleus, does not change. Again, the total number of nucleons does not change. We write this reaction as $^{14}_{6}C \rightarrow e^- + {}^{14}_{7}N$.

Having the same number of nucleons in the carbon and nitrogen nuclei does not mean they have the same kinds of nucleons. Here is where the drastic change occurred. The emitted beta (e^-) leaves the nucleus, catapulted into the empty space of the atom with an energy which is enormously large with respect to atomic-scale energy. The new nucleus has a charge of $7e^+$ since the electrons cannot balance the charge deficit. The number of nucleons remained the same, although the number of protons increased and the number of neutrons decreased. The new nucleus is now $^{14}_{7}N$. The upper number, A, is the number of nucleons and thus remains the same. Here is a case of spontaneous transmutation. One element, carbon, changed into another, nitrogen (what the alchemists always predicted but not in the way they hoped).

Had this carbon been in one of the long, flimsy, and complicated DNA molecules, which control genetic characteristics, the molecule might have been so weakened by the change of the chemical element that it would have spontaneously separated at this point. Mutations in genes can occur due to radiation damage, including via this process if the carbon 14 atom is substituted into a base or sugar that makes up the DNA macromolecule. Emission of the beta particle and the change of a carbon into a nitrogen atom can result in the deletion of a critical component of the gene; such an alteration may be deleterious. The new nucleus creates a new atom because the total charge of the atom *must* be zero. The number of orbital electrons also adjusts, picking up one of the free electrons that are always hanging around in solids. The important thing to note is that the element itself has changed both physically and chemically.

Since there are no electrons in the nucleus, where did the electron that emerged from the radioactive nucleus as a beta particle come from? Physicists think that it is generated at the instant of emission, that energy is converted to mass at this instant (remember the pair formation from the high-energy gamma). There is no reason to believe that the electron was hidden in the nucleus in some form before it was emitted. In fact, there is every theoretical reason to think that it was not. This electron has no relation to the orbital electrons. There is no indication that any beta particle was ever an atomic electron. Perhaps, it is best to characterize this "electron" as a particle necessary to make sense of the structure of neutrons. By this reasoning the neutron is then a proton plus an "electron" of dubious and somewhat mysterious origin.

A radioactive isotope can be identified by its emanations in an activity like instrumental neutron activation analysis. This measurement would yield 1–10 gamma rays, depending on the radioactive isotope, with specific energies depending on the isotope. From the number and energy of the gammas, the specific isotope can usually be identified. The half-life (which we discuss in the next section) is also used in this identification. To the question whether beta particles can be used, the answer is generally no. A beta particle emitted by a particular nucleus can have any energy up to some maximum, so only the maximum energy which the isotope can possibly emit is characteristic of the isotope. Further, electrons interact strongly with other electrons. You will remember that the space through which the emitted beta particle travels is full of electrons. The atom does not vary much in size throughout the periodic table, but the number of atomic electrons in the atomic volume does. Thus, for most high-Z materials, the emitted beta has a high probability of encountering another electron on its way out. It can be shown in physics that if a particle collides with a much heavier particle or a much lighter particle, it will not lose much energy. But if it collides with a particle about its own size, it is apt to lose as much as half the energy it has (recall the photoelectric effect). This figure is correct only to order of magnitude, but the beta, having a reasonable probability of losing energy in any solid (and a high probability of losing energy in a solid of high Z), will have its energy greatly modified on its way out of a solid material. Therefore, for these reasons, the energy profile of betas is not a good indicator to identify a particular radioactive isotope.

Dating Techniques Using Radioactivity: First Principles

In the transmutation of one isotope into another, the original isotope is called the parent and the resulting isotope the daughter. The transmutation takes place at a rate given by

$$N = N_o \exp(-\lambda t) \tag{5.1}$$

where λ is the decay constant and is related to the half-life, $t^{1/2}$ (the time needed for any amount of radioelement to decrease by one-half) by

$$\lambda = 0.693/T_{1/2}, \tag{5.2}$$

N_o is the original number of parents, N is the number of parents left undecayed, and time, t, is in the same units as the decay constant. The equation could be expressed as

$$N = N_o \exp(-0.693t/T_{1/2}) \tag{5.3}$$

where t and $T_{1/2}$ are in the same time units. The exponential notation means 2.73 (or e, the base of natural logarithms) raised to the power in the parentheses. Then, $N_o - N$ is the number of daughters created, called N_D. It is easy to see that

$$N_D = N_o - N = N_o [(1 - \exp(0.693t/T_{1/2})] \tag{5.4}$$

While it is not too important to be able to do the mathematics, it is good to be able to inspect an equation and tell what you can get from it. Someone can always solve the equation if all the information necessary is available. In any equation, if you have quantitative data for all the letters but one, you can solve the equation. What we wish to know for dating purposes is t in equation 5.4 (or 5.3). Age dating requires that we choose a particular radioactive isotope and obtain $T_{1/2}$ from tables. If there is some way we can measure the number of parents which were originally present, N_o, or the number of parents which are yet present, N, or the number of daughters, N_D, we will have all quantitative data for every number in equations 5.3 and 5.4 and can solve for the time t. Table 5.1 lists the battery of radioisotopes commonly used in geochronology today, with the list growing annually. Table 5.2 lists representative archaeological dates obtained by use of some of these isotopes.

Early Geological Dating

The first significant triumph in radioactive dating involved the determination of the age of the earth itself. This we shall recapitulate in the next few pages. The success was important in the development of radiogenic age determination, although from the standpoint of archaeological age determination, such a great age—billions of years—has little or no relevance. But as we so often hear in other matters seemingly tangential, "it is the principle of the thing" that is important. The principle *is* important. From Charles Lyell's day, geologists have speculated on the antiquity and origin of our planet. Likewise, paleontologists, naturalists, and, later, archaeologists speculated on the origin and antiquity of living organisms from bacteria to humans. The dating techniques based on radiogenic lead have allowed science to answer the full spectrum of questions about age, from the age of the earth to the antiquity of early humanity.

Alpha Particles

Alpha particles may also be emitted from isotopes. Alpha particles are helium nuclei (helium atoms minus the two orbital electrons). Helium nuclei quickly pick up electrons as they

Table 5.1 Radiometric Systems Used in Geochronology and Cosmochronology

Parent isotope	Half-life (yr)	Decay constant (yr^{-1})	Decay mode[a]	Daughter product(s)
^{1}H	12.33	5.62×10^{-2}	β^-	^{1}He
^{10}Be	1.6×10^6	4.33×10^{-7}	β^-	^{10}B
^{14}C	5730[b]	1.210×10^{-4}	β^-	^{14}N
^{26}Al	7.2×10^5	9.6×10^{-7}	β^-	^{26}Mg
^{40}K	1.25×10^9	4.962×10^{-10}	β^-	^{40}Ca (89.5%)
		0.581×10^{-10}	EC	^{20}Ar (10.5%)
^{87}Rb	4.88×10^{10}	1.42×10^{-11}	β^-	^{87}Sr
^{107}Pd	6.5×10^6	1.07×10^{-7}	β^-	^{107}Ag
^{129}I	1.6×10^7	4.33×10^{-8}	β^-	^{129}Xa
^{147}Sm	1.06×10^{11}	6.54×10^{-12}	α	^{141}Nd
^{176}Lu	3.53×10^{10}	1.96×10^{-11}	β^-	^{176}Hf
^{187}Re	4.3×10^{10}	1.61×10^{-11}	β^-	^{187}Os
^{210}Pb	22.26	3.11×10^{-2}	β^-	^{210}Bi
^{226}Ra	1.62×10^3	4.27×10^{-4}	α	^{222}Rn
^{230}Th	7.52×10^4	9.22×10^{-6}	α	^{226}Ra
^{231}Pa	3.25×10^4	2.134×10^{-5}	α	^{227}Ac
^{232}Th	1.40×10^{10}	4.9475×10^{-11}	α, β^-	^{208}Pb + 6^4He
^{234}U	2.45×10^5	2.794×10^{-6}	α	^{230}Th
^{235}U	7.04×10^8	9.8485×10^{-10}	α, β^-	^{207}Pb + 7^4He
^{238}U	4.47×10^9	1.5513×10^{-10}	α, β^-	^{206}Pb + 6^4He
		8.46×10^{-17}[c]	SF	Various
^{244}Pu	8.2×10^7	8.47×10^{-9}	α	^{232}Th + 3^4He
		1.06×10^{-11}	SF	Various

[a]Decay mode includes loss of electron (beta particle, β^-), positron (β^+), or alpha particle (α); electron capture (EC); or spontaneous fission (SF).
[b]Most accurate value. However, by convention most ^{14}C dating labs report ages based on a half-life of 5568 yr.
[c]Many workers use 6.85×10^{-18} for this decay constant.

move through a solid material. When they do, they become atoms and collide with the other atoms in the solid since all atoms are about the same size. The penetration of helium atoms in a solid material is only a few tenths of a millimeter. Beta particles can penetrate up to perhaps a centimeter or so. Gamma rays can penetrate to 4 or 5 cm. When a nucleus emits an alpha, it loses two protons and two neutrons. For example, $^{237}_{93}$Np $\rightarrow$ $^4_2$He + $^{233}_{91}$Pa. The element Np is neptunium and the element Pa is protactinium.

In the case of $^{238}_{92}$U chain decay, eight alpha particles are emitted. In the case of the final daughter of $^{235}_{92}$U chain decay, seven alphas are emitted. We talk about matter as being composed of long planes of atoms about 10^{-8} cm apart. The alpha particle is about 10^{-8} cm in diameter and it cannot easily diffuse through the atomic planes. Consequently, the alphas become trapped. The alpha particle is a helium nucleus and picks up four stray electrons to become a helium atom. Helium is an inert gas and does not easily form chemical compounds. If the sample in which it is generated is not too porous, it will be retained in the sample for long periods of time. For dating purposes, the rock is crushed and the helium gas is pumped off. The number of helium atoms can be calculated and will be seven or eight times the number of nuclei which have decayed depending on whether uranium 235 or uranium 238 is the parent nucleus. In samples where this technique is used, both uranium isotopes are present, so corrections have to be made on the basis of half-life for how much of

Table 5.2 Various Dates Obtained with Commonly Used Radioactive Techniques

Site and material dated	Method	Age (BP)
Iraq, skeleton of Meskalamdug (BM-64)	^{14}C	3920 ± 150 (2875–2095 BC)[a]
Iraq, skeleton of Queen Puabi (BM-76)	^{14}C	3990 ± 150 (2905–2160 BC)[a]
Egypt, Tarkhan, reign of King Djet	^{14}C	4150 ± 110 (3020–2420 BC)[a]
Egypt, Tarkhan, reign of King Djet	^{14}C	4160 ± 1100 (3030–2425 BC)[a]
Jericho, PPNB (BM-1320)	^{14}C	8540 ± 65
Jericho, PPNA (BM-252)	^{14}C	9582 ± 89
Jericho, Natufian (BM-1407)	^{14}C	11,090 ± 90
Stellmoor, Ahrensburgian (Y-159-2)	^{14}C	10,320 ± 250
Gönnersdorf, Magdalenian (Ly-768)	^{14}C	12,380 ± 230
Meadowcroft, Palaeo Indian (SI-2354)	^{14}C	16,175 ± 975
Meadowcroft, lowest occupation (SI-2060)	^{14}C	19,600 ± 2400
Laugerie-Haute, Middle Solutrean (GrN-4495)	^{14}C	19,740 ± 140
Abri Pataud, Perigordian IV (GrN-4477)	^{14}C	26,600 ± 200
Abri Pataud, early Aurignacian (GrN-4507)	^{14}C	34,250 ± 675
Lake Mungo 2, NSW (ANU-331)	^{14}C	32,750 ± 1250
La Quina, late Mousterian (GrN-2526)	^{14}C	35,250 ± 530
Border Cave, early LSA (Pta-446)	^{14}C	37,500 ± 1200
Istállóskö, early "Aurignacian"	^{14}C	39.800 ± 900
Jaua Fteah, Late Mousterian (GrN-2564)	^{14}C	43,400 ± 1300
Nahal Zin, Middle–Upper Paleolithic transition	^{14}C	45,000 ± 2400
	Uranium	46,500 ± 2900
Border Cave, MSA (Pta-1274)	^{14}C	47,200 ± 4200 −2750
Shanidar, level D, Mousterian (GrN-1495)	^{14}C	50,600 ± 3000
Tata, Hungary, Mousterian	Uranium	$101 ± 12 \times 10^3$
Tata, Hungary, Mousterian	Uranium	$109 ± 15 \times 10^3$
Pech de l'Azé, France	Uranium	$123 ± 15 \times 10^3$
Last Interglacial	Uranium (many dates)	ca. $127 ± 5 \times 10^3$
La Chaise, France Bed 11	Uranium	$145 ± 16 \times 10^3$
Zuttiyeh Cave, Jabrudian	Uranium	ca. 148×10^3
Gademotta, Ethiopia, early MSA	K/Ar	$149 ± 13 \times 10^3$
Gademotta, Ethiopia, early MSA	K/Ar	$181 ± 6 \times 10^3$
Kapthurin, Kenya, over Acheulean	K/Ar	$230 ± 8 \times 10^3$
La Cotte, Jersey, basal (OxTL222)	Thermoluminescence	$2388 ± 35 \times 10^3$
Tautavel, Arago, France	Thermoluminescence	ca. 350×10^3
Olorgesailie, Acheulean	K/Ar	$425 ± 9 \times 10^3$
Olorgesailie, Acheulean	K/Ar	ca. 486×10^3
Brunhes-Matuyama Reversal	K/Ar	ca. $0734 ± 10 \times 10^6$
Isernia, Italy	K/Ar	$0.73 ± 0.04 \times 10^6$
Kariandusi, Kenya	K/Ar	ca. $0.95-1.1 \times 10^6$
Karari Tuff, E. Turkana	K/Ar	$1.32 ± 0.10 \times 10^6$
Okote Tuf, E. Turkana	K/Ar	$1.56 ± 0.02 \times 10^6$
Chesowanja Basalt	K/Ar	$1.42 ± 0.07 \times 10^6$
Pliocene-Pleistocene boundary	K/Ar	ca. 1.6×10^6
KBS Tuff, E. Turkana	K/Ar	$2.8 ± 0.1 \times 10^6$
Olduvai Bed I (Tuff Ib)	K/Ar	$1.79 ± 0.03 \times 10^6$
Olduvai Bed (Basalt)	K/Ar	$1.96 ± 0.09 \times 10^6$
Omo, Shungura (Tuff F)	K/Ar	$2.06 ± 0.10 \times 10^6$

[a]Calibrated ^{14}C dates with 95% confidence range.

each was initially present. Disregarding this necessity and assuming that we are dealing with the uranium 238 isotope alone, the measured number of helium atoms would be 8 times the number of daughters. Therefore, the number of daughter nuclei, N_D, can be determined. From a measurement of the counting rate, R, and the use of the formula

$$R = \lambda N \tag{5.5}$$

and knowledge of the decay constant (through a knowledge of the half-life), the value N can be determined. We now have N_D, N, and $T_{1/2}$ (or λ). Since $N + N_D = N_o$ we can calculate N_o and use equation 5.4 to determine t. It is obvious that this technique is only good from the moment the rock has solidified. Helium released before solidification escapes. The range of helium (or alpha particle) movement in rocks is very small—much, much less than the dimensions of the rocks (on the order of 10^{-6} m), so not much helium escapes when the alpha particles are emitted unless the rock is very fine grained. However, over long periods of time, helium does have a tendency to diffuse, or gently drift, out of the rock, and errors are generated. The technique has been used to obtain ages of 2×10^9 years on granite samples.

The term *radiogenic lead* will be used for lead which occurs because it is a daughter of a radioactive chain. There is no way, however, of inspecting a particular nucleus and telling if it is radiogenic in origin. The naturally occurring isotopes of lead are ^{204}Pb, ^{206}Pb, ^{207}Pb, and ^{208}Pb. Of these, only ^{204}Pb *cannot* be radiogenic (i.e., could not be produced by a radioactive chain). The relative abundances of these isotopes in naturally occurring samples is 1.3%, 26%, 21%, and 52%, respectively. If there is no ^{204}Pb in a sample, it is assumed that all the lead occurring is radiogenic in origin. In any given sample, the amount of ^{206}Pb is measured and compared to the amount of ^{238}U as determined by counting rate. The same can be carried out for ^{208}Pb and ^{232}Th. The number of daughters would then be known (the number of lead isotopes), and the number of parents can be found by adding the N from the counting rate to the number of daughters (producing N_o), and equation 5.3 or 5.4 could be used to determine the time. Errors resulting from this technique occur because of the possibility of leaching or enriching selectively either the uranium or the lead from the sample or the loss of daughter elements in the decay chain. With some knowledge of geology, samples can be selected where the probability of these processes having occurred is small or at least accounted for by the analysis.

The population of the lead daughters can be determined by heavy-ion mass-spectrometric techniques. The samples are vaporized, charged, accelerated in a high vacuum electric field, and bent by a magnetic field. Different isotopes are bent in different amounts depending on their mass, and the desired isotopes can be selected out and their total charge measured to determine their total population. This type of measurement is routine and can be carried out for both the parent and the daughter if desired. The information obtained would be N and N_D. The N/N_D ratios are used for geochronology and provenance determination (see chapters 11 and 14).

U/Pb, Th/Pb, and Pb/Pb Ratios and Geochronology

Of the uranium and thorium chains geochronology has most used the two principal techniques involving (1) $^{238}U/^{206}Pb$ and $^{232}Th/^{208}Pb$ and (2) $^{206}Pb/^{207}Pb$. Both these techniques use mass-spectrographic analysis of Pb species. If ^{204}Pb is found, the correction for

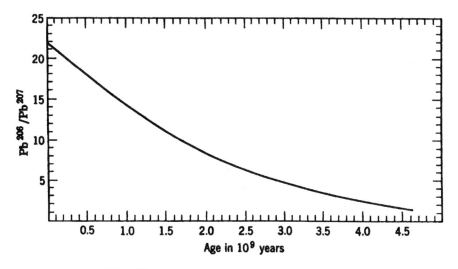

Figure 5.2 Ratio of ^{206}Pb/^{207}Pb vs. time.

contaminant ^{206}Pb/^{208}Pb must be made. ^{238}U/^{206}Pb ratios have proven more reliable than ^{232}Th/^{208}Pb ratios. Ages greater than 3×10^9 years are readily determined with lead ratios. The oldest rocks and meteorites have been dated to $4.5 \pm 0.1 \times 10^9$ years by lead ratios.

A. O. Nier was the first to examine ratios of Pb with the mass spectrograph. He noted that samples with more ^{206}Pb also contained more ^{207}Pb and ^{208}Pb. The oldest samples had the highest levels of "uncontaminated or primeval" leads. These leads existed when the earth's crust was formed. Joplin (Missouri) galena was used by Nier as well as Great Bear (Canada) lead. In 1938 Nier calculated an age of 2×10^9 to 5×10^8 years for the earth with a ratio of ^{206}Pb/^{207}Pb of 0.14 (figure 5.2).[2] Shortly thereafter, A. Holmes calculated an age for the earth of 3.35×10^9 years. F. G. Houtermans reported an age of the earth of $4.5 \pm 0.3 \times 10^9$ years based on the idea that the lithosphere's primeval lead isotopic composition would be very similar to values seen in the Canyon Diablo iron meteorite.

C. Patterson, using 10 meteorites as samples, in 1956 reported one uniform age from three independent techniques: Rb/Sr, K/Ar, and U/Pb. These results were indeed impressive. Patterson made four assumptions:

1. The meteorites formed at the *same* time as the earth.
2. They existed as *closed* and *isolated* systems.
3. They originally contained Pb of the same isotopic composition.
4. They contain U which has the same isotopic composition as that of the earth.

Using $K = {}^{238}U/{}^{235}U$ (today) $= 137.8$ and $\lambda_1\,({}^{235}U) = 9.72 \times 10^{-10}\,\mathrm{yr}^{-1}$ and $\lambda_2\,({}^{238}U) = 1.537 \times 10^{-10}\,\mathrm{yr}^{-1}$, Patterson[3] obtained an age (t) of $t = 4.55 \pm 0.07 \times 10^9$ yr.

The age of the earth was thus determined through the knowledge that the ratio of the isotopes of lead have been changing because of the constant addition of radiogenic lead. However, we do not know the original ratio of the isotopes of lead. The isotope ratios are taken with respect to ^{204}Pb, which is assumed not to have been radiogenic in origin. Then, the ratio of the lead isotopes is found as a function of the age of the sample. The ratio changes with time. See table 5.3.

Table 5.3 Pb/Pb Ratios

Time ($\times$ 10^6 yr)	Uranium 238: 206/204	Uranium 235: 207/204	Thorium 232: 208/204
0–50	18.5	15.6	38.4
1060	16.7	15.5	36.3
3500	11.4	13.5	31.1
Meteoritic	9.4	10.3	29.2

The meteoritic age may be close to the primordial ratios. By measuring these ratios in granite and galena where other techniques could not be used, the age of the earth was found. In Zircon, an accessory mineral in igneous rocks, lead ratios are most commonly determined today by use of a super high resolution ion–microprobe spectrometer (SHRIMPS), which can simultaneously measure a variety of isotopic ratios.

While lead isotopes are rarely used for archaeological dating, with the exception of early hominid sites, they have found great utility in determining provenance of ores and artifacts as discussed in chapter 14.

Uranium Dating

Ages can be determined by either the amount of decay of an initial excess of uranium daughters, chemically separated from the parent (daughter excess; D.E.), or the growth of an initially deficient daughter (D. D.) growing into equilibrium with the parent.[4] Figure 5.3 shows the three varieties of equilibrium in radioactive decay: transient, secular, and no equilibrium. Dating with the uranium-chain series involves measures of the equilibria in specific daughters, assuming (1) geochemical processes have been such to establish secular equilibrium in the sample. Other key assumptions are that (2) an initial state of material (the degree of excess or deficiency) can be inferred and (3) after deposition, material remained undisturbed from geochemical influences such as diagenesis, recrystallization, and dissolution, while parent and daughter slowly returned to secular equilibrium (i.e., a closed system).

Daughter Deficient Dating Methods

In solution, uranium dissolves and coprecipitates with $CaCO_3$, unlike thorium, which is generally adsorbed on clays. Uranium concentrations in carbonate minerals, as in corals, can reach hundreds of parts per million, though typically 0.1—10 ppm are more common. Over time, thorium daughters grow into equilibrium with their parent uranium. The presence of ^{232}Th is a good indicator that thorium contamination could have occurred and that the dates might be suspect.

U/Th dating is done by alpha-particle counting from ^{230}Th, ^{234}U, and ^{238}U. The activity ratios are calculated with a nomogram to graphically determine age (figure 5.4). With a precision of $\pm$1–5%, the technique is usable for samples less than 500,000 years old.

^{230}Th/^{234}U dating is routinely done for both geology and archaeology. In archaeology

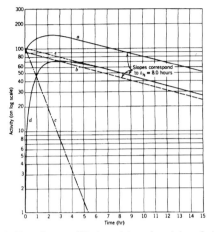

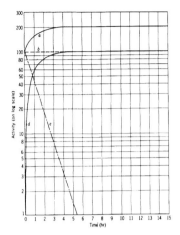

A. Transient equilibrium: (a) total activity of parent (b) activity of half-life = 8 hours c) decay of daughter fraction (d) in-growth of daughter (e) total daughter activity in parent + daughter mixture

B. Secular equilibrium: (a) total initial activity of parent (b) parent activity where half-life = ∞ c) decay of daughter fraction, half-life = 48 mins (d) in-growth of daughter

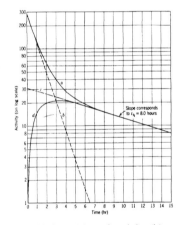

C. No equilibrium: (a) total activity (b) parent activity with half-life = 48 mins c) extrapolation of final decay curve to time zero (d) daughter activity

Figure 5.3 Varieties of radioactive equilibria.

and paleoclimate studies, cave deposits, specifically speleothems, have been the principal material studied.[5] The reliability of ages determined by this technique has been demonstrated by the concordance of the chronologies for both terrestrial and marine carbonates.[6] The calcite of the speleothems form from the $CaCO_3$ in groundwater, where the soluble ura-

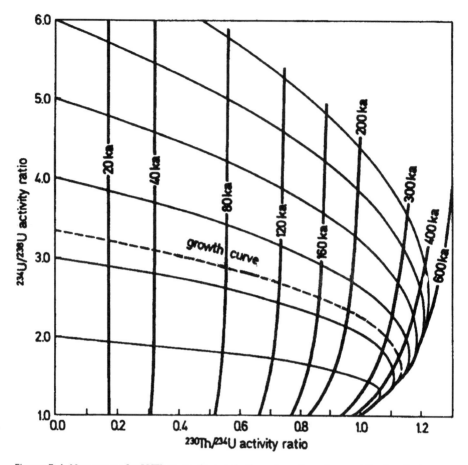

Figure 5.4 Nomogram for U/Th ages: changes in the ratios of uranium and thorium isotopes as a function of age. Growth curve has a constant ratio of 3.55.

nium is entrained. ^{230}Th will build up in the carbonate as secular equilibrium is attained for the activity of ^{238}U and ^{234}U or $\lambda_{238}N_{238} = \lambda_{234}N_{234}$. As in the case with all 4N chains, one must consider the "openness" of the geological system. If the parent, in this case ^{234}U, has been leached and the thorium is stable, any date obtained will be erroneously high. This can be checked by measuring the ^{238}U/^{234}U ratio ($\simeq$ 1.14:1).

Schwarcz and Blackwell discuss archaeological materials which can be dated by other techniques (see also table 5.2).[7] Compared to thermoluminescence and electron spin resonance techniques, environmental radiation has little impact on uranium-series dating.[8] In fact, uranium-series dates provide an independent check on the accuracy of these other techniques.

Bone and teeth are amenable to uranium dating in the sense that, like calcite, they take up the isotopes from groundwater. The amount of uranium can range up to 1,000 ppm, although a concentration of a few parts per million is enough to determine the age of fossil bone or teeth.[9] A major problem, however, with the use of bone or teeth is continued up-take from the burial environment. Uptake can be negligible or significant, depending on whether the system is closed or open. Since both teeth and bone are somewhat porous by nature, they emulate an open system and are of questionable value for uranium dating. Williams and Potts confirm this in the lack of homogeneity of trace elements in bone cross sections. The outer surfaces are elevated in uranium concentration, implying early uptake and fixing in the periphery of small and large bone, with values of 0.5 and 5 mm, respec-tively.[10] However, tooth enamel has been successfully dated by electron spin resonance. It is more dense than the tooth dentine and must approximate a closed system, suggesting that uranium-series dating might also be reliably applied here.[11]

Daughter-Excess Dating Methods

Uranium daughters are deposited in sediments of lakes and seas, then slowly decay away. As with Pb/Pb dating, an initial activity ratio is assumed. Generally, modern ratios are good analogues. These techniques are typically used for determining deposition rates and not for ages as with $^{230}Th/^{232}Th$ (see figure 5.4).

The ratios of $^{230}Th/^{231}Pa$, $^{234}U/^{238}U$, and $^{230}Th/^{234}U$ can be used for age determination, particularly the first, since ionium (^{230}Th) is insoluble. The half-life of $\sim 2.5 \times 10^5$ years is good for ages ranging from 100,000 to 1,000,000 years (figure 5.4). Th–Th dating is criti-cized because each isotope has a different origin.

The ^{210}Pb technique is short range and has obtained good results for low oxygen or anoxic lacustrine sediments. As valuable as ^{230}Th is in the study of deep ocean sedimenta-tion rates, for recent sedimentation, ^{210}Pb is a better gauge of very rapid rates such as those in the North Sea of 2 mm/year—1000 times more rapid than typical deep ocean basins. The $T_{1/2} = 22$ years, so a range of 200–300 years is typical maximum for dates. The decay se-quence is $^{210}_{82}Pb \rightarrow ^{210}_{83}Bi \rightarrow ^{210}_{84}Po \rightarrow ^{206}_{82}Pb$. It is used also for determining the antiquity of lead artifacts and pigment, where the ratio of $^{210}Pb/^{226}Ra$ is small in old artifacts and high in modern.

Potassium-Argon Dating

Potassium makes up about 28% of the earth's crust by weight. It has three isotopes: potas-sium 39 (93.2%), potassium 40 (0.119%), and potassium 41 (6.8%). Primordial potassium 40, the only radioactive isotope, probably once made up 0.2% of all potassium. Potassium 40 decays in three ways: electron capture, beta emission, and positron emission. About 99% of the time it decays by beta emission and K-electron capture:

$$^{40}_{19}K \rightarrow e^- + ^{40}_{20}Ca \text{ (beta emission: 89\% of all decays)}$$

$$^{40}_{19}K + e^- \rightarrow ^{40}_{18}Ar \text{ (K-capture: 11\% of all decays)}$$

K capture is named for the K-shell orbital electrons, which lie very close to the nucleus. Some actually go through the nucleus and, at times, an electron is captured by the nucleus, hence the name *K capture* (the *K* refers, not to potassium, but to the K shell). The mass of the nucleus is essentially unchanged by this small addition, but the number of protons is decreased by 1 and the number of neutrons increased by 1. The isotope $^{40}_{19}K$ decays by capture of an electron to $^{40}_{18}Ar$, an isotope of argon. Unfortunately, since 99.6% of all argon is ^{40}Ar, the radiogenic isotope cannot be distinguished easily from the normal isotope found in air. However, if the ^{40}Ar formed by decay of ^{40}K does not leave the sample, the age of the sample can be determined. Since ^{40}Ca comprises 96.% of all calcium, the radiogenic calcium cannot be distinguished from the original calcium in the sample.

Argon is a gas found primarily in the earth's atmosphere. It is a noble gas occurring in atomic form since it does not readily form chemical compounds, not even molecules with itself. All ^{40}Ar on earth probably came from decay of ^{40}K. Theoretically, one could estimate the age of the earth by estimating the amount of ^{40}Ar and the amount of ^{40}K and using the half-life. All helium found in the atmosphere probably also came from radioactive decay. But helium is a very light gas, which drifts to the top of the atmosphere and away from the earth. Argon is a heavy gas and thus retained in the atmosphere.

Ages from potassium decay can be estimated through the decay into calcium or argon. Decay into calcium is not used, because of the abundance of calcium in the crust of the earth. Most minerals with potassium probably had abundant calcium present as well when they first crystallized. Differentiating the new calcium formed from radioactive decay of ^{40}K from original calcium is an impossible task.

The date obtained by the $^{40}K/^{40}Ar$ technique is, of course, the time that the sample becomes a solid, since argon cannot be trapped except by a solid rock. This is termed *closure* of the rock system. Major errors could be produced if the sample were porous and some argon had diffused out of the sample. To estimate possible argon loss, the grain size and porosity of the rock, together with a knowledge of the diffusion rate of argon in the particular rock, must be known.

Potassium-argon dating is a very popular technique which has been substantially improved over the last 20 or so years, both in the precision of measurement and in the expansion of the range of dates for which it can be used. In the early 1970s it was considered good only for dates greater than 2 million years. In the middle of the decade this had been improved to 30,000 years, and now the method is used principally for early prehistory.

When potassium decays into argon in a crystal, the relatively large argon atoms cannot easily diffuse through the spaces between the atoms in the lattice structure. They remain trapped in the sample unless it is later heated or is porous. The argon atoms are the daughter isotopes of the parent ^{40}K. The beta-decay constant (β) is 4.72×10^{-10}/year, and the constant for K capture is 0.584×10^{-10}/year (K). The latter figure yields a half-life of 1.25×10^9 years. With potassium-argon dating, even though the calcium cannot be used, both decay constants must be known because the two modes of decay are competing. The previous law that we used in this chapter (eq. 5.4) must be modified in this case to

$$N_D/N = \lambda k/\lambda kt\lambda p[\exp{(\lambda\beta + \lambda k)t} - 1] \qquad (5.6)$$

The important idea to extract from this formula is that where the ratio of N_D/N is measured, both λk and λ must be known to determine the time, t. Here the symbol N_D is the number of argon daughters and N is the number of potassium parents remaining. Occasionally some references will give this formula in terms of "branching ratios" and a composite half-life λ_{Br} both of which are related to $(\lambda k + \lambda\beta)$.

To determine the amount of argon present, the sample is heated to liberate all gas, which is then trapped and analyzed. Some of the argon released will be from the atmosphere and some will be from the chamber, a result of atmospheric contamination in the laboratory apparatus. The argon is run through a mass spectrometer, where it is charged. The total amount of charge associated with each argon isotope is then measured to give the relative amount of each isotope present. A ratio of isotopic concentrations is obtained, which includes the atmospheric contamination and the radiogenic. From the known ratios of ^{38}Ar or ^{36}Ar to ^{40}Ar in the atmosphere, the total amount of ^{40}Ar atmospheric contamination can be calculated. The modern isotopic ratio of ^{40}Ar/^{36}Ar is 295.5. This is subtracted from the total amount of ^{40}Ar measured to give the radiogenic argon: ^{40}Ar $\times$ (^{40}Ar/^{36}Ar) = age.

If radiogenic argon is a very small fraction of the total ^{40}Ar, as it is in very young samples, then determining an accurate age is difficult. However, new developments in mass spectrometers and new vacuum techniques allow more accurate mass separation and reduce the residual amounts of argon in the system when the sample is melted. Determination of the amount of the parent ^{40}K is done chemically. Total potassium is determined and ^{40}K calculated from the known ratio of radiogenic vs. nonradiogenic isotopes.

From a crystallographic standpoint, ^{40}K is tightly held ionically in a mineral lattice, but ^{40}Ar is a gas and must be trapped by the lattice. This ability to trap argon varies with the mineral. For instance, feldspars have an open lattice, so dates are reported as minimum ages, whereas micas and hornblendes have tighter lattices and yield more reliable dates.

^{40}Ar/^{39}Ar Dating

A new Argon dating procedure, the argon-argon (or more specifically, single crystal, laser fusion) technique, has vastly improved argon dating.[12] The procedure and its differences with the older K-Ar technique are shown in table 5.4. In its most recent form it combines the use of heavy-ion lasers such as Ar with an activation analysis technique that uses "fast neutrons" (neutrons with energies above 0.25 MeV). The technique is useful for dating

Table 5.4 Differences in K-Ar and ^{40}Ar/^{39}Ar Techniques

K-Ar	^{40}Ar/^{39}Ar
1. Age determined by measuring *amount* (in moles) of K left in sample and *amount* of Ar produced by decay.	1. Age determined by measuring ratio of ^{40}Ar to ^{39}Ar in mass spectrometer
2. Chemical determination of total K concentration by flame photometry, atomic absorption spectrometry, isotopic dilution, or neutron activation. Amount of ^{40}K calculated by knowing its ratio to other K isotopes.	2. ^{39}Ar is produced from ^{39}K in nuclear reactor by fast-neutron flux. ^{39}Ar does not occur naturally.
3. Ar concentration determined by isotopic dilution (measure in mass spectrometer).	3. Total amount of K is calculated by knowing the ratio of ^{39}K produced in the reactor to ^{40}K. Chemical analysis not needed.
	4. Can liberate Ar from sample all at once and obtain age, or can liberate gas at a number of temperatures and produce an array of ages. The degree to which the ages correspond, "plateau ages," determines how meaningful the age is.

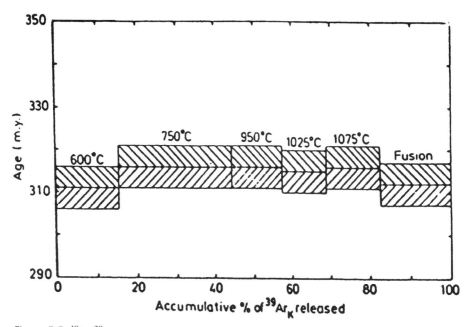

Figure 5.5 ^{40}Ar/^{39}Ar age spectrum for a biotite. Plateau age is 320,000,000 years.

grain-sized samples of minerals or volcanic glasses such as tephras. Multiple grains can be dated, yielding a suite of data which can be analyzed statistically to determine the most probable age range for the deposits or specimen. Even more important is the high precision in the determination of ^{40}K, which makes this procedure so attractive for geochronology.

In table 5.4 we see that Ar–Ar dating does not measure ^{40}K directly, thus eliminating chemical analysis, a step which can introduce other sources of error, as well as requiring larger samples in the determination. The technique uses the neutron activation of ^{39}K to ^{39}Ar: $^{39}_{19}$K $+ ^{1}_{0}$n $\rightarrow ^{39}_{18}$Ar $+ ^{1}_{1}$H.

The isotope ^{39}Ar is not a naturally occurring isotope of argon.[13] The fast-neutron flux necessary for the effective production of ^{39}Ar is produced by a relatively old reactor design called the TRIGA. A fast-neutron flux can be obtained in accelerators or, as the author Garrison has done, in a singular nuclear device—the fast-burst reactor—found at White Sands Proving Grounds, New Mexico. The fast neutrons produced are necessary for the neutron absorption cross section of ^{39}K. The daughter ^{39}Ar is a direct decay product of ^{39}K, which was produced by the neutron flux in a known ratio to ^{40}K. A mass spectrometer equipped for incremental heating releases the gases that are measured. The ^{40}Ar/^{39}Ar ratios "plateau" at characteristic temperatures (figure 5.5), the shape of the plateau being a statistical measure of the accuracy of the isochron. Most dates are averages of the isochrons of several analyses. Dates are determined as follows: $t = 1/\lambda_{total} \ln [(1 + J) (^{40}Ar/^{39}Ar)]$, where ^{40}Ar is radiogenic and J is a constant determined by use of a standard mineral.[14] Using what is called an isotope correlation diagram, shown in figure 5.6, the ^{40}Ar/^{39}Ar ratio can be calculated directly. This technique is better for "young" (thousands of years) sam-

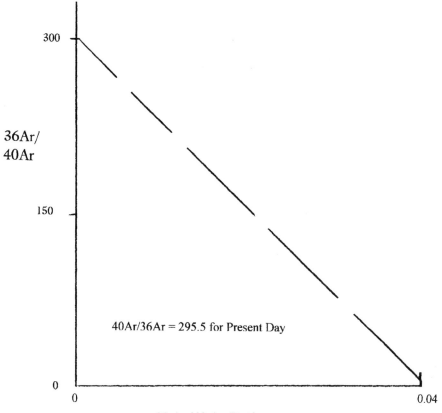

Figure 5.6 Isotope correlation diagram for Ar/Ar dating. A ratio of 295.5 assumes no radiogenic contribution. Extrapolation of the data to the axis yields the observed ratio and the age of the sample.

ples. The ratio of $^{36}Ar/^{40}Ar$ is also obtained by the irradiation procedure, wherein ^{40}Ca is activated to form ^{37}Ar and ^{36}Ar. The advantages of Ar–Ar dating over the conventional K-Ar method lie in increased accuracy and lower sample weight. Accuracy is better than 1% and only microgram quantities are required for most samples.

In 1994 two papers on the Ar–Ar dating of early hominid sites produced a stir in the scientific community. One was a study of the early australopithecine "Lucy" and the other of the Trinil locality in Java, site of the oldest finds of *Homo erectus*.[15] Both studies reinforced the usefulness of the Ar–Ar technique.

Lucy, found by Donald Johanson's expedition to the Hadar region of Ethiopia in 1973, is the most complete skeleton of the australopithecine subspecies *A. afarensis*.[16] While the antiquity of *A. afarensis,* a Pliocene hominid, is not challenged, its precise temporal placement was questioned.[17] The principal difficulty in ascertaining a reliable date for the Hadar hominid's geological provenance stems from the diagenetic reworking of volcanic ash and rock members of the Hadar Formation.[18] Weathering and chemical alteration of the tephras at Hadar have removed the feldspars, a favorite medium for K/Ar dating.[19]

Lucy and the 13 individuals represented in the early Hadar finds have been called "the First Family." These discoveries, coupled with Mary Leakey's serendipitous "footprints" from Laetoli, Tanzania, provide a remarkably human picture of our early ancestors.[20]

6

Radiation-Damage, Cosmogenic, and Atom-Counting Methods

Fission-Track Dating

Fission-track dating, one of the more recent techniques involving the use of radioactivity, has developed one of the widest ranges of applications.[1] Dates of objects have been obtained ranging from 6 months to 10^9 years BP. Volcanic tephra, obsidian, man-made and basaltic glass, meteorites, and mica have been dated.[2] A more apt term is *nuclear-track dating* because fissionable elements do not have to be present in the material. Fission, which produces one form of nuclear track, is a rare mode of radioactive decay. A more common decay is alpha decay, which produces a different type of track.[3] Uranium 238 fissions spontaneously and has a well-defined half-life. It also fissions in the presence of neutrons such as are produced by reactors, accelerators, or neutron "howitzers." About 99.27% of all uranium is uranium 238. Robert L. Fleischer, Paul B. Price, and Robert M. Walker, who have done most of the original work in this field, have determined that most minerals contain this isotope in amounts from a few parts per billion (ppb) to many parts per million (ppm).[4] These researchers devised a chart which characterizes the ease of use of this technique as a function of the uranium concentration (see table 6.1). A high uranium concentration allows an "easily measured" age where the observer spends an hour at the microscope counting chemically etched fission tracks. For "considerable labor," 40 hours of such work is assumed.

Ancient synthetic glass typically contains 1–2 ppm of uranium, so most glasses older than 8,000 years are datable. Most pottery clay contains about 5 ppm of uranium in either the clay itself or other minerals that occur as inclusions. It is very probable that some pottery clays or the mineral inclusions, such as zircon, might contain higher concentrations than this, which would make the age measurement lie between "easily" and "with considerable labor." It is important to point out that mineral inclusions such as zircons or micas act as solid-state detectors in that they register fissions as a track on the surface in contact

Table 6.1 Ease of Age Determination by Fission Tracks
as a Function of Uranium Concentration

Youngest age measured easily	Uranium concentration by weight	Youngest age measured by considerable labor
300,000,000 yr	1/1000 ppm	8,000,000 yr
30,000,000 yr	1/100 ppm	800,000 yr
3,000,000 yr	1/10 ppm	80,000 yr
300,000 yr	1 ppm	8000 yr
30,000 yr	10 ppm	800 yr
3000 yr	100 ppm	80 yr
300 yr	1/100 of 1%	8 yr
30 yr	1%	10 months
3 yr	10%	1 month
4 months	100%	3 days

with the pottery clay. Both fission and alpha events can do this. Alpha-recoil track dating will be discussed later in the chapter.

Fission-track dating results from observations made of the damage done in insulating solids by the two nuclei which are produced in fission. In a mineral, the principal damage process resulting from fission is ionization, as first described by Bohr.[5] Fission occurs when a nucleus splits into two or more nuclei (usually two), such as krypton 92 and barium 141. Uranium 238 fissions naturally, with a definite half-life. The heavy Kr and Ba nuclei are generated with kinetic energy; they recoil and move through the material in which the fission occurred. The fragments are large with respect to the spaces between atoms, and they have a tendency to remove large numbers of electrons from the atoms which they pass. This is ionization. The highly charged atoms which they leave in their wakes repel one another with considerable electrostatic force and move away from the path of the fission fragments. In insulator solids, atoms so repelled form dislocations or damage areas in the crystal lattice, and these areas do not recombine or "heal" without significant new inputs of energy to the lattice. Etching of surfaces of insulators such as muscovite mica with hydrofluoric acid causes a track to appear on the surface as long as 10 μm which can easily be seen with an optical or atomic microscope. For plastics, such as Lexan, which can serve as detectors of radioactive decay, the etching agents used may be common household bleaches.[6] These detectors are essentially without cost compared to the more conventional methods of particle detection—photovoltaic, GeLi, NaI crystal detectors—which cost thousands of dollars.

To date using fission tracks, a phase-contrast optical microscope and access to a nuclear reactor or some other neutron source are required. No counting-time difficulties are involved such as with neutron activation, where the sample must be inspected immediately after irradiation. The fission-track dating sample can be sent off, for example, to a nuclear (research) reactor laboratory where neutron irradiation is offered as a service. The irradiated sample is returned and, under almost all conditions, inspected at the leisure of the microscopist. In most cases, the technique is relatively nondestructive to the sample. For example, any induced radioactivity in mica subsides, in most cases, within days.

Almost all minerals are formed with uranium within them, and almost all minerals that have the characteristics of insulators—glasses, micas, zircons, and quartz—will contain fission tracks. The number of fission tracks will be proportional to the amount of uranium in the mineral grain or crystal and the length of time the mineral has existed in its present form. Thus, it can be shown intuitively (with that intuition heavily bolstered by the previous discussion on radioactivity) that the time the insulating mineral has been in its present form can be determined. If a mineral has been exposed to high heat after its formation—such as in metamorphism—the fission tracks existing before thermal exposure will be erased, and the method can be used to determine the time since heating. For determination of dates, the decay constant for spontaneous fission of ^{238}U is used, since it constitutes 99.27% of all uranium $-6.85 \pm 0.20 \times 10^{-17}$ yr^{-1}. When the ^{238}U fissions, it generally splits into two heavy elements, with atomic numbers ranging from zinc (30) to terbium (65). After the ^{238}U atom fissions, the two elements travel through the material and produce the damage seen as fission tracks. The total amount of fissioning is determined by counting the number of tracks. This total is a result of the age of the sample and the amount of ^{238}U originally present. To find the amount of ^{238}U, the amount of the isotope ^{235}U is determined. ^{235}U fissions readily under neutron bombardment whereas ^{238}U does not (thus the atom bomb consists primarily of ^{235}U). Since the ratio of the two uranium isotopes is known, the amount of parent ^{238}U can be calculated. The procedure followed is to first count the number of spontaneous fission tracks in the sample. Then the sample is annealed by heat causing the fission tracks to fade and disappear. The sample is next exposed to a neutron flux, which produces induced fissioning of ^{235}U and a new track population. Counting the new fission tracks and knowing the total amount of the neutron flux, the amount of ^{235}U in the sample can be determined. We can now calculate an age for the sample since we know the amount of each uranium isotope present and the amount of decay that has taken place. The standard fission-track age equation used today is as follows:[7]

$$A = \frac{1}{\lambda_D} \ln\left(1 + \frac{P_s g \lambda_D \sigma I \phi}{P_i \lambda_f}\right) \tag{6.1}$$

where P_s = spontaneous track density from ^{238}U (tracks/cm^2), P_i = neutron-induced track density from ^{235}U (tracks/cm^2), ϕ = neutron flux (neutrons/cm^2), $\lambda_D \cdot \lambda_f$ = total decay constant for ^{238}U (1.551×10^{-10} yr^{-1}) and decay constant for spontaneous fission of ^{238}U ($6.85 \pm 0.20 \times 10^{-17}$ yr^{-1}), σ = cross section for thermal neutron-induced fission of ^{235}U (580×10^{-24} cm^2/atom), I = isotopic ratio for ^{235}U/^{238}U (7.252×10^{-3}), g = geometry factor, and A = age in years (more correctly "neutron-years")

For natural glasses, such as obsidian or rhyolite, the date would be that when the lava cooled. For synthetic glasses, the date would represent the time that the manufactured object—bottle, jar, etc.—finally cooled. It is important to note that the addition of uranium has been used in the past to color glass, so some glasses have very high contents of uranium. If glass artifacts or flakes were subjected to fire after their manufacture, the fission tracks will yield the date of this event. Archaeologically, this has been a fairly infrequent type of find.

It is possible that stones, such as granites, that contain flecks of mica or other minerals near their surface and that have been exposed to high heat might be used for dating. Inspection of most potsherds under the microscope will show crystalline minerals appearing in the freshly broken sherd edge suggesting that the fission-track technique could be used for dating potsherds if the amount of uranium present is sufficient. At the present time, fission dating is not commonly used in the dating of pottery but rather thermoluminescence and electron spin resonance dating.

Alpha-Recoil Tracks

A nuclear-track technique that showed some promise but, for reasons given below, has never been actively pursued is alpha-recoil track counting. Fission is relatively rare even in nature, but alpha-particle emission is quite common. There are about 2 million alpha particles emitted for every fission event. In some materials, such as high-molecular-weight cellulose,[8] alpha particles themselves can produce detectable tracks, but in most instances the track forms when the nucleus recoils upon emitting an alpha particle in the conservation of momentum. The tracks generated are only about 1/2,000 of the length of the fission track on the average (200 Å). Because the frequency of alpha tracks is higher, the ability to date young samples is theoretically greatly enhanced.

Both the uranium and the thorium decay chains are dominated by alpha decay. As a result, lower concentrations of uranium and thorium will produce enough tracks so that a smaller area can be searched. Using mineral inclusions, it seemed possible to date pottery through the use of alpha-recoil tracks. This proved to be more difficult than thought, but led to the detecting of alpha damage by other techniques such as electron spin resonance spectroscopy.[9]

The alpha-recoil track technique was first described by W. H. Huang and R. M. Walker.[10] Again, in any radioactive technique, the decay constant, the number of parent nuclei, and the number of daughters must be determined. The number of daughters is determined by counting the number of recoil tracks. Each alpha emitter of the three radioactive chains ^{238}U, ^{235}U, and ^{232}Th produce tracks making the parentage of the daughters uncertain. In addition, a "linear assumption" for the decay process cannot be made in most mineral systems, because radioactive daughters such as radon are gases and daughters such as ^{234}U are commonly lost.

D. Wolfman and T. M. Rolniak touched on another problem, that of registration efficiency in the use of alpha-recoil tracks in materials that have been heated.[11] Unlike fission-track formation, where the correlation of fission event to fission track is assumed to be 1:1, alpha-recoil track formation may require four or more alpha decay events. Additionally, the density of alpha-recoil tracks in mica taken from archaeological ceramics is not easily reconciled to the number of such tracks expected from the decay of uranium and thorium impurities.[12] This disparity is disturbing and can be as great as an order of magnitude. The early promise of alpha-recoil tracks has not been fulfilled by the development of any reliable dating technique.

Thermoluminescence

Theory of Thermoluminescence Dating

In a solid, such as a crystal, the electrons of the constituent atoms occupy certain energy levels, or quantum states, which are split into closely spaced levels called energy bands. Using the hydrogen atom as our example, only one orbital electron surrounds a nucleus with a single proton. Atoms in electronically excited states undergo spontaneous transition to lower energy states by means of radiation emission. Quantum mechanics gives the expression for allowed electronic energy levels of a hydrogen atom in terms of a single quantum number, n:

$$E_n = \frac{2\pi^2\mu e^4}{h^2} \quad \frac{1}{n^2} \quad (n = 1, 2, 3, \ldots, n_1) \tag{6.2}$$

where e is the charge of the electron, and h is Planck's constant divided by 2π. reduced mass, μ is in terms of the mass of the electron, m and the mass of the proton, M:

$$\mu = \frac{mM}{m + M} \tag{6.3}$$

Energy states converge toward a limit at $n \to \infty$ where $n \to \infty$ indicates a continuum of energy states corresponding to a complete separation of the electron and the proton (i.e., the electron leaves the atom). Energy levels have a small dependence on the quantum number, which is 1. The most tightly bound inner electrons (K, L, and M shells) have the lowest total energies. Each of these electrons is anchored to its own atom, and for identical atoms each has essentially the same energy.

For the less tightly bound atomic electrons, the close spacing gives rise to a phenomenon known as *tunneling,* whereby particles such as electrons may appear in regions forbidden by classical physical principles. The branch of physics known as quantum mechanics deals with this type of phenomenon. The electron is considered a quantized energy wave packet that exists in discrete energy states but is allowed to penetrate to higher or lower energy states on the basis of probability considerations. Max Planck theoretically predicted this quantum nature of subatomic particles when he was working with the emission of light from an incandescent surface. In this work he observed, in agreement with his assumptions, that light is emitted in discrete quanta of energy corresponding to $h\lambda$, where h is Planck's constant divided by 2π and λ is the frequency of the light emitted.

Electrons, which exist in discrete energy levels, may be raised to higher states either by the "spontaneous" mechanism discussed or by the more common mechanism whereby energy is input into the atomic system causing a loosely bound valence electron to be boosted up to the conduction level, where the particle behaves much like a free electron. Hence, electrons in solids exist in band levels of discrete quantized states determined by the number, N, of atoms in the solid; that is, each discrete state for an isolated atom becomes part of a band with N states in a solid. This is the case with the valence band in crystals.

In the conduction band, as mentioned, electrons are in configurations that allow them to be freer than their valence counterparts but are constrained by the regularity of the crystal's atomic lattice, which modifies the wave functions and creates successive bands of conduction energy states. The most important attribute of a solid in determining its electrical properties is the extent to which the conduction and valence bands are filled. In the ideal case, the valence bands are exactly full, with as many valence electrons as there are allowed states. However, two important factors modify this situation and are important in understanding the principles of thermoluminescence (TL) dating.

First, a valence electron can be broken loose from its normal place in the crystal by giving it enough energy by collision or other processes to raise it to the conduction band. Second, within a real crystal, there are imperfections which give rise to vacant states in the valence band. For instance, an incident photon can break a covalent bond connecting sili-

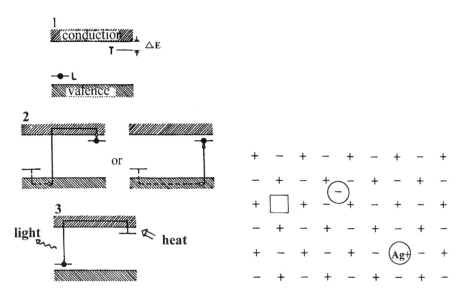

Figure 6.1 *Left:* Band model for thermoluminescence (after Zimmerman). *Right:* Simple defects in a crystal lattice: negative-ion vacancy; negative-ion interstitial, and substitutional impurity (Ag$^+$) (modified from Aitken, 1974).

con atoms by giving, upon collision, to the valence electron enough energy to rise to the conduction band. A vacant site is created in the valence band and, hence, the creation of what is termed a *hole-electron pair.*

In general, electrons raised to the conduction band return by a process termed *population inversion,* whereby the hole-electron pair recombines with the excess energy of the stimulated electron given off as emitted light.[13] Due to the effects of vacancies, interstitials, and impurity atoms within the crystal lattice, termed defects (figure 6.1), not all the hole-electron pairs created recombine. These defects trap the electron in metastable states. These states require the input of a great amount of energy to raise the trapped electron to a level where it can recombine with a hole. The electron will remain trapped until sufficient energy is available to free it.

Defects are created within crystals by radioactive processes before and during crystallization. Radioactive impurities produce ionizing radiation within and outside the crystal, which in turn produces hole-electron pairs. Typical damage centers that may be produced in crystals by high-energy irradiation can be classified as follows:

1. F centers are produced when a negative ion is missing from the crystal lattice and an unpaired electron is trapped in the Coulomb vacancy.
2. V centers are the opposite of F centers and are produced when a positive ion is missing from the lattice and an electron is also missing from one of the neighboring ions. This is a "hole" trapped in the vacancy.
3. Interstitial atoms form centers such as the U center, which is a negative-ion vacancy

containing a hydrogen or deuterium ion (negative). The unpaired ion is localized on the hydrogen ion.

These damage centers are stabilized within the crystal if recombination of the radicals is prevented by the solid. This is particularly so if the incident radiation has sufficient energy to cause the broken bonds to move away at high speed relative to the time necessary for recombination ($>10^{-8}$ s). Imperfections in the crystal perform the same functions as the damage centers in forming "traps". These centers, interstitials, and impurity atoms form localized energy levels within the energy gap between the valence and conduction bonds that are known as metastable levels. Storage of the metastable electrons is additive over time, and release is possible by suitable heating whereby the induced thermal energy frees the trapped electrons and thus measurable light is emitted. These are the principles on which TL depends.

Most minerals contain radioactive elements such as uranium, thorium, and potassium measured in parts per million (or μg/g) up to percentage amounts. A typical clay contains 3 ppm uranium, 12 ppm thorium, and 1% potassium.[14] Over time, their decay will create hole-electron pairs within a crystal. The number of metastable pairs formed is a function of the intensity of radiation and time. To empty the trapped electron states, temperatures in the range of 500–1200°C are necessary. Hence, it is immediately obvious that if such temperatures were achieved by the thermal pretreatment of lithic materials, then the time of firing will be measured by TL dating, in which the material is reheated and the light emitted is measured, which in turn provides a direct measure of the age of the specimen.

It was once assumed that the radiation dose per year is more or less constant over time. This seemed a reasonable assumption when dating crystal inclusions in pottery by TL but is somewhat untenable for the, now, wide range of archaeologically interesting materials. With pottery, the material with which the technique was explicated, a linear increase in the stored TL was expected. This assumed that no saturation of available "conduction" levels occurred within the artifact. The assumptions of linearity and saturation are questionable in regard to other materials, such as stone and bone, as well as pottery. In the case of stone, increasing antiquity of the artifact raises the question of sublinearity wherein the natural TL has approached a saturation value for the material. This can be visualized as a change in slope of the stored response of the material over time, with a plateau region being formed in the curve's slope, which (the slope) decreases with increasing radiation dose and produces an upper limit for age determination of the material.

Materials, Environment, and Paleodose

The TL observed is the result of the buildup of metastable traps in a particular material's matrix. In the main these materials are mineralogical in nature. For pottery dating, quartz has been the favored material because it is routinely present in clay fabrics.[15] The radiation producing the TL in a specimen is resident in two principal locations: (1) internal, or within the material, and (2) external, or within the environment within which the specimen has been resident. In general, the internal radiation dose, called the internal dose, is less important than the external, or environmental, dose.[16] Again, this relation is highly specific to the material and the environment.

The principal radiation types are alpha, beta, and gamma particles. A less important but prevalent type, for nonburied or shallowly buried materials, is cosmic radiation. Again, in most soils and minerals the sources of the primary types are the uranium and thorium decay chains—^{238}U, ^{232}Th—and potassium. Gamma radiation is lightly ionizing, being able to interact only through the Compton effect (i.e., ejection of an electron by a reduced-energy gamma ray). Beta and alpha radiation dominate the paleodose.

Much has been written on the vicissitudes of radiation disequilibria in the environment, as well as the partitioning of the source nuclides (radioactive isotopes) in minerals and artifacts.[17] Suffice it to say, it is important to understand that most geochemical systems are radiologically "open" systems in which migration of radioelements into or out of the systems is to be expected. The crux of most discussions on this subject centers on how this impacts the calculated ages as error.

Another important factor is the efficiency of trap production by the various radiations; the most efficient is beta, followed by alpha and then gamma. The efficiency of trap production is still a difficult question in TL studies owing in large part to the basic physics of the process itself, which is complex to the point of being called "hydra-headed" by one of TL's pioneering figures, M. J. Aitken.[18] Much of our understanding is due to early work by D. W. Zimmerman, whose contribution was cut short by his untimely death.[19] The range of trapped electron centers and their energy distribution make up the envelope of the TL glow curves (figure 6.2), which in their elegant measurement simplicity mask the number and variety of trap sites within a mineral.

The material most used in TL studies has been quartz, first examined by S. J. Fleming for pottery dating.[20] This has been called the large-grain technique as opposed to the fine-grain technique à la Zimmerman, which has become the most commonly applied technique.[21] The advantage of the latter lies in the recognition of alpha-particle ranges in soils and clays (ca. 25 μm). By removing 2 mm of the sherd exteriors, one effectively eliminates the alpha contribution of the soil or burial matrix, leaving only the clay. The next step is the separation of quartz grains less than 10 μm in size, which simplifies dose rate considerations by effectively eliminating the "$K\alpha$" correction, where K is a value for the attenuation of alpha radiation. The larger-grain technique attempts to reduce K by etching away the exterior 20 μm of larger quartz inclusions. The total error for each procedure is ±5% and ±7%, respectively. For minerals such as zircons and apatites, the higher concentration of U and Th effectively reduces the environmental component to negligible values.[22] The error is reported to be ±4%. Feldspars hold promise for TL dating of ceramics, metamorphics, and granites because the alkali feldspars and plagioclase have a TL sensitivity greater than quartz, with ages of 5×10^8 yr^{-1}. G. Guérin has dated lava flows by feldspar TL in Auvergne, France.[23] Flints have received extensive attention, notably those that have been heated.[24]

To date lithic materials by the TL technique, the following formula can be solved for A (sample age):

$$\frac{A}{Y} = \frac{O-B}{IR-B} \qquad (6.4)$$

where O = observed TL, B = background TL, IR = irradiated or induced Tl, and Y = years represented by induced irradiation (a, g, b). Since many materials are nonlinear in response, the age determination is most accurate if

$$\frac{O-B}{IR-B} \simeq 1 \qquad (6.5)$$

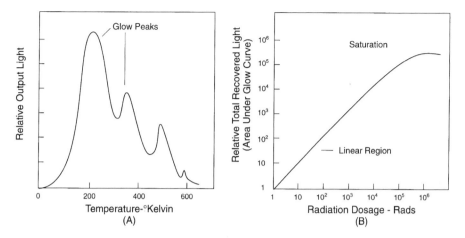

Figure 6.2 TL glow curves: (A) as a function of temperature; (B) as a function of radiation dose.

The error factor for the derived age is calculated as $= A + B + C$ where $A =$ standard deviation of background as percentage of the TL readings of O and IR (i.e., s.d. $= 0.5$ of 3.00 nanocoulomb [nC]; thus, a 16% error), $B =$ percentage of error in irradiation (10 krads $=$ 10%; 1 krad and 0.1 krad are negligible, i.e., $\approx 1\%$), and $C = 1\%$ error for National Bureau of Standards calibration source.

Sediment dating, by TL, is reviewed by A. G. Wintle.[25] Pioneering research in TL age determinations of loess deposits was begun in the Ukraine by V. N. Shelkophyas and others.[26] Table 6.2 provides a series of TL dates and reflects the growing importance of this age determination technique, particularly in the realm of early hominid evolution.

Electron Spin Resonance

Basic Theory

Stern and Gerlach, in a landmark experiment, demonstrated that the electron has the property of spin.[27] As such, it has a component known as the spin angular momentum. The allowed values for this quantity are given by $[S(S + 1)]^{1/2}h$, where S is the spin quantum number and h is Planck's constant divided by 2π. The components of the spin angular momentum vector in a space-fixed direction are limited to $M_s h$, with $M_s = S, S - 1, \ldots, -S$. A single electron has the value $S = 1/2$. For systems of two or more unpaired electrons the maximum value of M_s is 1, 3/2, 2, etc. The quantity S is equal to the maximum value of M_s. The allowed values of M_s range from $-S$ to $+S$, giving $2S + 1$ components along an arbitrary direction. Examples of spin angular momentum vectors for $S = 1/2$, $S = 1$, and $S = 3/2$ are shown in figure 6.3.

The spinning electron behaves like a small bar magnet or magnetic dipole in that it has a magnetic moment defined as

$$W = -\mu \cdot H \tag{6.6}$$

Table 6.2 Recent TL Dates of Archaeological Interest

Sites and material dated	Age	Investigators
Tautavel, Arago, France; teeth	ca. 350×10^3	Grün et al. 1987
Kebara, Israel; flint	$60–65 \times 10^3$	Valladas et al. 1987
Tabun, Israel; burnt flint	171×10^3	Mercier et al. 1995
Qafzeh, Israel; burnt flint	$92 \pm 5 \times 10^3$	Valladas et al. 1988
Qafzeh, Israel; mammal teeth	$115 \pm 15 \times 10^3$	Valladas et al. 1988
Skhul, Israel; teeth	$101 \pm 13 \times 10^3$	Mercier et al. 1993
Middle Band Keramik, Germany	4950 ± 490	Wagner, 1979, (*Archaeo Physika*, 10)
Skhul, Israel; flint	$119 \pm 18 \times 10^3$	Mercier et al. 1993 (JAS)
Hummal Well, Syria; flint	$104 \pm 9 \times 10^3$	Mercier and Valladas 1995
Yabroud I, Syria; flint	$195 \pm 15 \times 10^3$	Mercier and Valladas 1995
Kebara, Israel; flint (Neanderthal level)	$48.3–59.9 \pm 3.5 \times 10^3$	Mercier and Valladas 1995

Sources: R. Grun, H. P. Schwarcz, and S. Zymela, 1987, ESR dating of tooth enamel, Canadian Journal of Earth Sciences, 24:1022–1037.

H. Valladas, J.-L. Joron, G. Valladas, B. Ahrensburg, O. Bar-Yosef, A. Belfer-Cohen, P. Goldberg, H. Laville, L. Meignen, Y. Rak, E. Tchernov, A. M. Tillier and B. Vandermeersch, 1987, Thermoluminescence dates for the Neanderthal burial site at Kebara in Israel, Nature, 330:159–160.

N. Mercier, H. Valladas, G. Valladas, J.-L. Reyss, A. Jelinek, L. Meignen and J.-L. Joron, 1995, TL dates of burnt flints from Jelinek's excavations at Tabun and their implications. Journal of Archaeological Science, 22:495–509.

H. Valladas, J.-L. Reyss, J.-L. Joron, G. Valladas, O. Bar-Yosef, and B. Vandermeersch, 1988, Thermoluminescence dating of Mousterian "Proto-Cro-Magnon" remains from Israel and the origin of modern man, Nature, 331:614–616.

N. Mercier, H. Valladas, O. Bar-Yosef, B. Vandermeersch, C. Stringer and J.-L. Joron, 1993, Thermoluminescence date for the Mousterian burial site of Es-Skhul, Mt. Carmel, Journal of Archaeological Science, 20:169–174.

N. Mercier, H. Valladas and G. Valladas, 1995, Flint-thermoluminescence dates from the CFR laboratory at Gif: contributions to the study of the chronology of the Middle Paleolithic, Quaternary Science Reviews (Quaternary Geochronology) 14:351–364.

where μ and H are vector quantities. The magnetic moment and angular momentum are proportional.

Only electrons outside closed atomic shells give rise to angular momentum and thus magnetic moments. Shells are filled according to the Pauli exclusion principle, which specifies that electrons occupying the same level must differ in spin, and Hund's rule, which states that when orbits are occupied by single electrons, they are most stable when the electron spins are parallel. If a shell is unfilled, it will have unpaired electrons in any subshell for which $L \neq 0$. A free atom or ion will have no orbital magnetic moment if it has no resultant spin ($S = 0$).

Except for the noble gases, few atoms or ions exist in the free state. They are usually bonded, ionically or covalently, so that the momentum components are zero and accordingly, are not magnetic. If spins are paired so the resultant spin, S, is zero, the atom has *diamagnetic* properties. If S is not zero, then the atom will have a magnetic moment aligning with the external magnetic field. Substances with this property are termed *paramagnetic*.

Since μ and S are proportional, it is conventional to write $\bar{\mu} = g\beta S$, where β is the Bohr magneton (9.27×10^{-21} ergs gauss^{-1}), g is a dimensionless constant called the spectroscopic splitting factor ($g = 2.00232$), m_e is the electron mass, and c is the speed of light. Placing the electron in a magnetic field gives it an energy $H = -\mu \cdot H$ classically, and inserting the expression above for μ, we get magnetic energies $h\nu = g\beta S \cdot H = g\beta H S_z$, and so our quantum mechanical prescription for S_z says there are two possible energy states: $E_+ = g\beta H(+\frac{1}{2})$ and $E_- = g\beta H(-\frac{1}{2})$, with a net separation $\delta E = g\beta H$. Inserting rea-

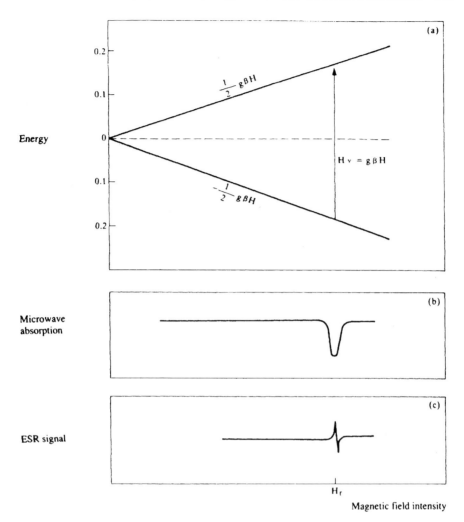

Figure 6.3 (a) Allowed values for spin angular momentum in spin = 1/2 system; (b) ESR absorption spectra for $+/-$ transition; (c) ESR derivative signal (modified from Poupeau and Rossi, fig. 1 in E. Roth & B. Poty, eds., *Nuclear Methods of Dating*, 1989).

sonable numbers for a field of an electromagnet, say $H \approx 3000$ gauss, we find $\delta E = h\nu$ with $\nu = 10^{10}$ Hz (roughly a wavelength of 1 cm).

If a sample of free electrons is placed in a magnet and illuminated with microwave photons such that $h\nu = g\beta H$, we have a resonance absorption of energy from the microwave beam as electrons are excited from E_- to E_+. Transitions between these Zeeman levels involve a change in the orientation of the electrons' magnetic moment. These transitions can occur only if the electromagnetic radiation brings about such a reorientation.[28] Assuming that the electromagnetic radiation is polarized such that the oscillating magnetic field is parallel to the static magnetic field, then the effect of the radiation will be to cause oscillations in the energies of the Zeeman levels according to

$$h\nu = g\beta H \cdot S \tag{6.7}$$

However, no reorientation of the electron magnetic moment will occur. For this to occur the oscillating magnetic field must have a component perpendicular to the static magnetic field. The characteristic feature of electron spin resonance (ESR) spectroscopy is the variation of the energy-level separation by variation of the applied magnetic field.

Complications are introduced when the electron is bound in an atom or, for example, at a crystal defect site. Then the current produced by the motion of the electron in its orbit gives rise to a magnetic dipole moment of its own, given by $\mu_L = g_L\beta L$, where L is the orbital angular momentum and $g_L = 1$. Then we have a contribution to the energy given by $H = +g_L\beta L \cdot H$. This contribution is not trivially additive with the spin contribution, because the spin ($g = 2$) and orbit ($g = 1$) interact with each other in a complex way. The net effect (if the orbital ground state is nondegenerate) is to give an effective interaction (the spin hamiltonian) in the form

$$H = g_{eff}\beta S \cdot H \tag{6.8}$$

where g_{eff} is not quite equal to 2.

The difference $g - g_{eff}$ depends on the expectation value of orbital angular momentum in the ground state $\langle L_z \rangle$, which can only be found with a detailed microscopic model of the defect center. This is actually a blessing. Inside a crystal g_{eff} will even depend on the orientation of the field with respect to the local axes of the defect, and the measured values of $g_{eff}(e)$ provide a stringent test of any microscopic model. Many paramagnetic single crystals have ESR absorption lines which occur at magnetic field strengths that depend on the orientation of the crystal with respect to the magnetic field direction.[29]

It can be shown that equation 6.7 can be rewritten as

$$H = \beta S \cdot g \cdot H \tag{6.9}$$

where g is a symmetric second-rank tensor fixed with respect to the axes of the defect. In our case g contains all the microscopic information that can be obtained from an analysis of the experimental data.

The value of H for resonance will depend upon the orientation of the crystal in accordance with expression 6.7, with

$$g_{eff} = (g_1^2 \cos^2 \theta_1 + g_2^2 \cos^2 \theta_2 + g_3^2 \cos^2 \theta_3)^{1/2} \tag{6.10}$$

Here θ are the angles between the direction of the magnetic field and the coordinate axes in which the g tensor is diagonal.

Two other features of an ESR spectrum can also give detailed information on the electron wave function and its environment:

1. *Fine structure:* If the total spin of the system involves several interacting electrons, we may have $|S_z| > H/2$ (but always integral or half integral). Then the interaction with microscopic electric fields can split some energy levels even in the absence of a magnetic field. These electric field splittings can also be measured by ESR and serve, for example, to define the local symmetry.

2. *Hyperfine structure:* If a nucleus with a magnetic moment is in the vicinity of an un-

paired electron, the effective magnetic field acting on the electron is the sum of nuclear and applied fields $H = H_0 + H_{nucleus}$. The quantity $H_{nucleus}$ depends on the possible quantized orientations of the nucleus, S, that each ESR line splits into several components. This can be a valuable diagnostic tool (e.g., in studies of ancient marbles).

When the paramagnetic substance is a powder or an amorphous solid, the absorption line will be asymmetric due to the random orientation of the substance's molecules.[30] The resonance line for flint behaves this way, in that the shape is determined by a random distribution of orientations.

ESR is a relatively recent dating technique first suggested by E. J. Zeller, P. W. Levy, and P. L. Mattern in 1967.[31] It is based on the direct measurement of the radiation-induced paramagnetic electrons trapped in crystal defects. These "free," or unpaired, electrons are generated, over time, by alpha, beta, and gamma radiation on minerals such as quartz and apatite. An ESR age is obtained by dividing the total accumulated dose within an incident in grays (1 Gy = 100 rads) by the annual dose:

$$\text{ESR age} = \frac{\text{accumulated dose (Gy)}}{\text{annual dose}}$$

It can be used to identify heat treatment of flint artifacts. The accumulated dose is measured by counting the number of unpaired electrons in a sample, which is directly proportional to the amount of radiation incident on the sample.

The ESR spectrometer "reads" the number of unpaired electrons (or "spins"), typically 10^{11} or more, and produces a characteristic absorption line, which is plotted as a derivative (I) versus the magnetic field strength (H) of the spectrometer's magnet. Points of inflection or turning points on the spectrum are characteristic of radical species in the sample. The characteristic parameter of an ESR signal is the g value, which is referenced to the free-spin value of the single electron ($g = 2.00232$).

Free electrons are formed by being knocked from their parent atom in the valence band and diffuse through a crystal's conduction band. These bands of electronic states, or energy levels, m, are separated by a band gap in which traps or defects exist that prevent the negative electron's recombination with "positive" holes in the valence band. This physical process underlies the more familiar dating technique of TL, as we have seen.

TL and ESR differ in that it is rarely possible to determine the particular trap and center responsible for the TL signal. The TL glow curve is the result of recombination of electrons and holes of widespread energies (trap depth) in a crystal. With ESR it is possible to characterize the particular defect and its energy level.

TL and ESR both have ranges for mineral dating in excess of 10^6 years. ESR has been called "a child among two parents (^{14}C and U-series dating) and an older sister (TL)." ESR's utility stems from its temporal range; sample variety (flint, bone, teeth, stalagtites, glass, zircon, and wood), specificity as to radiation defect, small sample size, and ability to date a specimen without destroying it.[32] See table 6.3.

ESR Dating of Flint

In the 70's and early 80's, studies were directed at characterizing the ESR signals of flint for geological dating purposes.[33] Three principal hole centers in quartz or flint include: two

Table 6.3 Materials Dated by ESR

Material	Location	Age ($\times 10^3$ yr)
Bone	Heidelberg, Germany (Mauer)	164–240
	Tautavel, France	250–90; 435
Teeth	Choukoutien, China	88–179
	Steinheim, Germany	216
	Petralona, Greece	430
	Tautavel, France	215
	Kabuh, Java	550
	Pucangan, Java	810
	Alberta mammoth (Canada)	90–120
Carbonates	Tautavel, France	147, 242–313
	Petralona, Greece	205 ± 20%
	Grotte du Prince, Italy	211
	El Kown, Syria	80–160
	Vertezollos, Hungary	127
	Tata, Hungary	81–127
Flint	Devonian (U.S.)	336,000
	Combe Grenal, France	375
	Kebara, Israel	65

created by the removal of an oxygen atom and electron from the SiO_2 quartz lattice and the other by removal of an Si atom. The oxygen centers are termed *E centers,* and the lost silicon creates the *nonbridging oxygen* (NBO) center. An ionized oxygen radical, O_2^-, bonded to a silicon atom produces the *peroxy* hole center.[34] These centers have characteristic *g* values in silicate minerals and have been found to anneal and reconstitute in a predictable fashion.

E centers typically result from gamma or beta radiation, whereas the NBO-peroxy centers result principally from ionizing processes, including direct "knock-on" collisions. The behavior of the E centers in flints in annealing and reconstitution studies with gamma radiation shows a thermal stability up to temperatures where annealing of crystal defects occur. The thermal stability of a defect determines the maximum age that can be obtained. Annealing studies imply high stability or deep trap depths for quartz of 10^8 or greater years.

ESR Dating of Bone

The assumption of a linear uptake of uranium by bone has some problems because bone must be viewed as a chemically "open" system where elements such as uranium or thorium can be gained or lost over time due to mineralogical and geochemical processes.[35] G. J. Hennig and R. Grün have examined the role of water content and have noted that the higher the water content, the greater its influence on ESR ages.[36] As the annual dose rate determination is critical to ESR ages, it is very important to rigorously determine the influence of water and other factors that either enrich or leach radioisotopes from mineral systems. R. Grün and H. P. Schwarcz suggest that it may not be possible to obtain a reliable ESR signal from bone but only from the enamel of teeth.[37] However the dating of ancient hominid bones form Tautavel, France, and Mauer/Heidelberg, Germany has produced reasonable ages (see table 6.2).[38]

ESR Dating of Teeth

Teeth have been the subject of the most extensive studies of ESR dating in Quaternary research.[39] The three components of teeth are cementine, dentine, and enamel. Enamel, in particular, can be composed of up to 98% HA $(Ca_5(PO4)_3(OH)_1)$. Living tissues rarely have uranium-thorium contents of more than a few parts per billion, whereas mineralized fossils may have up to 1,000 ppm.[40] The mechanism for this increase in the radiogenic fraction is uptake from the soil, presumably under less-than-saturated conditions. There appears to be differential uptake between the components of teeth, with the enamel differing significantly from the dentine.[41] This has resulted in two models for uranium uptake rate: early-uptake (EU) and linear- or late-uptake (LU) models. In the EU model, uranium is assumed to have been absorbed soon after deposition; the LU model assumes a linear uptake at a constant rate.[42] The 1987 paper by Grün, Schwarcz, and Zymela presents an excellent discussion of the technique and attendant problems.[43]

ESR Dating of Speleothems

Until the recent proliferation of ESR studies of teeth, the most highly developed application of ESR to archaeological problems was the dating of stalagmitic calcite and other speleothems.[44] The transparent speleothem crystals of calcite have produced up to seven different paramagnetic centers giving rise to absorption signals.[45] Most are associated with the well-known CO_3^{3-} and CO_2^- groups.

Ikeya began this line of inquiry in 1975 with his studies of calcite from Akiyoshi Cave, Japan.[46] The interest in speleothems such as at Akoyoshi Cave arose from the observation that these carbonates are precipitated in caves as stalagmites, stalactites, and flowstones and are frequently associated with Paleolithic occupation levels. Paleolithic occupations of a number of caves have been dated using travertines (Arago, France; Lazaret, France; Vallonet, France; Abri Pie Lombard, Italy; Petralona, Greece; Skhul, Israel; El Kown, Syria; Ehringsdorf and Bilzingsleben, Germany).[47]

Cosmogenic Techniques

Radiocarbon Dating

Carbon 14 (radiocarbon) was produced in 1940 when S. Ruben and M. D. Kamen exposed a graphite target in the 37-inch cyclotron for one month at the University of California at Berkeley.[48] The nuclear reactions that took place were $^{12,13}C$ ($_1^2H$, $_1^1H$) ^{13}C, ^{14}C. The graphite was burned to CO_2 and precipitated as $CaCO_3$. Ruben and Kamen observed a weak activity inside a screen-wall counter of a type designed by W. F. Libby. They estimated the half-life as about 4,000 years. They then tried the reaction $_7^{14}N$, $_7^{15}N$ ($_0^1n$, $_1^1H$) $_6^{14}C$, $_6^{15}C$, with

2- to 5-gallon containers of saturated ammonium hydroxide (NH_4NO_3) near a neutron source (the deflector region of the 60-inch cyclotron). They were astonished to obtain several microcuries of ^{14}C. Their interest in the ^{13}C (^2_1H, ^1_1H) reaction vanished!

Neutrons are formed in the upper atmosphere by energetic cosmic rays of 5–10 MeV which collide with air molecules (N_2). The energy of the cosmic rays is reduced to thermal levels. These collisions are called "spallations" and result in the emission of nuclei, protons, neutrons, and particles called mesons. All this debris is generated within a few millionths of a second after the energetic particle strikes the original nucleus.

When a neutron from such an event strikes a $^{14}_7N$ nucleus in the atmosphere, the latter absorbs the neutron and immediately emits a proton. The number of nucleons remains constant and the number of protons decreases so that $^{14}_6C$, radioactive carbon, is generated in the atmosphere: $^1_0n + {}^{14}_7N \rightarrow {}^1_1H + {}^{14}_6C$ Other reactions that take place are $^{14}_7N + {}^1_0n \rightarrow {}^{11}_4B + {}^4_2He$ and $^{14}_7N + {}^1_0n \rightarrow {}^{12}_6C + {}^3_1H$. Libby reported that neutrons produced by cosmic radiation may produce ^{14}C and ^3H.[49] The ^{14}C decays to nitrogen as follows: $^{14}_6C \rightarrow \beta^- + {}^{14}_7N$. Libby estimated a neutron flux of $2.6/cm^2$, which on the earth's surface equals 5.1×10^{18} neutrons/cm^2. Therefore, by equilibrium, there are 1.3×10^{19} beta decays/s. If the specific disintegration rate is $^{14}C = 1.6 \times 10^{11}$ (decays/s)/g, then the ^{14}C inventory is

$$\frac{1.3 \times 10^{19}}{1.6 \times 10^{11}} = 8.1 \times 10^7 \text{ g}$$

The ^{14}C is converted chemically to CO_2, which is incorporated into plants and animals. Libby estimated an activity/g of 16.1 ± 0.5 (decays/s)/g with the Nier effect value of 1.06 ratio for the abundance of ^{14}C of inorganic to biological ^{14}C, the activity is 15.3 ± 0.1 (decays/min)/g. Radiocarbon from this atmospheric reservoir is taken up by living organisms. When an animal or plant dies, the amount of $^{14}_6C$ at death is N_0. This amount diminishes by $N_0/2$ every 5,730 years (half-life).

Ages are calculated from the equation

$$t = \frac{5730}{0.693} \ln \frac{I_0}{I} \tag{6.11}$$

where I_0 = activity of 1 g of "living" carbon[50] and I = activity of 1 g of "dead" carbon. In modern carbon on 15 decays/min = I_0 while in ancient carbon, I, decreases to 1 decays/min at 23,000 years.

Any living organism is in a state of equilibrium between cosmic radiation and the natural rate of decay of radiocarbon as long as it lives. At death, assimilation stops, but the disintegration continues according to

$$I = 15.3 \exp(-0.693t/5568) \tag{6.11a}$$

or

$$I = 15.3 \exp(-0.693t/5730) \tag{6.11b}$$

where I is the specific activity and t the age of the organic material in years.

Libby developed the thin-walled Libby counter (a cylindrical Geiger counter) to measure solid carbon's ^{14}C. Hydrocarbon emissions dilute atmospheric ^{14}C; nuclear tests increase it. By 1963 the concordance between real time and ^{14}C time was beginning to diverge. Libby was fortunate to have begun by dating events in the 800–2400 BP range, where tree-ring and ^{14}C dates agree well.

Counting time is important to the precision of a radiocarbon date. With a *zero* background rate in the counter's detector, it would require 12 hr of measurement on 1 g of modern $^{14}_{6}C$ to achieve a 1% accuracy. Background rate and, hence, detectability set a practical limit of 50,000 years on conventional ^{14}C dates. As we shall see, new techniques such as the combined use of accelerators and mass spectroscopy—atom counting—can extend this limit to about 75,000 years.

Libby's basic assumption of *uniform* production rate was clearly in error. We now know that the atmospheric reservoir fluctuates due in large part to solar activity and magnetic field strength. Solar activity and radiocarbon production are linked as follows: if solar activity is up, radiocarbon production is down; if geomagnetic field is up, radiocarbon production is down. This was confirmed by comparison of tree-ring ages and radiocarbon ages between 3000 and 4000 BP which are consistently older than ^{14}C ages. By 1970 factors such as solar activity, the geomagnetic field, and cosmic-ray flux were proven to be the source of significant fluctuations in ^{14}C and hence the ages so determined. Additionally, H. E. Suess[51] had also discovered a 400-year periodicity in ^{14}C short-term fluctuations. R. E. Lingenfelter[52] demonstrated a clear relationship between sunspot activity and ^{14}C fluctuations. Suess also reported long-term fluctuations on a scale of 10,000 years based on 300 measurements of bristle cone pine from the past several millenia had a 10% amplitude. Its explanation requires a 50% change in ^{14}C production, which is in line with a factor of 2 change in the intensity of the geomagnetic field.

Radiocarbon dating's impact on archaeological age determination cannot be overstated. It was the first chronometric technique to be able to date anthropogenic materials directly. Moreover, it dated materials within the Quaternary, that period when fully modern humans began their floruit and supplanted Neanderthals, both advanced and classic forms. The use of radiocarbon dating has been termed a "revolution" by the noted British archaeologist Colin Renfrew.[53] Indeed, he identifies two radiocarbon revolutions: the first with its seminal impact on archaeological dating and the second with the recalibration of the technique using dendrochronology and the correction factors identified by Suess, Stuiver, and others. The accuracy of the dates improved, as well as the precision, so that by 1978 it was possible to determine a date to within 0.2%, or 16 years.[54]

Another important correction factor in modern radiocarbon dating is carbon isotope fractionation. The impact of variation in the ratio of carbon isotopes can be significant, as seen in the following relationship where isotopic fractionization is corrected by normalizing the sample to $-25‰$:

$$A_{norm} = A_{spl} \frac{1 - 2(25 + \delta^{13}C)}{1000}$$

If a radiocarbon age is calculated using A_{norm}, then a 1% shift from the standard $-25‰$ value equals about 16 years. For a more detailed discussion of stable isotopes in general see chapter 14.

Atom Counting and Radiocarbon: The Third Radiocarbon Revolution

Radiocarbon research and the advances made into the 1970s were fundamentally the province of chemists, albeit radiochemists and geochemists. It was the physicists, however, who made the next significant advance in the use of radiocarbon for age determination: atom counting.[55] Atom counting was not new; Nier and others had used it in their studies of the lead isotopes (see chapter 5). Indeed, Aston had invented the mass spectrograph—the device that actually counts atoms—in 1905. What was new was the resolution achieved in today's mass spectrographs. Until the 1970s the resolution of earlier devices was not sufficient to distinguish ^{14}C from other ions of the same mass, such as $^{13}CH^{2-}$, $^{12}CH^{2-}$, $^{11}BH_3^-$, ^{14}N. Of these, ^{14}N is the most difficult to separate from ^{14}C. An approach to the problem other than traditional "mass-only" spectroscopy was needed.

The technique for ^{14}C atom counting is most correctly termed *mass and charge spectroscopy* (MACS), but it has been given the less accurate title *accelerator mass spectroscopy* (AMS). The latter name is correct insofar as it identifies the use of accelerators and mass spectrographs but the key term is *charge*. Without the contribution of our current understanding of the physics of charged particles, the newest radiocarbon dating technique would not be possible.

The trajectory of a charged particle (or ion in flight) through a magnetic field is characterized by the charge of that ion, but that charge is not its elemental or nuclear charge, Z. The accelerator raises the energy of the ion in its path to the mass detector. Along its flight, the ion can gain or lose electrons. In AMS dating the carbon ion is totally ionized (e.g., $^{14}C^{4+}$), with the loss of all its electrons, giving it an unambiguous trajectory and impact position at the spectrograph's counter.

This approach to radiocarbon detection is the most sensitive way to detect and count the amount of radiocarbon in a sample—at present it can detect three ^{14}C atoms in a concentration of 10^{16} ^{12}C atoms. Such a concentration of $^{14}C/^{12}C$ is equal to an age of about 70,000 years.[56] Compare this sensitivity to the efficiency of conventional radiocarbon counting of a contemporary sample at 15 decays/min out of a potentially available 60×10^9 atoms of ^{14}C in a gram of modern carbon. The advantage of direct atom counting becomes immediately obvious. Equally obvious is the reduction in the amount of carbon needed for determination of a date. Now investigators can reduce their sample size to a milligram (10^{-3} g) or less.

Accelerator power output is measured as a current, or *beam current*. For accelerator dating, beam events are typically 20–30 microamperes (μA). Each microampere of "current" of ^{14}C ions produces about 450 ^{14}C ions in a contemporary sample, a gain of 450:15 over conventional detection procedures. Since the first accelerator technique was only about 25% efficient, the gain was 100:15, or roughly 7:1. Today, with familiarity and improvement in technique, that efficiency has nearly doubled. The improvements in sensitivity and the reduction in sample size have given the archaeologist and earth scientist a powerful new tool to investigate questions previously too difficult due to those limitations. A signal of the acceptance and reliability of the new technique was its use in dating the famous Shroud of Turin in 1989.[57] This important Christian relic, reputed to be the burial shroud of the Christ, had been off-limits to conventional radiocarbon dating because of the large piece needed to ascertain a reliable date. The atom-counting date was determined using a few strands of the cloth, and the shroud was found to be a medieval fabrication. This was a coming-of-age moment for the AMS (MACS) technique.

In 1991 a new archaeological discovery—the "Iceman"—led to yet another application of the atom-counting technique. This extraordinarily well preserved mummy, found in the Tyrolian glacier at an altitude of 3000 m, was dated to the 5th millennium before present (roughly 4 millennia "before Shroud") on the basis of a few milligrams of material from his apparel and his own tissue.[58]

Paleo-art and Decoration: Dating the Magdalenian

The discovery in the last century of Upper Paleolithic painting, sculpture, and engraving was a watershed in the recognition, by the modern world, of the intellectual fecundity of this late Glacial period. Long the subject of archaeological and art historical interest, "cave art" is known principally from the karstic limestone regions of France and Spain (figure 6.4). The antiquity of the archaeological cultures—Châtelperronian, Aurignacian, Perigordian, and Magdalenian—has been established by traditional radiocarbon dating over the past 30 years. Yet because of the nature of Upper Paleolithic art—for example, painted images and small, portable sculpture pieces—we have had little direct dating of the art itself until the advent of AMS technique.

Cave paintings and etchings are very widespread among Paleolithic cultures in both the New and the Old Worlds. They have been discovered in Australia, East Asia, Europe, and North America. Any comparative study of cave art must include time correlations among the various sites. The dating of cave art has proven to be a difficult task. During the past 10 years, however, absolute dating techniques have sparked new interest in the study of Paleolithic cave art.[59] Absolute dating of cave paintings often becomes a search for organic material. A positive yield will of course facilitate radiocarbon dating, but this technique when applied to cave paintings has several drawbacks.

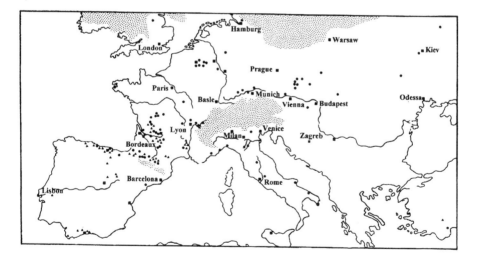

· ▲ Art Sites ▪ Modern Cities
▫ Maximum extent of last glaciation

Figure 6.4 Distribution map of Upper Paleolithic cave art sites (modified from Marshack 1979).

Cave Paintings   Archaeologists and anyone studying art distinguish two functional elements in paint: its pigment which gives the paint its color and its binder. Binders are usually required to give the pigment the cohesive and textural qualities needed for it to act like a paint. Paleolithic cultures had many paint recipes, each unique to the region and time of occupation and the resources available. Luckily for modern-day researchers, humans have established a tradition of using natural and organic materials to produce pigments and binders. Organic pigment materials include blood, charcoal, and plant substances. Inorganic substances include ochres and umbers. Lipids from both plants and animals were often used as binders.[60] Two of the most common pigments are ochre and charcoal. The charcoal is very easy to date using the radiocarbon method, but until recently large samples had to be collected in order to get an accurate date. Recently, however, use of accelerator mass spectrometers, which separate and count individual carbon isotopes and require a much smaller sample size, have been employed. Paintings composed of ochres and clays cannot be directly dated unless they contain substances such as organic binders or charcoal. Organic binders include blood, urine, honey, and wax. Much to the disappointment of many archaeologists, few cave paintings have been found to contain organic binders or charcoal. However, Tom Loy of Australian National University has found human blood mixed with ochre at several sites in Australia. He has done this using tests which involve monoclonal antibodies which recognize a sequence of amino acids that form part of the human serum albumin molecule. Trace amounts of the blood extracted from the paintings were enough to analyze with the mass spectrometer. This device is capable of dating a sample smaller than 100 μg as far back as 40,000 years with a margin of error of only 10–15%.[61]

Intensive studies were also performed at the caves of Altamira, El Castillo, and Niaux by H. H. Valladas et al.[62] These sites contain some of the finest examples of Franco-Cantabrian cave art. The following will describe the procedure followed by Valladas and his colleagues to establish dates for similar paintings of bison in three different regions.

First, in order to obtain the amount of carbon needed for dating, several sections were scraped with a nickel-plated scalpel. About 20–40 mg were collected. Next, to free the scrapings of calcite contamination, each was placed in an ultrasonic bath of 0.2 N HCl. Any material left undissolved after this was collected on a precleaned quartz fiber filter. This was then washed with a weakly alkaline solution in order to get the humic acid fraction. The filter and remaining residue were then washed with 0.2 N HCL to remove any CO_2 that might have been absorbed during the alkaline wash. This was followed by heating under a controlled oxygen flow for 1 hr at 300°C to remove any modern organic pollutants. Optical microscopy, which had been used initially to determine the presence of the charcoal in the paintings, was then used to determine that the charcoal came from a coniferous tree.

Direct radiocarbon dates have been obtained for paintings at Altamira, El Castillo, and Niaux Caves of the French-Spanish Pyrenees. The paintings date to the middle and late Magdalenian.[63] From Niaux, Valladas et al. dated charcoal from one of the bison paintings to 12,890 ± 160 BP. From nearby La Vache, the direct dates are from portable art: of 12,450 ± 105 BP, 12,850 ± 60 BP, and 11,650 ± 200 BP. Paintings at Fontavet have produced a date of 12,770 ± 42 BP.

At the Altamira site, three painted bisons were sampled, with the high water, clay, and limestone content of the cave making the separation and purification very difficult. This difficulty may explain in part the range of dates from 13,600 to 14,300 years BP for stylistically similar drawings. To test for contamination with modern organic matter, the humic acid alkali-soluble fraction were dated separately and a mean of 14,450 years BP was obtained. Comparison with bones from the Solutrean and Cantabrian Early Magdalenian strata excavated at the cave entrance indicated that the bison were the work of Magdalenian occupants.

At the site of El Castillo, two adjacent bisons were dated to 13,060 ± 200 and 12,910 ± 180 BP. This evidence also favored that the paintings were done during the Cantabrian Early Magdalenian period. Although the bisons from El Castillo appeared stylistically similar to those sampled at Altamira, the dates obtained show the former to have been painted by people living much earlier. Recent dates for the spectacular paintings at Vallon Pont d'Arc (Ardèche) of 30,000 BP have provided further evidence of both the utility of direct dating of the art and the great antiquity of these works.[64]

Problems in Dating Cave Paintings   Most caves are limestone, of which carbon is a chief component. A difficulty arises in separating the carbon in the paint from the carbon in the limestone, which invariably contaminates the paint. In 1988 Shafer and Rowe used oxygen plasma, which reacts only with organic materials.[65] When mixed with organic materials, the oxygen plasma reacts and produces water and carbon dioxide. The carbon can be extracted and dated. At the time of their publication (1990) the results of this procedure were not in.

Another problem is that much organic material in paints is from blood or urine. These proteins are quite complex, their stability in mixtures on rock surfaces is not known; and they are more prone to contamination than proteins found in bone or teeth. Contamination almost always results in dates younger than actual.[66] Methods are now lacking to purify the samples.

Paleo-Indians in Beringia

Archaeologists have sought convincing evidence of Paleo-Indian occupation in the New World in the late Pleistocene. The Bering Strait region is also known as Beringia (figure 6.5).[67] The sites most often mentioned as of an Upper Paleolithic age are not in the immediate vicinity of the land bridge. These are Old Crow Flats near the Alaska-Yukon border and Bluefish Caves, 65 km to the southwest. Both sites have been dated using conventional radiocarbon techniques, with a date of about 27,000 BP for Old Crow Flats and dates of 13,000–24,800 BP for mammoth bones at Bluefish Caves. Utilization of the AMS technique to redate the bone from Old Crow Flats yielded a date of less than 2000 BP. No redating of the Bluefish Caves materials has been carried out.

Paleo-Indian-like projectile points have been found in Alaska since the 1930s, but most have come from undatable contexts.[68] M. C. Kunz and R. E. Reamier report 11 AMS dates for lanceolate points in association with nine shallowly buried hearths from the Mesa site.[69] The lanceolate points are most similar in form to those known from midcontinental North America (Wyoming), such as Agate Basin and Hell Gap. None are the classic fluted Paleo-Indian clovis form with the distinctive channel-flute scar. AMS dating clearly places the Mesa site assemblage well within the Paleo-Indian period, with dates ranging from 9730 ± 80 BP to 11,660 ± 80 BP. A Paleo-Indian tool assemblage from the Putu site, 290 km to the east of Mesa, yielded a date of 11,660 ± 80 BP.[70] Together with revisions in radiocarbon dates for the Clovis site, in New Mexico, placing it between 10,900 and 11,200 BP,[71] the AMS dates for the Mesa and Putu assemblages support a Beringia origin for the continental Paleo-Indian technocomplexes.

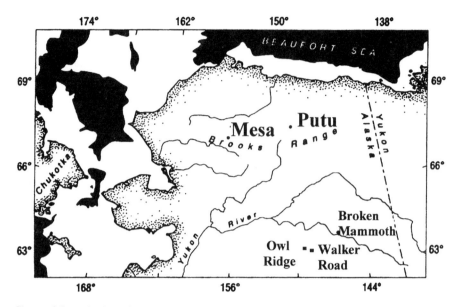

Figure 6.5 Beringia and locations of Paleo-Indian sites (modified from Kunz and Reamier, 1994).

Chlorine-36 Dating of Rock Surfaces and Artifacts

Chlorine-36 is produced by cosmic ray bombardment of the upper atmosphere in a manner not unlike that of its more famous cosmogenic cousin, radiocarbon. Likewise, on the earth's surface cosmic radiation can spallate potassium ($^{39}K(n,2n2p)^{36}Cl$) and calcium ($^{40}Ca(n,2n3p)^{36}Cl$) atoms. Less spectacular is the absorption of neutrons by potassium (n,α) and chlorine-35 (n,γ). Secondary cosmic ray muons (μ^-,μ) are subatomic particles of mass $207m_e$, where m_e is the mass of the electron. They are produced along with the neutrons in the atmosphere and lithosphere. Muons, particularly μ^-, can fall into an atom's K-shell then be captured by the nucleus. The reaction, $^{40}Ca(\mu^-,\alpha)^{36}Cl$, takes a greater role in the production of ^{36}Cl as neutrons are attenuated in the lower atmosphere and surface rock.

The production of chlorine-36 is then a function of altitude and latitude because of the moderation of the cosmic ray flux by the atmosphere.[72] While it is reasonable to think that the earth's cosmic ray flux is tied to our sun's activity, it really isn't. The production of in situ cosmogenic isotopes is dominated by the galactic cosmic ray flux from outside our solar system. This galactic flux is moderated by the magnetic field of the sun which varies with solar activity.[73]

The obvious utility of ^{36}Cl dating of surface-exposed lithology is to give Quaternary geologists and archaeologists a reliable means to date landforms. In particular, the study of glacial deposits provide direct dating of Quaternary climatic oscillations. Erosional modeling and the dating of paleosol surfaces are potential areas for the use of this "new" technique. Volcanism, alluviation, and catastrophic flood events are, likewise, important geological processes important to archaeology.[74]

7

Other Chronological Methods

Dendrochronology

Dendrochronology relies on the seasonal changes in the wood growth of trees that result in the annual production of rings; each ring starts with large cell elements associated with spring and ends with small cell elements associated with summer and autumn growth. The age of the tree is known by counting these rings. The sequence of rings produced over the years is distinctive and shared by trees of the same species over a broad region. In the western and southwestern United States, the bristlecone pine from the White Mountains of California and the eastern Great Basin[1] has allowed the establishment of a tree-ring chronology of 10,000 years.

The California bristlecone pines are found west of the Sierra escarpment's White Mountains, on the Trans-Sierra Valley slopes. The oldest groves of the trees are at an altitude of 13,000 ft (3936 m), with a few hundred trees. The oldest living tree is "Methuselah," at 4,700 years, while some of the dead trees have ages of 8,000 years. Shaped by the wind, their silvery trunks have tightly packed ring sequences (figure 7.1).

The growth of trees, which occurs from spring to autumn, is marked each year by the formation of a new ring of wood cells. The thickness of the rings is a function of the temperature and precipitation at the time of their formation. The trees of a region experience the same variations in climate and, therefore, present the same series of growth rings for the same data (period) sequence.

In 1911, an astronomer, A. E. Douglass, was studying tree rings to correlate them with sunspots and climatic changes. He succeeded in establishing one of the most precise dating techniques used in archaeology.[2] In order for the technique to be used, the tree rings must contain an arrangement of both narrow and wide rings that vary considerably in width. Each of the rings found within the cross section is called an annual ring. A wide annual ring signifies plentiful moisture in the soil, whereas a narrow ring signifies insufficient moisture in

Figure 7.1 Dendrochronological dating of a Bronze Age dugout boat from Twann, Switzerland. (Courtesy Cantonal Museum of Archaeology, Neuchâtel; photograph by Béat Arnold.)

the soil for robust growth. An annual ring may not form if there has been severe dryness, whereas constant amounts of rainfall will produce rings with relatively constant widths. Some tree ring patterns do not show any climatic influence and are termed *complacent*. If the tree rings are complacent, they cannot be used for dendrochronology.[3]

Problems of absolute accuracy do exist. Some trees form "false" rings, which result from either extreme climatic changes (e.g., frost) or from insect defoliation. These two processes may also partially deform a ring or cause the annual ring to be absent, therefore preventing the construction of an accurate chronology. These ring characteristics may, nevertheless, be useful in studies of paleoclimate.[4]

Tree Species and Dating

In U.S. tree-ring chronologies, three distinct species of trees were first used, and two still play a large role: (1) the giant sequoia (*Sequoia gigantea*), a species of the Taxodiaceae family, which flourished during the Tertiary period; (2) the yellow pine (*Pinus ponderosa*); and (3) the bristlecone pine (*Pinus longaeva*). The giant sequoia is found at very high elevations, 1500–2000 m, and may live for 3,000 years. These giant trees are found along the slopes of the southern Sierra Nevada of California. Although these trees have an extremely long life, they are not found in many of the most important archaeological sites. For that reason, pine species have formed the bases for absolute chronologies in the United States.

Oak trees (*Quercus robur* and *Q. petraea*) have been used in Europe for many years, and the tree-ring absolute chronology goes back over 9,000 years.[5] Using oak in Europe has advantages: for example, clear annual ring character, long tree life, and very distinguishable bark. Within the British Isles, oak is very common in wetland sites. Because of its hardness and durability, oak has been used as a building material since prehistoric times. Samples date from the last seven millennia.

When dating living trees, the bark ring corresponds to the present day and counting in from the bark toward the pith will determine the tree's age. One main concern is that an ancient piece of wood will most likely not contain bark due to natural or intentional removal after felling. Cross-linking can be done using different portions of a timber to correlate the ring-width structure of each portion of the same timber. This technique is known as the *bridging process* (figure 7.2).

The dendrochronologist uses many techniques in collecting samples, most commonly by coring. For example, the Swedish increment borer is used for living softwoods and provides a pencillike cross section. Power-driven extractors are used for both softwoods and hardwoods. One aim of collecting wood samples from an archaeological site is to make a dendrochronological correlation. These correlations are used along with other analyses (e.g., radiocarbon) of materials that are also found within or around the surrounding site. It is possible to do a radiocarbon analysis within a single tree ring and produce an extremely precise date.

Cross-dating, where ring patterns are cross-matched with others, was developed in 1904 by Douglass, who first dated an unknown specimen by dendrochronology. The study was done in Flagstaff, Arizona, when Douglass found that a certain group of narrow rings occurred in each tree around the year 1880.

The annual tree ring is formed by the cambium, which is located between the bark and the actual ring. During the spring, or when the growing season begins, thinly walled cells are added to the wood and produce the annual ring. Following this period, when summer begins, the cell walls become thicker and then the cells cease to be produced. This results in a distinct line or demarcation between the summer wood and the next spring growth.

To detect this natural demarcation, careful sample preparation is followed in the dating laboratory. Smoothing of the surface is usually done with a razor or by polishing, and certain liquids may be used to enhance the visibility of the rings. The "skeleton plot" (figure 7.2), first derived by Douglass, is then made by plotting the variation of the rings. For example, every tenth ring is marked. The rings that have ordinary thickness are only counted on the plot, but rings with highly different thicknesses from their neighbor rings are plotted on a vertical line. The letter *B* is used for larger rings.

Cross-correlation Techniques beyond Skeleton Plots

The skeleton, or *signature,* plot technique used successfully in the American Southwest does not transfer well to temperate tree species.[6] In Europe early workers realized that (*a*) skeleton plots ignored too much information available in the ring patterns and (*b*) correlation of whole ring patterns worked better for cross-dating purposes. Thus in Europe, where the trees are more prone to relative complacency in ring series, dendrochronologists have developed statistical techniques that rely on the directional or absolute covariation in measured ring series to establish cross-dating.[7] While sounding somewhat daunting, these analyses can be as straightforward as averaging ring widths of five rings and expressing individual ring widths relative to this "running average."[8] Combined with the removal of

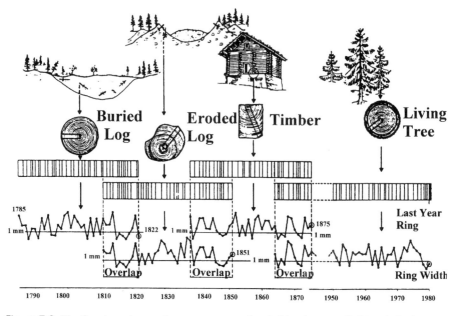

Figure 7.2 The four tree-ring sections are age correlated either by cross-linking their ring patterns or by matching their ring widths.

growth trends—some trees growing faster in one region than trees in another location—one obtains an index that results in a better master sequence (figure 7.2). Correlations of ring widths between two series can be done with simple statistics.

Densitometric, or ring-density, cross-dating has been much discussed and holds great promise for complacent series that resist matching on the basis of ring-width variation.[9] Ring-density data require more rigorous data collection and analyses than simple signature plots. For one thing, the sample must be of standard thickness and absolutely perpendicular to the plane of growth; precision borings are necessary and cross sections are even better.

Once in the laboratory, the samples are passed at a uniform rate under an x-ray beam exactly parallel to the vertical plane of the wood.[10] The image formed on the photo negative is analyzed with a scanning microdensitometer, yielding the density structure of each ring. Various aspects of the density spectrum—minimum, maximum, mean of whole or partial ring variation—can be used. Density varies from ring to ring even when size does not, making it a valuable index for both dating and dendroclimatic purposes.

The Bristlecone Pine Chronology

Douglass began his studies with living specimens. His trees were not the "Methuselah-like" bristlecones but were mostly yellow pines from around his Flagstaff location. Although Douglass sought, and located, older living examples, he was forced to turn to dead trees not unlike that shown in Figure 7.1 to build his tree-ring chronology. Still, his chronology only went back 500 years.[11] The answer, fortuitously for Douglass, lay in two related phenomena: prehistoric archaeological sites and the aridity of the American Southwest.

For tree-ring chronologies to be constructed that can be used for absolute or calendrical dating, wood preservation is sine qua non. In either desiccated or waterlogged conditions, dead wood can survive virtually intact. The American Southwest is hardly a true desert, but its dry climate is perfect for preserving wood, particularly the wood used by ancient people for building the now-abandoned dwellings that dot the landscape. Douglass turned to these ruins for successively older samples to construct a "floating" prehistoric chronology. Floating chronologies can span millennia but are not linked to modern samples. To establish the 10,000-year chronology used today, Douglass had to connect the prehistoric sequence to his sequoia chronology, which by 1919 extended to 1300 BC.[12] To accomplish this task he turned to the ancient bristlecone pines, hence the name "bristlecone chronology."

In that year, anthropologist-archaeologist Clark Wissler of the American Museum of Natural History sent Douglass sections of wooden beams from the Aztec Pueblo ruin, New Mexico. Douglass soon had a short, cross-dated sequence. He established that the larger Pueblo Bonito (Chaco Canyon, New Mexico), 80 km south of Aztec, was begun 40–50 years earlier than Aztec, but he could not tell in what year; thus, the frustration of a floating chronology. Further expeditions (1923, 1928) to Aztec and Pueblo Bonito collected specimens which extended the Flagstaff chronology to AD 1260 and built a 585-year prehistoric sequence but no overlap. A third "Beam Expedition" in 1929 closed the gap.[13] From Show Low, Arizona, a charred and rotted pine wood specimen, HH-39, was excavated from a large ruin. HH-39's inner rings overlapped the outer end of the floating chronology, and the outer rings cross-dated with the inner rings of the Flagstaff chronology (figure 7.2). The last year-ring of the prehistoric chronology was AD 1284.

The European Oak Chronology

Beginning with Kapteyn's (1880s) and Huber's (1930s) research, German dendrochronology has been in the forefront of the development of a chronology based on oak.[14] Combined with the Irish bog oak chronology developed by the Belfast Paleoecology Centre, the European oak chronology reliably spans 7,272 years.[15] By bridging the latter part of the floating sequences of the oak chronology with the very early bristlecone pine chronology (extending 1,500 years beyond the master oak sequence), it has been possible to date back to 11,000 BP. Linking the floating, subregional ring sequences from Bronze and Neolithic Age sites in south Germany and Switzerland (Cortaillod-Est, Hauterive-Champréveyres; see chapter 3) has produced a chronology of 6,000 years.[16]

Other Chronologies

Beyond the southwestern United States, the prospects for other chronologies have slowly developed. By the 1970s a midwestern chronology had been constructed back to the 12th century.[17] Since then, there has been little effort to extend the sequence due largely to the rarity of preserved wood in older archaeological contexts on the Great Plains.

The use of bald cypress (*T. distichum*) in Arkansas has raised hopes for a subregional chronology based on this species, which is highly resistant to degradation, particularly in wet environments.[18] Tree dating in the Arctic shows significant promise based on activity of the 1940s and 1950s.[19] The local chronologies established at that time extend back to the 10th century. The potential for dendrochronology in this area is buttressed by the likelihood of preserved wood in early Eskimo (Inuit) sites and in earlier prehistoric sites in the interior.

In Russia and Anatolia, dendrochronology has been productive—used primarily for paleoclimatic reconstruction in the former and in archaeological dating in the latter.[20] At Novgorod (near St. Petersburg), dendrochronology applied to pine logs formed in superimposed medieval woods led to the development of a 579-year floating chronology that ultimately was tied to wood from five historically dated churches.[21] Juniper logs from the site of Gordion in Turkey led to the development of an 806-year floating sequence. Later historic dating of Tumulus MM to 725 BC established that the Gordion chronology extended to 1600 BC.[22] Dendrochronology continues to flourish in Turkey and the circum-Aegean through the work of such laboratories as the Weiner Dendrochronological Laboratory at Cornell University.

Dendrochronology and Radiocarbon Dating

While providing the most accurate absolute age determination available to archaeology and geology, the tree-ring date is that of the outermost growth ring of the particular wood being examined. In the happy situation where that outer ring coincides with the final year of a tree's life, one can with some assurance date when the tree either died or was cut, called the *cutting date*. A *cutting date,* while calendric, does not necessarily coincide with the last year of a tree's life. Many things can remove outer rings of a tree, such as burning, trimming, and decay. In structural wood one should expect to find a non–cutting date rather than the felling age. The point is that this technique dates the tree, not the archaeological event or phenomenon of interest. When the last year of the tree coincides with the archaeological event, one generally finds a pleased archaeologist.

As seen in the preceding chapter, the radiocarbon technique in both its guises—conventional and AMS—has come to rely on the accuracy, precision, and resolution of dendrochronology as a standard for ages so derived. Very early in the history of radiocarbon dating H. L. DeVries and others realized that there was an appreciable divergence between radiocarbon ages and calendar ages.[23] Subsequent research has shown the effect is due either to dilution by terrestrial carbon or to the differential production of atmospheric radionuclide carbon 14.

By comparing radiocarbon chronologies to well-established dendrochronological sequences, radiocarbon researchers such as E. Ralph have revealed a systematic increase in the difference between the two beyond 2,000 years.[24] This difference is expressed as radiocarbon ages being progressively too young; for example, radiocarbon dates around 5000 BC are some 750 years too young (figure 7.3). Suess found that shorter-term fluctuations are superimposed on this longer trend.[25] Using the bristlecone pine sequence dates as a correction has resulted in the publication of a series of correction factors up to the latest "high-precision" calibration of 1985 presented at the Twelfth International Radiocarbon Conference at Trondheim, Norway.

Shown at the bottom of figure 7.3 is the radiocarbon age determination of samples from the destruction layer caused by the Thera eruption in the Bronze Age. What is important to note is, first, the slope of the long-term radiocarbon trend; second, the short-term fluctuations (called "wiggles"); and, third, the effect these fluctuations have on determining a cal-

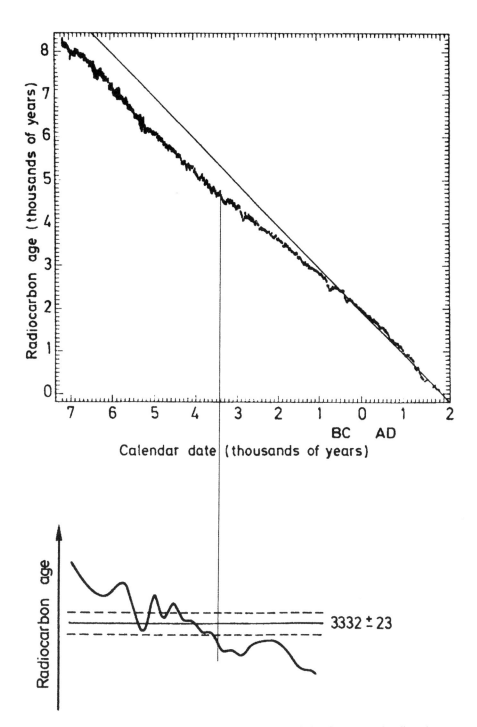

Figure 7.3 Dendrochronology and radiocarbon dating: variation in an actual radiocarbon age (*jagged line*) from expected age (*straight line*). The lower figure illustrates the range in the age introduced by this variation. Note there are three possible ages for the sample.

endar age. The central line for the age bisects the wiggles at three different points, leading to the possibility of three ages based on the same data. Still, most archaeologists today is first use radiocarbon when given the opportunity to obtain a date in a Quaternary context. This is good because most modern workers truly understand the reality that such dates represent: a closely determined estimate of age leavened with over four decades of research into all the ways that age can or cannot be "real." Dendrochronology has contributed immensely to the present understanding and reliability of radiocarbon ages.

Archaeomagnetic Dating

Archaeomagnetism is one of the fields where the interaction between a natural science and archaeology has been mutually beneficial, rather than one primarily serving the needs of the other. Geophysically oriented archaeomagnetism uses dated archaeological materials to delineate the long-term variation of magnetic field changes, while chronometric archaeomagnetism uses these master curves to infer dates for other features.

Remanent Magnetism

Basic Concepts

All fired clays and rocks are weakly magnetized. This permanent magnetism arises from the paramagnetic, ferromagnetic, or diamagnetic (no net moment) nature of their minerals. Paramagnets are random and have no real net moment. Iron has four unpaired electrons (4+) in uncompleted shells. Within 1 μm domains (e.g., Fe grains), the moments align or are parallel, forming ferromagnets in Fe_3O_4 (magnetite). The ferromagnetic iron compounds have antiparallel moments, so the net moment is a vector sum of the parallel and antiparallel moments. Each grain of αFe_2O_3 is a magnetic domain (a dipole) that is spontaneously magnetized along a direction dictated by the shape and the crystal anisotrophy of a grain. Fe_3O_4 can only exist in single domains less than or equal to 0.03 mm, while αFe_2O_3 (hematite) can be as large as 1 mm. For single-grain domains of αFe_2O_3, the theory is straightforward. In unfired clays the direction of magnetization of the grains is vectorally zero due to random orientations.

X is the magnetic susceptibility where $m = 1$ where μ_r is the permeability (Vs/Am) where $M = \chi^{\circ} H_0$ and H_0 = field strength in amperes. M is called magnetization. The total magnetization of a clay or rock is thus

$$M = \frac{\Sigma m_1}{V}$$

where V = the volume of the sample and m_i = vector sum of magnetic moments of all domains (e.g., Fe grains) in a sample.

Thermoremanent Magnetism

The alignment of magnetic domains is dependent on temperature. The quantity X decreases with increasing temperature to a critical temperature called the Curie point, T_C, where χ

equals zero. At this temperature all domains unalign and the sample becomes paramagnetic (i.e., random). Upon cooling, the magnetization of a clay or rock can change with a component parallel to the earth's magnetic field. The temperature where reversal is probable is termed the *blocking temperature,* or T_B. Magnetic domains become parallel to the earth's field and are "frozen" in the mineral, giving it a net magnetic moment, with the excess of parallel domains proportional to the earth's magnetic intensity. In 1899 Folgheraiter measured the direction of thermoremanent magnetism (TRM) in Etruscan vases and bricks.[26] TRM is associated with iron oxides (αFe_2O_3 or Fe_3O_4) present in clays in the form of ferromagnetic minerals. The Curie point for hematite (αFe_2O_3), is 675°C, and that for magnetite is 565°C. Upon cooling from the Curie point, the rocks' or clays' domains will follow the earth's magnetic field, H_0. With cooling, the relaxation time, t, reaches seconds, minutes, and years until the magnetization is fixed (i.e., the TRM is parallel and proportional to that magnetic field, H_0, at that time, t).

Most of the TRM is acquired at 200°C below the Curie point. The TRM gained upon cooling from $T_1 \rightarrow T_2$ is exactly the same as that TRM lost upon heating from $T_2 \rightarrow T_1$. Viscous remanent magnetization (VRM) occurs when domains that follow slow alterations in the geomagnetic field typically are unstable above 150°C. Chemical remanent magnetization (CRM) is due to oxidation and dehydration that alters some of the magnetic character of a sample.

The earth's dipole forms a field that is aligned along the axis of the planet's rotation (i.e., H_0 is orthogonal to the current of the earth's core). It is inclined at an angle (Θ) of 11.5 to the axis of rotation, intersecting the earth at the north and south geomagnetic poles, 78.5° north latitude, 69° west longitude, and 78.5 south latitude, 111.5° east longitude. The number of degrees of variation from true north and compass north is the angle of declination (D). The number of degrees the compass needle varies from horizontal is the angle of inclination or dip (I) which increases with latitude. Inclination is zero at the equator. Field strength (intensity) varies across the earth increasing with latitude and is 0.3 gauss at the equator and 0.6 gauss at the poles. These values are constantly changing: short-term fluctuations are *transient variations,* long-term fluctuations are *secular variations.*

Field reversals in the past give rise to polarity epochs, typically dated by K-Ar methods using volcanic lavas. These epochs can last from 5,000 to 200,000 years. Examples are the Bruhe's Polarity Epoch (normal), at 780,000 BP; the Laschamp Reversal, at 12,000 BP; the Blake Reversal, at 110,000 BP, and the Matuyama Polarity Epoch (reversed).

In this century the field, H_0 is moving westward at 0.2° of longitude per year. The direction of the earth's geomagnetic field is described at a point on the geoid in terms of D and I which is called the virtual geomagnetic pole (VGP). The VGP at latitude A' and longitude B' is given by

$$cot(p) = 1/2 \tan(I)$$

$$sin(A') = \sin(A) \cos(P) + \cos(A) \sin(P) \cos(D)$$

$$sin(B' - B) = \sin(P) \sin(D)/\cos(A')$$

A and B are the latitude and longitude of the point where the measurements D and I were made; P is the angular distance along a great circle from the point where D and I were measured to the VGP.

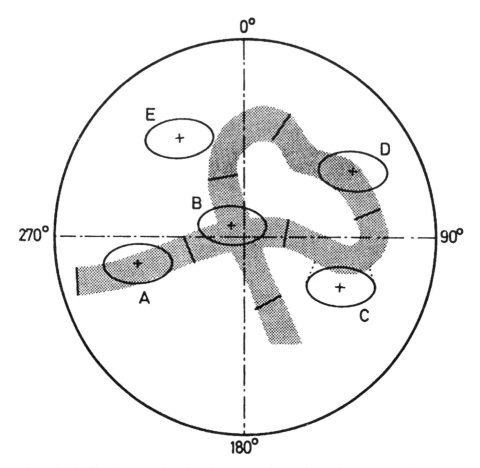

Figure 7.4 VGP path for the American Southwest with confidence limits (*shaded band*). The ellipses represent specific archaeological ages and their confidence limits (after Sternberg and McGuire 1990).

Accuracy and Precision of Archaeomagnetic Dates

The change of VGP can be shown in an area roughly 500 miles across as a path called the polar data representation curve (PDRC). This curve illustrates the movement of the VGP over time. In the American Southwest the VGP position has an accuracy of $\pm10 - \pm30$ years at the 95% confidence level (2σ) (figure 7.4). Since the dipole field makes a significant contribution to the direction of H, the PDRC for one area *cannot* be used for another area. No records of secular variation before AD 1540 exist, so a curve of D versus I must be obtained from archaeological evidence, calibrated by other means of dating.

For the virtual pole position, the apparent polar wander (APW) path (used in paleomagnetic dates) as well as VGP plots (figure 7.4) have continued to refine the precision of the age determined by reference to dates obtained from either geological or historical phenomena. The problem of accurately and precisely inferring the date of remanence acquisition is an important limiting factor in the reconstruction of secular variation.[27] Curves mapping secular variation in a region are constructed for remanence values, which are assigned dates. The better the independent age, the more accurate the resulting plot.

R. S. Sternberg and R. H. McGuire treat ages for remanence values as uniform probability distributions, with the true age of the remanence event equally likely anywhere in that age range and having a zero probability outside that range.[28] Error in the archaeomagnetic date comes from two principal sources: (1) error in the independent variable of time and (2) error in the dependent variable of magnetic direction. At each point along the secular variation curve, confidence limits are assigned, resulting in more of a band than a line (figure 7.4). Initially, archaeomagnetic workers used a "moving-window" method of smoothing the secular variation values from point to point on the curves. This can be visualized as a simple running average of consecutive points—say, four—with the average value plotted at the midpoint. As a result, short-term variation, such as weekly or monthly, is suppressed, and instead, annual variability is emphasized.

The sample plot in figure 7.4 resulted from a modified moving-window procedure suggested by Sternberg and McGuire.[29] The weighing of both age and magnetic direction errors, data density, window size, and the interval between successive windows all are converged in this smoothing procedure. Window size is an important aspect: too large, and information that may be chronometrically meaningful is lost; too small, and too little information is contained. Window length in our example is 100 years (indicated by the transverse bars). This works well where data are sparse ($N = 2.74$), whereas a 50-year window length is better where data are more dense ($N = 5.74$). Each VGP along the plot has a confidence interval at the 95% level (A95) represented by an elliptical area. When directional data are used, the confidence value is written in angular notation, α_{95}.

What then is the best number of points or data density for the best precision in archaeomagnetic dates? Following the above discussion but not trivializing it, it should be apparent that the A95 (or α_{95}) value depends inversely on N and, from experimentally determined results, reaches a fairly constant value of A95 = $1.5°$–$2°$ for $N > 7$. A recent study of an extensive set of archaeomagnetic data from Pueblo Grande indicates that this number may be somewhat optimistic. J. L. Eighmy and D. R. Mitchell observe that an N of 8–12 can reduce the α_{95} to the $2°$–$5°$ range.[30]

Archaeological Geology and Archaeomagnetism

Writing in 1990, Dan Wolfman discussed the progress of archaeomagnetic dating to the present day.[31] Along with the editors of the authoritative volume *Archaeomagnetic Dating* (1990), he sees the method entering a second generation. As with other important dating techniques such as dendrochronology and radiocarbon, archaeomagnetic dating has undergone an intensive and somewhat extensive gestation period where basic principles and techniques were established, leading to the routine reporting of archaeomagnetic dates today.

Like dendrochronology, archaeomagnetic dating is fundamentally a pattern-matching method.[32] Instead of skeleton or density plots one compares the traces of VGP curves such as the first such plot developed by the pioneer N. Watanabe in the 1950s.[33] Again, as with dendrochronology, archaeomagnetic data apply best to regional scales. Where paleomagnetic phenomena, such as the Bruhnes-Matayama Reversal (figure 7.5), are globally recognized, archaeomagnetic features covary less over large areas. Paleomagnetic temporal scales are large in a relative sense when compared to the variation seen in archaeomagnetic

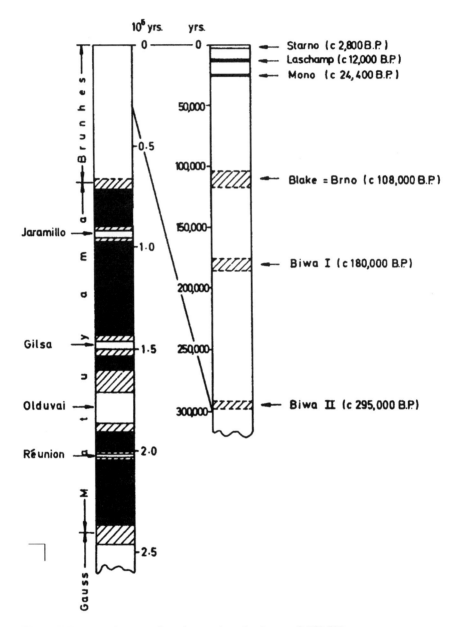

Figure 7.5 Magnetic reversals and excursions for the past 2,500,000 years.

master curves. While mainly the province of archaeologists, archaeomagnetic studies are becoming more relevant to less expansive levels of geological scale. After the dynamic 20-year period roughly from 1950 to 1970, when paleomagnetism flourished in the epochal inquiries into continental drift and field reversals, there was a return to geological questions that bear on archaeological issues, such as sedimentary sequences, ash falls, and lava flows.[34] The recognition of the less-than-optimal magnetic stability in sediment cores has led to a renewed interest in secular variation studies in sediments using archaeomagnetic correlations.[35] Additionally, archaeomagnetic dating of volcanic events such as ash falls and lava flows shows promise for the reconstruction of past secular variation, particularly when combined with ancillary age determinations by radiocarbon and tephrachronology.[36]

Tephra Chronology

Volcanic ash and pyroclastic ejecta are termed *tephra*.[37] The origin of tephra can be either air fall or pyroclastic flow. Whatever its source, tephra is composed of phenocrysts and glasses which reflect the parent magmas of felsic volcanoes. As such, they become markers of volcanic eruptions whose specificity relies principally on their varying geochemistry both within and between grains.

In geological terms, tephra is composed of (*a*) volcanic glass, (*b*) phenocrysts, and (*c*) small amounts of xenocrysts and xenoliths.[38] Individual tephra events are characterized by their petrography and geochemistry including heavy minerals and trace element content.

Volcanic eruptions are often quite dramatic and remain in the cultural memory of many peoples for centuries. In the Aegean world of the Late Bronze Age, Thera's eruption had major effects on the eastern Mediterranean, with ashfalls from the blasted remains of Santorini recorded as far east as the Nile River delta.[39] Vesuvius in 79 BC, Krakatoa in 1815, Mount Saint Helens in 1980, and Pinatubo in 1991 immediately come to mind as major eruptions that left their trace on both land and memory. From a geological and archaeological perspective these eruptions mark a precise moment in time. The mantle of ash covers several thousand square kilometers in the largest events to only a few hundred in more localized eruptions such as at Ceren (El Savador). While not as regular in nature as tree rings or varves, the tephra layers are ideal time markers. For the Pliocene and Pleistocene epochs, tephras provide chronological relations among hominid-bearing sediments, as in Ethiopia.[40]

Since the 1970s, tephrachronology has developed as a valuable aid in both archaeology and geology. Correlation between tephra exposures has been aided by the use of reference samples;[41] better geochemical characterization using instrumental analyses such as by microprobes; advances in the chronometric dating of tephras;[42] and the use of paleontological analyses where possible.[43] Volcanology has contributed to this continued development, as recent major eruptions around the globe (Mount Saint Helens, Pinatubo, Rabaul [1994]) have sparked heightened interest in all aspects of their geology.

Archaeological sites suggest a link between air-fall tephra and environmental impact. One of the most dramatic examples of this is the record of the Laki Fissure eruption in 1783–84, which devastated Iceland; approximately 50% of the cattle, 79% of the horses, and 24% of the human population perished (figure 7.6).[44] While the local effects of such eruptions are severe, extralocal impacts are less clear. Another massive Icelandic eruption, Hekla 4, ca. 4000 BP, has been linked to a "synchronous" decline in Scots pine in north

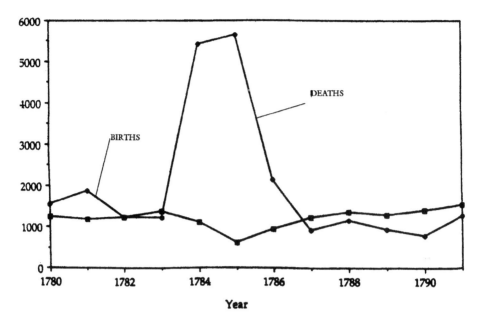

Figure 7.6 Impact of the 1783–84 Laki Fissure eruption on Iceland (modified from Halfdanarson, 1984).

Ireland, but detailed examinations of Hekla tephras in Scottish archaeological sites have revealed that this is unlikely.[45] Hekla, or the "Gateway to Hell," as it was known in the lore of Europe, has been a regular producer of tephra.[46] Figure 7.1 illustrates the frequency of Hekla and another volcano, Katla, in a stratigraphic section from Icelandic sites. Hekla 4, the eruption in the 5th millenium, has been identified by use of x-ray radiography in peats of Scotland, where the thin ash layers resist easy comparison to those much closer to the vent.[47] Analysis of biotite mica from the Baily ash (western United States) found titanium, iron, and magnesium as the discriminatory elements.[48]

   Age determination of tephras can be done by use of paleomagnetism, radiocarbon, potassium-argon, and fission-track methods. The tephra's paleomagnetic character in acquiring and retaining remanent magnetism combined with a radiometric age greatly enhances its use as a stratigraphic marker.[49] K-Ar/Ar-Ar radiometric methods have been used to date phenocrysts, although the potential for atmospheric argon contamination in mica is great.[50] Sanidine has proven to be the best mineral material for dating Pleistocene tephras, while glasses may produce ages that are too young because of replacement of sodium by potassium through natural hydration processes.[51] Fission-track dates of glasses are considered to be minimum ages due to spontaneous annealing of the tracks.[52] Zircons have proven excellent solid-state recorders of fission tracks because stability is not a problem, and uranium is sufficient to generate a statistically reliable number of tracks (see table 6.1). Zircon, however, does not occur in all felsic eruptions and becomes finer grained with distance from the vent.[53]

Ceren, El Salvador

The Ceren site, located 20 km northwest of the capital of San Salvador, has been the subject of extensive archaeological study since the late 1970s.[54] Since its discovery, it has been recognized as a World Heritage Site by UNESCO. It is buried under 5 m of ash and scoria bombs from the nearby Loma Caldera (figure 7.7). The detailed eruptive sequence at Ceren is well known from the research of Miller and others.[55] Ceren's well-preserved Mayan settlement was founded on the reworked tephra of the cataclysmic Ilopanga eruption of AD 260, known as TBJ ash or Tierra Blanca Joven tephra. This ashfall was 50 cm thick at 50 km from the vent and was detectable as far as 2000 km away. By contrast, the Loma Caldera eruption which buried Ceren was a highly localized event which was catastrophic for the inhabitants of Ceren but certainly not of the scale of Ilopanga. Based on radiocarbon dates from charred roof thatch, the Loma Caldera eruption occurred around AD 500.

Gönnersdorf, Germany

The Upper Palaeolithic site of Gönnersdorf lies in the Neuwieder basin of the middle Rhine. It is archaeologically well known, with numerous examples of Paleolithic art in the form of carved stones.[56] Gönnersdorf is, like Ceren, stratigraphically bounded by volcanic facies: the habitation level lies on tuff and the overburden is ash (figure 7.8).

In the West Eifel region, of which the Neuwieder basin is part, the last 600,000 years have seen several volcanic eruptions. Three of these eruptions were massive: Riedener Kessel volcano, ca. 400,000 BP; Wehrer Kessel volcano, ca. 220,000 BP; and Laacher See volcano, ca. 11,000 BP. Over 100 other, smaller eruptions are known for the region as well.[57] Around 18,000 BC at Gönnersdorf, one of these smaller eruptions produced an ashfall known as the Eltviller Tuff (figure 7.8).

Around 10,500 BC, Magdalenian hunters settled at Gönnersdorf. These settlers built large, round, tentlike shelters with central hearths. Found beside the hearths were stones with bowllike depressions interpreted as stone lamps.[58] Noteworthy as these remains are, it is, as with most other Magdalenian sites, the art that impresses. Found at Gönnersdorf and the neighboring settlement of Andernach were carved and polished stones. The variety of subjects includes human and animal figures. Gönnersdorf's inhabitants created memorable beauty in a Postglacial world of fire and ice. While the ice was probably a recent memory of the Magdalenian groups, the "fire" was a contemporary reality.

With the beginning of the Alleröd, ca. 10,000 BC (12,000 BP) the Laacher See volcano exploded. In less than a week, over 5 km^3 of lava was spewed forth. Described as a plinian eruption, great fire clouds and falls of ash were blown to the northeast, resulting in layers up to 1 m thick. This layer plus a second eruption, around 9000 BC (11,000 BP) are important markers for tephra chronology in central Europe. Near the Laacher See volcano, the ash is over 20 m thick.[59]

Figure 7.7 Ash layers, structure 11, Ceren site, El Salvador. (Photograph courtesy Payson Sheets and Larry Conyers.)

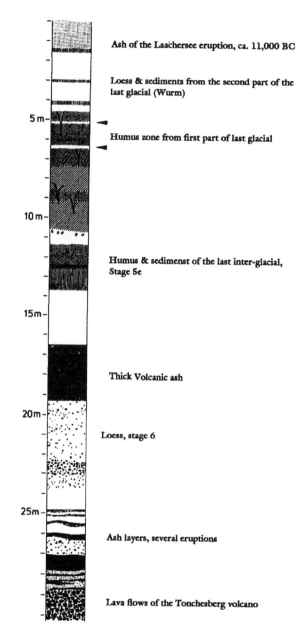

Figure 7.8 Ash strata in the vicinity of the Upper Paleolithic site of Gönnersdorf, Germany. (Courtesy Gerhard Bosinski.)

PART III

ARCHAEOLOGICAL SITE EXPLORATION

8

Archaeogeophysical Exploration

Geophysical techniques are a commonplace tool in today's archaeology as a result of an extensive collaboration between scientists and archaeologists on both sides of the Atlantic. This "cross-fertilization" has produced growing subdisciplines, of which archaeological geophysics is one example. As may be recalled from our introductory chapter, K. Butzer defined geoarchaeology as archaeology done using a geological methodology.[1] G. Rapp and J. A. Gifford describe archaeological geology as the use of geological techniques to solve archaeological problems.[2] Fagan has called geoarchaeology a "far wider enterprise than geology," involving (1) geochemical and geophysical techniques to locate sites and features; (2) studies of site formation and spatial context; (3) geomorphology, palynology, paleobotany; (4) absolute and relative dating procedures; and (5) taphonomic studies.[3] Archaeological geophysics is a major aspect of archaeological geology.

The application of geophysical exploration techniques in archaeology is also known as archaeogeophysics. Geophysical methods of potential usefulness to archaeological geology fall within the following classes:

1. seismic: reflection/refraction
2. electrical & electromagnetic: resistivity and conductivity
3. magnetic
4. radar
5. microgravity
6. thermography

All have been used on a variety of archaeological problems. The application of geophysical techniques has grown as (1) the access to the instruments and (2) the methodological understanding of the users have increased. Access to geophysical instrumentation has been made easier by the steady development in solid-state design and computerization, which has reduced size and costs as it has in almost every technical field. The beneficiaries are the geologists and archaeologists. The first to recognize the applicability of geophysical methods to archaeology were the geologists—more specifically, the geophysicists. Working in association with their archaeological colleagues, the earth scientists translated the objectives of the archaeologists into practice.

Such cooperation was very productive but suffered from the same kinds of problems that dogged the early usage and acceptance of radiocarbon dating. The archaeologists' untutored enthusiasm, coupled with their lack of a true understanding of the physics and atmospheric chemistry inherent in that technique, led to a backlash of skepticism when dates reported by the first radiocarbon researchers were found to be in error.[4] Conversely, the scientific specialists' zeal to apply their new techniques led to a good many well-meaning but often banal applications. A common occurrence in the first instances of joint cooperation between archaeologist and scientist were papers long on the description of the technique and short on archeological problem solving. This still occurs today but an appreciation of the balance between the valuable exposition of a technique and a substantive demonstration of its merit in the study of archaeological questions is more often the rule than the exception. However, archaeologists are often naive in their expectations about what geophysical prospecting can do for them. In many instances the expectations far outstrip the results, and archaeologists lose interest when, with a bit more practical an outlook, real insights into the specific archaeological problem can be obtained.

In a physical sense geophysical techniques only measure properties of the earth—soil and rock—or natural and man-made alterations in these components. One has to determine whether these geophysical properties have any archaeological meaning. For the traditional geologist, application of geophysical techniques is generally directed at deep lithological or sedimentological facies in stratigraphic relationships. For the archaeological geologist the application is generally nearer the earth's surface, seldom deeper than a few meters. In many instances, the objects or features to be studied are found in nonstratigraphic associations. The shallow depth range has given this branch of geophysics the name *shallow geophysics*. The reason for the difference in depth scale lies in the temporal nature of the phenomena under investigation: the deeper the geology, the older the features and the less likelihood of an anthropogenic connection. While the deeper geology can be of profound interest in itself—for instance, the placement of anthropogenic features can be and is determined by it—for our purposes we shall emphasize the shallower horizons most likely to have been modified or created by human agents.

Geophysical techniques can be differentiated as either active or passive systems. Active systems are those which introduce a perturbation into the subsurface and measure the physical result. Electromagnetic and seismic techniques are examples of active techniques. Passive techniques measure magnetism, temperature, and gravity. Passive techniques use sensors that measure ambient levels with the aim of detecting any perturbation within these. Both varieties have great utility in archaeological geology.

Active Techniques

Seismic

Seismic techniques have the longest history in geophysics, having been developed for the prospection of petroleum and gas deposits.[5] They are discussed first not only because of this historical priority but also because of their phenomenological importance in guiding intellectual inquiry into the use of subsequent techniques for archaeological prospection. Seismic techniques had little direct use in the detection of terrestrial archaeological phenomena due to the frequencies and power levels utilized, which completely missed archaeological features. Seismic exploration has had success in the detection of underwater archaeological sites, however. We shall discuss this field of research in some detail, but first we shall examine the physical properties of the original application of seismic techniques.

Seismic prospection of archaeological sites on land still suffers from the difference in the properties of sound and its absorption when compared to those in water. The physical principle involved in seismic techniques, notably seismic reflection and refraction, is of importance to the discussion. The introduction of sound into the ground or water at sufficient levels to produce a return echo is the simple principle used in seismic reflection and refraction. Measurement of the velocity of the returning echo is used in refraction surveys, and the frequency of the returning pulse is used in reflection studies. Neither technique has had great utility in terrestrial archaeological survey, except for seismic reflection of the seafloor.[6]

The seismic pulse is a pressure phenomenon and its principal component is termed the P-, or pressure, wave. The secondary component, the S-wave, is that part of the wave train that radiates orthogonally to the direction of the P-wave. Other components have lesser importance to seismic studies, such as the Rayleigh wave, but most models are built around the physical behavior of the P and S components. The measurable strength of the reflected wave or pressure component and the change in the phase or frequency comprise the S or secondary component. Changes in the P and S components can tell the geophysicist much about the nature and depth of geological strata, each of which can be characterized by the seismic data and termed *reflectors*. A typical seismic profile is shown in figure 8.1.

The limiting factor in the use of seismic techniques in terrestrial archaeological prospection has been the depth of the typical archaeological feature. Seismic reflections and refractions from these shallow features tend to be directly proportional to the strength of the incident wave as well as its wavelength. The typical seismic wavelength is in the range of a few hertz (Hz) to a kilohertz (kHz). The "seismic band" is usually thought of as 0 or 3 to 125 Hz. Waves of this size have the advantage of penetrating deeply into the earth but resolving only features coincident with their size. It follows from this that an unconformity of a specific thickness will only be resolved by an incident wave of equal or lesser size. The accuracy in determining depth to an unconformity is termed *vertical resolution. Lateral resolution* is determined by the geophone array (a series of acoustic receivers). For instance, the vertical resolution of a 20-Hz wave will be on the order of a few meters (~5 m). The resolution of a 3-kHz wave, commonly used in shallow marine surveys, is only about 0.5 m. While there are archaeological features of these sizes, such as building remains, walls, ditchs, roads, and subterranean constructions, their burial depth is often within the initial part of the seismic wave and reflections of same are difficult to interpret. Additionally, the burial medium itself adds to the problem of discrimination by scattering the wave in such a manner as to make a seemingly simple process unduly complicated in interpretation.

In terrestrial circumstances the seismic wave is generated by an energy source such as a metal plate struck by a hammer. The detection of the returning sound energy is accomplished by the use of geophone arrays. In the seismic refraction technique the incident wave encounters a subsurface with varying elastic properties, as with reflection, the wave's energy is partitioned into reflected *and* refracted components. At some particular angle of in-

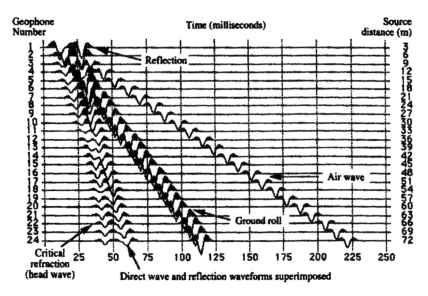

Figure 8.1 *Top:* schematic of a typical seismic profile with principal elements indicated: head wave; ground roll; direct arrival; air wave; *right:* a seismic profile (image is rotated 90° with respect to upper image) from adjacent to the Thames River (modified from EASEDROPPER(c), Kansas Geological Survey).

cidence, the downward wave front propagates along the surface of a subsurface medium or reflector at the velocity of that medium.[8] A portion of this wave is received by the geophone array, and the depth and refractor velocity can be calculated using the arrival times at the various, equally spaced geophones. The geophones are typically equally spaced, such as 10 feet (3.3 m) apart. The first wave to arrive is called the "first or direct arrival" and stops a timer started by the sound source. Generally the direct arrival has traveled the shortest distance, just under the ground surface, to the nearest geophone. Should the wave travel the 10 feet in 10 milliseconds, then the velocity is 1,000 ft/s (330 m/s).

Typical seismic velocities for sediments near the surface are 1,200–2,200 ft/s; velocities in bedrock strata range from 5,000 to 16,000 ft/s. By measuring the arrival time, T_x, and the distance, X_n, for different sensor/source pairs we can, by a simple time-distance plot, determine the depth of strata. One can determine the respective line-segment slopes of the example in figure 8.1 and determine velocities. Here the 0.4 ms/ft yields a reciprocal of 2.5 ft/s or 2,500 ft/s for the first line segment, which represents the shallow surface layer. The deeper bedrock layer, indicated by the steeper line segment, corresponds to a reciprocal slope of 10,000 ft/s (3300 m/s). It can be seen that a keystone assumption of seismic studies is the increase in velocity with increasing depth.

The first arrival at the second geophone, 20 ft (6.7 m) distant, stops the timer and its velocity is calculated. If the velocity is the same, then we are still measuring the surface wave. At the more distant, third geophone, we measure a time of 26 ms, which produces a difference in the expected velocity. Instead of 1,000 ft/s we now calculate a velocity of 1,150 ft/s. At the fourth geophone, 40 ft (13 m) away, the sound arrives in 28 s. It traveled the addi-

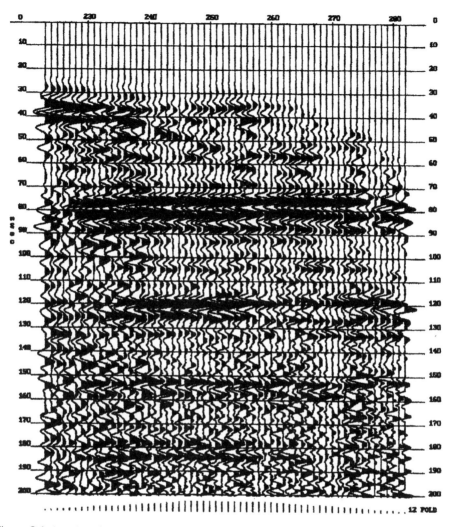

Figure 8.1 (*continued*)

tional 10 ft between the third and fourth geophones in only 2 s, or 5,000 ft/s (1650 m/s).

The following formula can be used to calculate the depth of the high-velocity section, which is most probably bedrock: $D = C/2 \sqrt{(V_1 - V_2)/(V_2 + V_1)}$, where V_1 is the velocity at or near the surface, V_2 is the velocity in the bedrock, and C is the "critical distance" that is found graphically in a time-distance plot. Two lines are formed: one for the velocity of 1,000 ft/s and one for 5,000 ft/s. Where the lines intersect is the critical point, and the distance to it is the critical distance, or depth C.

Differences in velocities can also indicate weathering and competency (intactness) of the bedrock and variation in compactness and continuity of soil horizons. The best returns are obtained from bedrock surfaces due to the low contrast in refractive velocities.

Table 8.1 Seismo-stratigraphic Layer Model for the Burial Mound at Rabenschwand

Seismic horizon	Velocity (m/s)	Lithic sediment description
V1	235 ± 15	Humic overburden
V2	385 ± 25	Sandy gravel horizon with prehistoric debris
V3	630 ± 50	Gravel-sand, probably lined with layered stone 2–3 m deep
V4	1040 ± 90	Morainic till gravel in natural layers
V5	2360 ± 80	Compact, moraine

Relatively spatially contiguous features such as paving or floors, can refract the incident seismic wave as in the case of bedrock. Discrimination of subsurface features is heightened by the contrast in the velocity layers, with the best results obtained where velocity differences are large. Resolution in terrestrial seismic survey is further limited by geologic noise, defined as variations in subsurface conditions not related to the exploration objective.[9]

In 1978 seismic refraction was used in concert with resistivity and radar methods at the Mayan archaeological site of Ceren, El Salvador.[10] The seismic results at Ceren were disappointing; radar less so. Elsewhere, such as at the capital of Roman Switzerland, Aventicum, the results were more positive. At Aventicum, refraction was able to successfully detect and follow the buried course of the city's long circuit wall.[11] In the former East Germany, seismic methods were in use in 1987.[12] Perhaps the most impressive demonstration of the refraction method in archaeology was in the determination of the stratigraphy of the Hallstatt period (ca. 1250 to 850 BC) barrow at Rabenschwand, Austria.[13] Along a 192-m profile geophones were set at 2-m intervals using a 12-channel system with a resolution of 0.0005 s. Table 8.1 and figure 8.2 illustrate the results of the seismic sounding of the mound, with the clear demonstration of five horizons which correlate to pedological and geological strata. Two collapsed chambers were also detected in the mound (LTI and II; figure 8.2).

Most recently, A. J. Witten et al. have demonstrated the newest shallow seismic refraction technique (termed geophysical diffraction tomography, or GDT) in the characterization of subsurface tunnels and rooms at the Chalcolithic site of Shiqmim in the Negev Desert. Combined with imaging software it was possible to display a portion of the large site (>10 hectares) using the GDT data.[14] The technique holds great promise for archaeology although it shares some of the logistical problems common to both seismic and resistivity techniques: extensive cables and ground contact probes.

In the marine environment the use of direct seismic or sonar reflection has proven to be a valuable technique for detecting surfaces and unconformities which aid in the detection of anthropogenic features. Individual cultural features are resolved in these marine strata as parabolas. At the seafloor sonar can define objects with great detail. Archaeologists searching for the remains of prehistoric cultures on the sea or lake floor rely on a hypothesis that posits a high probability of co-occurence of archaeological facies with specific landforms such as Pleistocene terraces or inundated shorelines.[15]

In marine and lacustrine surveys, subbottom profiling, combined with side-looking (side-scan) sonar, can survey large areas with both bottom and subbottom returns produc-

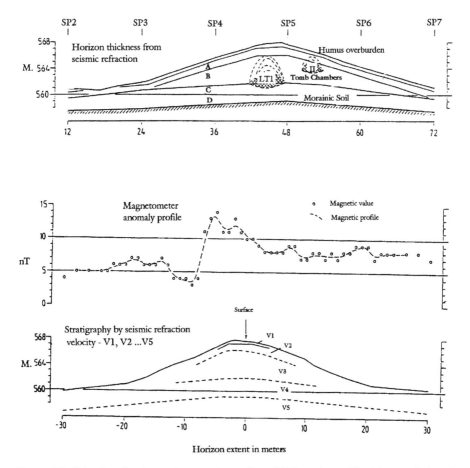

Figure 8.2 Seismic refraction and magnetic profiles of Rabenschwand burial mound, Austria (after Wolach, 1987, and Neubauer, 1990).

ing unambiguous records of topographical and subsurface features, as illustrated in figure 8.3 (see chapter 3 also). The high definition of these sediments is accomplished by the use of sound frequencies higher than those used in deep seismic studies. The frequency range is small, from a few to 1000 kHz, but these frequencies can profile the upper marine sediments to depths of up to 30 m with ease. The first use of these "high-resolution" techniques was for the detection of what are termed "geological hazards or geohazards," such as unconsolidated biogenic sediments and near-surface gas pockets, which can cause problems for the placement and operation of offshore petroleum structures, from slumpage to catastrophic blowouts.

M. J. Stright recognized the archaeological potential of geophysical data collected during geohazard surveys. Beginning in the late 1970s she evolved a methodology for the recognition of potential archaeological sites embedded in marine sediments of the Gulf of Mexico and elsewhere.[16] Combined with archaeological geological techniques first used on

Figure 8.3 *Top:* Side scan sonar mosaic image of 81 km² of seafloor showing sediment hardness differences (darker = harder; lighter = softer); *right:* Graphic interpretation of top figure. Note the paleoshoreline along the 70 m isobath. This feature corresponds to a late Pleistocene stillstand.

terrestrial sites, this methodology reached its maturity in an ambitious attempt to locate and identify prehistoric habitation sites in the now drowned lower sections of the Sabine River valley on the northern Gulf of Mexico continental shelf of Louisiana and Texas.

Sherwood Gagliano, Charles Pearson, and their coworkers used elements of spatial geography, archaeological geology, and geophysics in a unique attempt to detect and identify buried archaeological sites on the continental shelf. Gagliano had used geomorphic attributes to predict the occurrence of prehistoric sites in the lower Mississippi River valley and in coastal Louisiana.[17] In 1982 Gagliano and Pearson collaborated on a study to characterize archaeological sites by the use of sedimentological attributes.[18]

The expectation of finding offshore archaeological sites is a direct consequence of the cycles of dramatically lowered sea levels as a result of the Pleistocene glaciations (figure

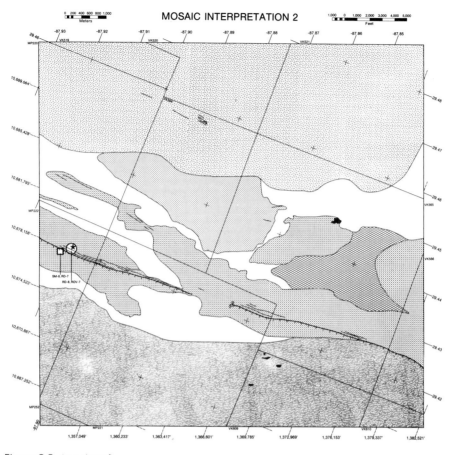

Figure 8.3 (*continued*)

3.3). Knowing that Pleistocene sea levels were as much as 200 m lower than today's, archaeologists speculated on the variety of archaeological sites that must exist on the now drowned continental shelves of the world's landmasses.[19]

Coupled with this intellectual interest were the more practical concerns of U.S. environmental policy related to the development of offshore oil and gas deposits. Included under the umbrella of U.S. environmental policy was a concern for the protection of archaeological sites. In its original form this policy was more readily applicable to historic shipwrecks, but M. J. Stright—a federal archaeologist—realized their legal extension would include prehistoric archaeological sites.

In the 1970s not all archaeologists were convinced of the existence of archaeological sites on the continental shelf. Those that did accept their reality were still not readily convinced as to their merit as objects of serious scientific study.[20] One question concerned the preservation of these prehistoric sites in the face of the marine transgressions that followed the end of the glacial periods. Indeed, there were serious questions even about the nature of the rise of Postglacial sea levels. The early sea-level curves of Fairbanks and others were

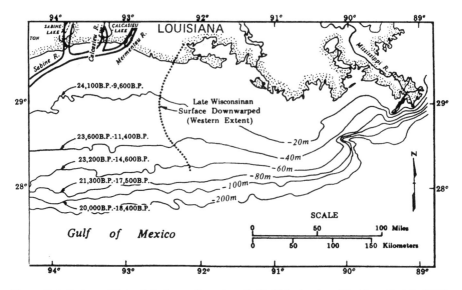

Figure 8.4   Various stillstands in the northwestern Gulf of Mexico (modified from Stright 1986).

based on a paucity of data and relied more on intuition than reliable curve fitting.[21] Research since the 1960s has uncovered direct evidence of the behavior of the sea-level rise and has made more reliable estimates of the dating of the sea's low- and stillstands using new techniques such as [230]Th and TL on shell and coral.[22]

As we noted in our chapter on geomorphology, sea level is a complex issue tied to variables as disparate as the shape of the geoid, isostatic rebound and the mass-wasting behavior of the polar ice sheets, to name but three of the more important.[23] What is sure is that the sea-level rise was anything but slow and steady. On the Gulf of Mexico continental shelf alone the sea has retreated and returned in a start-and-stop fashion over the last 20,000 years (figure 8.4). There was adequate time for the gradual processes of alluvial and colluvial action to build up deposits that would protect prehistoric habitation sites from the returning sea's erosive force, particularly if the transgression was a rapid one and if the site was in a felicitous location such as a lagoonal deposit behind a coastal barrier island chain.

Gagliano and Pearson's models for settlement patterns and therefore prehistoric site distribution on the Gulf of Mexico's continental shelf were extrapolations based on patterns observed for coastal Texas and Louisiana. Early Holocene sites were known to cluster on terraces at or near confluences of main channels and distributaries of the Sabine River valley. Using seismic signatures for these paleolandforms derived from Stright's work, the study proposed to sample these loci using deep (30-m) vibracoring. The recovered core samples were then analyzed using the attributes found from their earlier work: point count analyses of fine fractions, grain size, soil chemistry, and radiometric age. The numerical and statistical parameters of the various data were then compared to their predictive model based on the coastal site research.

On the basis of their analyses Pearson and his colleagues were able to locate two loci where the indicator variables were in good enough agreement with those of the coastal sample to conclude that they were potential archaeological sites.[24] Figure 8.5 shows their geo-

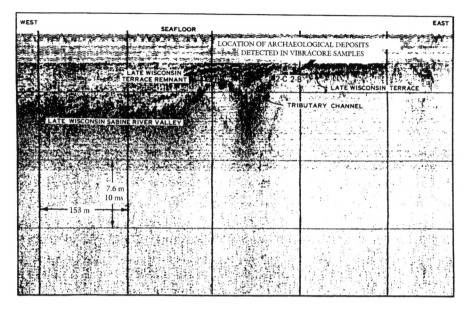

Figure 8.5 3.5 kHz subbottom profiler record across a paleochannel.

graphic and vertical location below the sediments of the drowned Sabine River valley, as well as the core sections through the anthropogenically produced strata.

Are they really prehistoric sites? Good science demands healthy skepticism. Clearly, the potential sites meet the experimentally defined criteria for human disturbance of the physical and chemical makeup of the sediments. But can other processes mimic the range of values for those attributes chosen to define human presence on a landscape long submerged? In truth, we cannot be sure as Pearson and his colleagues would agree. Nonetheless, the results are compelling support for the proposition that early human presence can indeed be found if one approaches the problem with good scientific tests.

Digital Sonars

New seismic systems such as swept frequency modulated devices or CHIRP sonars, which use digital processing of the returning signals, can produce dramatic increases in the resolution of thin sediment layers, as demonstrated in figure 8.6, a CHIRP sonograph of Narragansett Bay.[25] The resolution of the sediment bedding and the sharp delineation of unconformities as thin as a few centimeters are indicative of the power of these new systems. Coupled with graphic processing of the seismic image, the CHIRP sonar offers exciting opportunities for the archaeological geologist in the characterization of inundated geomorphologies.

Electromagnetic

In most discussions of geophysical survey methods based on electromagnetism, techniques are discussed as "electrical" methods and "electromagnetic" methods. There is some merit

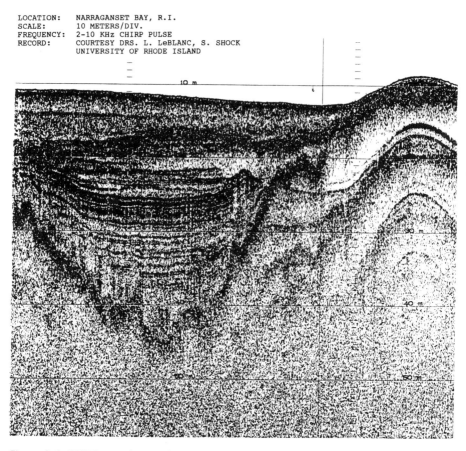

```
LOCATION:    NARRAGANSET BAY, R.I.
SCALE:       10 METERS/DIV.
FREQUENCY:   2-10 KHz CHIRP PULSE
RECORD:      COURTESY DRS. L. LeBLANC, S. SHOCK
             UNIVERSITY OF RHODE ISLAND
```

Figure 8.6 CHIRP sonar image of Narragansett Bay, Rhode Island. (Courtesy datasonics, Inc.)

for this in a phenomenological sense, because the former relies solely on electrical fields, E, whereas electromagnetic and strictly magnetic methods must consider both E and the magnetic field, H. Nonetheless, in a physical sense, the phenomenon of electromagnetism is treated uniformly using equations such as Maxwell's:

$$\nabla \cdot (\mu H) = 0, \qquad \nabla \cdot (\epsilon E) = q$$

where E and H are the electric and magnetic fields; and μ, ϵ, and q are magnetic permeability, dielectric permitivity, and the volume charge density, and E and H are related through the Maxwell-Ampère equation, $\nabla \times H = \sigma E$. The operator ∇, nabla, accounts for the local divergence of the magnetic field; σ is the symbol for conductivity.

Resistivity Methods

The resistance to the passage of electrical current in a volume is simply termed a material's resistance and is generally determined by the bulk electrical properties of the material.

Resistance is highest in insulators like glass and lowest in conductors like metals. Resistivity is represented by the Greek letter ρ in the following relation: $\rho = E/i$. Here E is the electrical field and i is the current density. The reciprocal of resistance is conductivity, σ. In soil and rocks, conductivity is enhanced by porosity and the salinity of water in these media. Resistance is measured in units of Ohm-meters, Ωm, while conductivity is measured in siemans, S.

To measure resistance, an electrical potential, V, must be introduced into a medium like soil. The electrical field is equal to the gradient between two electrodes placed in the soil. If I is the electric intensity through a medium in unit time, t, the current is $i = dI/ds$.

It can be further shown that $i = \sigma E$ or $1/\rho$. If V is used as a measure of E, then the current between the electrodes is proportional to the potential divided by the resistance, or $I = V/R$, which yields $R = V/I$. To measure resistance, one simply measures the potential between electrodes, or ΔV.

If the current traverses a surface separating two conductors of diffferent resistances, ρ_1 and ρ_2, then it can be shown that the trajectory of that current is broken and follows the relation $\rho_1 \cot\theta_1 = \rho_2 \cot\theta_2$. Here the angles θ_1 and θ_2 are formed in each conductor by lines of current normal to the surface of the interface. When R is large, the line of current refracts, and where R is low or equipotential, the line of current is orthogonal or nearly so to the interface. These cases are shown in figure 8.7.

The resistivity of the soil is termed the apparent resistivity, ρ_a, since it will vary relative to the local conditions found in the soil at a particular locus. In practice soil resistivity is measured using multiple electrodes in spaced arrays. The most common is the Wenner array, in which equally spaced electrodes are arranged linearly. The electrodes at the ends act as poles between which the current passes. The two middle electrodes measure the voltage. Other common arrays are the Schlumberger array, with unequal distance between the electrodes; the dipole-dipole (double-dipole) array, with the first two electrodes acting as the measurement electrodes and the second set being the poles for the current; and the three-electrode (pole-dipole) array, much like the dipole-dipole arrangement except using only one electrode current source (figure 8.8). Of these four arrays, the dipole-dipole arrangement is the most sensitive to lateral variations in resistivity because it measures the second derivative of the potential, V, from a simple current source. Resistance varies widely in materials. Table 8.2 gives R values for typical rocks and sediments.

One can look for resistivity contrasts to distinguish the separate layers with either the Wenner or the double-dipole array, the resistivity will also vary as a function of the depth. To evaluate this, the electrode spacing can be varied. For small electrode spacing, the apparent resistivity is generally equal to the true value of the upper or surface layer. Increasing the spacing, ρ_a will, in theory, approach that of the basement layer, with the depth to the top of that layer linearly proportional to the interprobe spacing.[26]

European archaeologists were the first to exploit resistivity and conductivity to rapidly, when compared to the drudgery of spadework, find sites and features.[27] First attempts often used far too much of the wrong kind of voltage, several ohms of direct current, which made life hazardous for the surveyor as well as sharply reducing the effectiveness of the electrodes themselves. This was particularly true with the simple two-electrode arrays. Experimentation and the advice of colleagues from the physical and engineering fields soon rectified these early problems. Resistivity and noncontact techniques using conductivity measurements rapidly proved effective, in appropriate soil conditions, in location and delineation of archaeological features from fire hearths to fortification ditchs. In such a man-

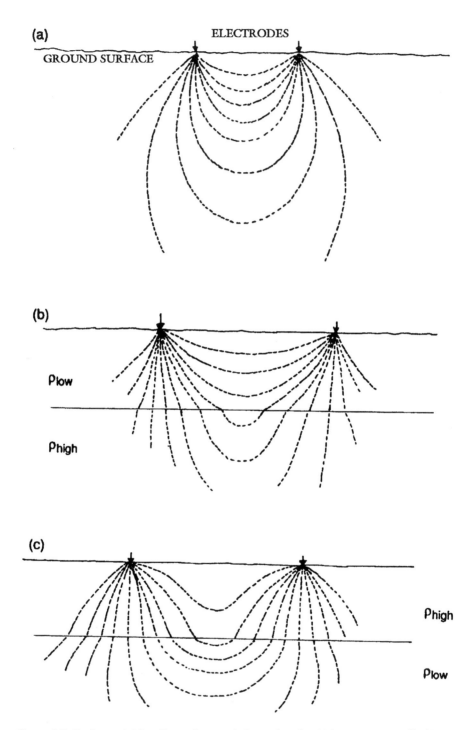

Figure 8.7 Equipotential flow lines of current in layered earths: (*a*) homogeneous soil; (*b*) two-layer soil, with lower soil more resistive; (*c*) two-layer soil, with upper soil more resistive.

ner archaeological geophysics was born. Resistivity studies of archaeological sites now take advantage of advances in digital processing, as well as the acquisition of sounding data, to enable the creation of vertical and horizontal resistivity sections.[28]

Conductivity Methods

Development and experimentation has continued in the use of archeological remote-sensing procedures using electromagnetism. The principal aims have been to make the resistivity and its reciprocal, conductivity, easier to use. Conductivity techniques evolved to a large degree in response to the cumbersome and slow nature of early surveys using resistivity meters with large lead-acid batteries and confusing arrays of electrodes and wires. The main advantage of time and frequency domain conductivity methods lies in the rapidity with which a survey can be conducted with reasonably the same results as can be achieved with resistivity.

The conductivity, or EM, meter is much like the familiar metal detector in that it has an active transmitter (search) and receiver coil circuitry in which voltage generated in the search coil produces eddy currents (voltage) by electromagnetic induction in sediments or rock. In contrast to the metal detector, the EM device uses separate coils that are often widely spaced. The weak eddy currents, so generated, in turn produce magnetic fields which reproduce currents in the detection coils of the device. A voltage meter, built into the circuit, measures the strength of electrical permitivity of the soil or rock as a function of the induced and measured voltages in the coils.

I. Scollar and others have pointed out the importance of distinguishing the phase components of the secondary magnetic field at the receiver coil.[29] All waves—electromagnetic, sound, or whatever—have the two properties of phase and amplitude. Although we do not ordinarily think about it, the FM and AM bands of our radios are direct expressions of these properties. The strength of the received radio signal can be modulated (i.e., controlled) by using either the phase (through the frequency, FM) or the amplitude (AM).

A simple elastic wave can be described in one dimension, x, as follows: $\Phi = \Phi \sin(\lambda x - \Omega t)$, where Φ is the amplitude, λ the wavelength, Ω the frequency ($2\pi f$), t the time, and x the direction. The wave varies as a sine function. A wave can vary in amplitude and phase. The change in phase, or phase shift, reflects the time difference between the waveform at the transmitter coil and the waveform at the receiver coil. This phase shift is measured as an angular measure in degrees—45, 90, etc.—of the offset in the phase of the wave.

M. S. Tite and C. Mullins determined that EM meters do not measure electrical conductivity directly unless they use a property of the wave's phase called the "quadrature" component rather than the "in-phase" (180-degree) component.[30] The quadrature component is a direct property of the sinusoidal waveform. The quadrature component of the wave at the receiver coil is exactly 90 degrees out of phase with that of the transmitter coil. As such, the magnitude of this property of the secondary magnetic field at the receiver is directly proportional to the ground conductivity. Instruments such as EM meters can measure both the in-phase and the quadrature components at a point by simply turning a switch. The in-phase measurement is particularly sensitive to highly conductive, buried objects such as metal. In this configuration the device acts as a very effective metal detector. The principal difference between the ordinary metal detector and an EM device operating in an in-phase mode is that the EM device typically operates at a frequency at or below 10 kHz. At fre-

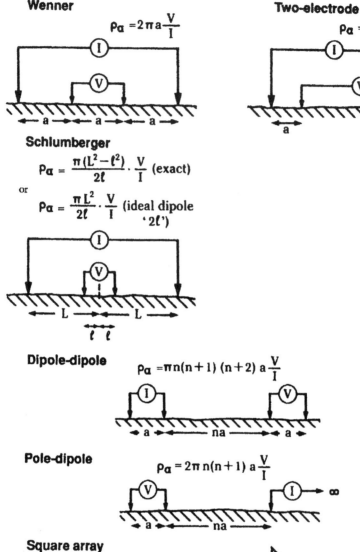

Wenner

$$\rho_a = 2\pi a \frac{V}{I}$$

Two-electrode

$$\rho_a = 2\pi a \frac{V}{I}$$

Schlumberger

$$\rho_a = \frac{\pi(L^2 - \ell^2)}{2\ell} \cdot \frac{V}{I} \text{ (exact)}$$

or

$$\rho_a = \frac{\pi L^2}{2\ell} \cdot \frac{V}{I} \text{ (ideal dipole '} 2\ell\text{')}$$

Dipole-dipole

$$\rho_a = \pi n(n+1)(n+2) a \frac{V}{I}$$

Pole-dipole

$$\rho_a = 2\pi n(n+1) a \frac{V}{I}$$

Square array

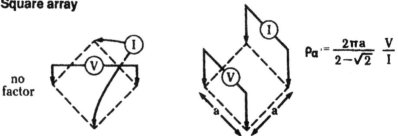

no factor

$$\rho_a' = \frac{2\pi a}{2 - \sqrt{2}} \frac{V}{I}$$

Figure 8.8 Various resistivity electrode array configurations.

Table 8.2 Resistivities of Common Rocks and Sediments

Material	Frequency at measurement Hz (if not direct current)	Resistivity, ohms/cm
Galena	—	0.5–5.0
Pyrite	—	0.1
Magnetite	—	0.6–1.0
Graphite	—	0.03
Rock salt (impure)	—	3×10^3–5×10^5
Serpentine	—	2×10^4
Siderite	—	7×10^3
Igneous rocks		
Granite	—	10^8
Granite	16	5×10^5
Diorite	—	10^6
Gabbro	—	10^7–1.4×10^9
Diabase	—	3.1×10^5
Metamorphic rocks		
Garnet gneiss	—	2×10^7
Mica schist	16	1.3×10^5
Biotite gneiss	—	10^8–6×10^8
Slate	—	6.4×10^4–6.5×10^6
Sedimentary rocks		
Chattanooga shale	50	2×10^3–1.4×10^5
Michigan shale	60	2×10^5
Calument and Hecla conglomerates	60	7×10^3
Muschelkalk sandstone	16	7×10^5
Ferruginous sandstone	—	
Muschelkalk limestone	16	1.8×10^4
Sediments		
Unconsolidated wet clay		20
Marls		3–70
Clays		1–100
Alluvium and sands		10–800
Glacial till	—	5×10^4
Oil sand	—	4×10^2–2.2×10^4

quencies below 10 kHz, the in-phase component is practically independent of the quadrature component. These properties can be used to determine information about a buried feature in that a low-conductivity feature has relatively the same shape and response in both in-phase and quadrature but the in-phase component reacts more sharply to higher conductivity.

Most EM devices are highly transportable. By fixing coil separations to less than 4 m, a device can be constructed that one person can carry. Some models have coil separations that maximize deep penetration of the subsurface. Coil separation plays an important role in the detection of small, nonconductive objects or features because the larger the separation, the less the sensitivity. As a rule, in in-phase operation, the EM device can detect objects as large as or larger than 1.5 times the coil separation. Devices used to measure the magnetic field amplitude and phase are said to be operating as "frequency domain EM." The other form is termed "time domain EM." In physical principle, frequency domain EM is a simple Fourier transform of time domain EM.[31] The practical difference, from an archaeogeophysical standpoint, is the frequency domain EM's greater sensitivity to near-surface features. Time domain EM methods do not typically measure the conductivity of the first few meters of the subsurface. This is a result of the continuous nature of the primary field current in frequency domain EM as opposed to the discontinuous field signal of the time domain EM instrument. The strength of the secondary field in this situation is directly proportional to the exploration depth.[32]

Increasingly, both EM techniques are being used with other instruments (magnetometers and radar) with good success. When used with radar, the effectiveness of the radar can be estimated directly from the measurement of the magnetic susceptibility the EM device can provide. We typically use the term *skin depth* to describe the relationship of frequency to sounding depth. Skin depth refers to the fact that electrical fields tend to concentrate near the ground surface. Lower frequencies tend to have greater penetration. With most radar, the frequencies are so high that a strongly conductive soil limits the penetration. A rough rule of thumb is 35/conductivity = skin depth for radar (in meters), where, for example, a σ of 35 millisiemans (mS) would imply a skin depth of 1 m.

Other Electromagnetic Techniques and Very Low Frequency

For the detection of very shallow (~1 m) ferromagnetic materials, a variety of instruments, termed "metal detectors," based on electromagnetic principles are available.[33] These devices operate according to several principles, of which (1) beat frequency, (2) induction balance, (3) pulsed induction, and (4) very low frequency (VLF) are the most common.[34] Many, if not most, of the models bought are in the hands of the public, many of whom use them on archaeological sites to "hunt" with metal detectors, resulting in the loss of artifacts and much information. However, their use by archaeologists, such as at the Little Bighorn and Palo Alto battlefields, has produced important data on the spatial distribution and type of ammunition and metal artifacts.[35] This information has resulted in the reinterpretation and, subsequently, a fuller understanding of events.

The VLF method has the advantage of being either an active or a passive device. In the active mode, the device generates a low-frequency (2–25 Hz) pulse which is conducted along the strike or axis of an object or feature.[36] Differences in the phase of the induced current are measured rather than the amplitude. For instance, a ferromagnetic material will increase the magnetic field while a diamagnetic object will decrease the field. This allows the VLF device to "discriminate" between metals such as iron, bronze, and gold. Some VLF instruments are passive devices relying on the induction of electrical currents in conductive bodies by powerful radio transmitters. These VLF transmitters exist at a variety of locations around the globe. A metallic feature whose axis is orthogonal to the VLF field will not be detected. Interpretation is simple: the conductor is located at the point of positive/negative inflection and maximum field strength.[37]

Passive Techniques

Magnetic

Speed, due in part to their noncontact measurement system, and sensitivity to small-scale variation in local magnetic fields have moved magnetometric instruments to the forefront of archaeological geophysics. Magnetic devices rely on the fundamental nature of the earth itself. Because of its structure, the earth behaves as a dynamo, with the metallic nickel-iron core in contact with a viscous lower mantle generating electrical fields which in turn produce a magnetic field. This field behaves as a huge magnet with north and south magnetic poles and varies in intensity from the poles to the equator, roughly from 70,000 to 35,000 nanoteslas (nT). The magnitude of the magnetic field observed at any point is a function of a feature's magnetic susceptibility, remanent magnetization, size, and distance from the measurement device.

Soils, features, and objects vary in their ability to be magnetized largely due to the amount of iron present. The degree to which a material becomes magnetized in a magnetic field is termed its *magnetic susceptibility*. Remanent magnetism is a product of an object's composition and its thermal, depositional, and diagenetic history.

Magnetic materials are "susceptible" to both electrical and magnetic polarization, hence their utility for resistivity, EM, and magnetic geophysical methods. This polarization results in part from the displacement of the electron clouds surrounding atomic nuclei. Due to their orbital motion, electrons have magnetic moments, so their displacement produces a net dipolar configuration in the atom or molecule much like that of a bar magnet, with a $+/-$, or "north-south," arrangement. Some atoms have permanent magnetic moments due to a surplus of electrons in unfilled orbits (shells) and are called paramagnetic. Atoms with no net moment are called diamagnetic. Most paramagnetic materials have little net magnetism. Ferromagnetism is in essence the same as paramagnetism except atoms of this type behave very differently individually and in volume. Ferromagnetic atoms, such as iron with four unpaired electrons, have a strong moment and these electrons can act together in what are termed "domains" of 1 μm in size, giving the material a large net magnetic moment per unit volume.

The magnetic moment of an atom, μ, is determined, as we have said, by the number of unpaired electron spins, n: $n = \sqrt{[n(n + 2)\mu_B]}$, where μ_B is the Bohr magneton, 0.927×10^{-20} ergs gauss^{-1}. Iron, Fe^{2+}, has a spin-only μ of 4.9.

In the absence of a strong applied magnetic field, H, the net magnetic moment is essentially random in direction due to thermal motion. In the presence of an applied field, along with that of the earth, there is partial alignment of these domains, although the net alignment is slight. This magnetic induction due to the applied field allows the direct measurement of magnetic strength in a magnetometer. The simplest form of magnetometer is the compass needle, which responds nicely to fields of several thousand nanoteslas. Another early instrument is the Schmidt magnetometer, which is a bar magnet with its center of gravity offset from its center of magnetism. In a strong local field, the bar will rotate proportionally to the strength of small fields (<100nT).

The first practical electronic magnetometers used the fluxgate principle. A coil is wrapped in magnetic metal, and an alternating current sent through the coil is distorted proportionally to the local magnetic field. The fluxgate instrument measures only that component of the field that is parallel with the coil axis, and it has the advantage of combining a Schmidt device's sensitivity with the ability to move freely about. As a single-axis device, a fluxgate magnetometer gives information on the interaction of the local field with that of the earth (ambient field). This interaction takes the form of a vectorial phenomenon where magnitude and direction of the two fields can significantly alter the total field reading. If H and H represent the ambient and local fields, respectively, and if the two are orthogonal, then the resultant, R, is the vectorial sum of the two. Likewise, in the unrealistic instance of equal but opposite field strengths and directions, the resultant must be zero.

Perhaps the most common magnetometer used in archaeological surveys is the type based on a measurement of the precession frequency of the proton. This frequency is roughly 2000 Hz. To measure this quantity, an electrical current of a few amperes is impressed on a volume of proton-rich paramagnetic fluid like water or kerosene. When the current is turned off, a strong local magnetic field, H, is created which causes the protons to align momentarily and precess about the lines of magnetic force of the ambient field, H. The precession frequency is determined by H and departs from the "free-spin" value. A microvoltage is generated in a coil surrounding the container of fluid proportional to the precession frequency and amplified and converted to an analog or digital signal. Various paramagnetic fluids have different decay times for the precession signal (the decay time of water is about 3 s), which in turn determines the rate for the pulse/read cycle for the instrument. The faster the decay, the faster the instrument. Most proton magnetometers can operate at cycle times of 1 s at a 1 nT sensitivity. A 1 nT sensitivity means the ability to measure a 0.04 Hz variation in the precession frequency.

"Continous" reading magnetometers have been created using a property of electron behavior in volume known as "optical pumping," which relies on a phenomenon termed the Zeeman effect. The Zeeman effect simply refers to the characteristic of orbital-electron energy levels to separate by some value of δE in the presence of a magnetic field (recall our discussion of ESR, chapter 6). This value is equal to Planck's constant, h, times some frequency, v, or: $\delta E = hv$. When energy, such as heat or light, is absorbed by one of these electrons, it can move to another energy level provided it is allowed by the spin value of the particular electron and the frequency of the energy absorbed. An electron with a spin of $+1/2$ can only move to a level with a $-1/2$ and vice versa. The energy difference between levels is affected by a magnetic field, as discovered by Zeeman. When this difference is measured, one finds a value for the local magnetic field. Electrons that are raised to different levels are said to have been "pumped" to those levels.

Spectral lines produced by some atoms, such as sodium, have opposite electrical vectors such that they are polarized and the light absorbed versus that emitted when the pumped electron returns to its original energy level is opposite in sign. Optically pumped magnetometers use absorption of a specific frequency in a volume of sodium or other gas (often cesium) to measure the magnetic field strength. Only so many electrons can be excited to a finite number of available energy levels (states). When this occurs, the energy levels are "filled" and no more absorption is possible. The vapor becomes transparent to that frequency of light. To cause the electrons to move back to their original levels, energy normally in the form of a high-frequency radio wave is applied to the vapor. When the vapor absorbs again, the frequency is measured, and this value is directly proportional to the magnetic field just as the change in the proton frequency is a measure of magnetic strength. Optical magnetometer systems have such rapid response in the absorption/depumping cycle they are termed "continous" in their operation. They also have very high sensitivities, in the range of 0.01 nT. While it is nice to have the ability to rapidly read the field strength, one does not ordinarily need a three-order level of resolution in archaeological situations.

Another form of magnetometer that also relies on Zeeman splitting (Zeeman Effect) of energy levels is the Overhauser-Abragam effect magnetometer.[38] This device uses a solvent like methanol in which free radicals such as triacetone-amine-nitroxide are dissolved. These ions have available free electrons which couple with the protons to raise the net polarization of the fluid by a factor of 4000–5000 when compared to that of a typical proton magnetometer. The spin-coupled media now can have as many energy levels as optical systems, again being dependent on the spin state (e.g., 1/2, 3/2). There are $2(2N + 1)$ levels available, where N is equal to the spin value. The impression of a high-frequency field "pumps" the system to saturation of the allowed electronic transitions as determined by the spin value, resulting in highly polarized media available to precess about the magnetic field. The Overhauser-Abragam magnetometer has a 30% improvement in the signal-to-noise ratio, and its sensitivity is 0.01 nT, approaching that of the optical devices. The device looks and operates like a standard proton magnetometer but, because of its greater efficiency, uses much less power in the polarization cycle of its smaller sensors.

Differential and Gradiometer Configurations

In archaeogeophysical surveys, the magnetometer, no matter what the operating principle, with perhaps the exception of fluxgate devices, measures the vectorial sum of the local and the earth's magnetic fields. In theory, the magnetometer is simply measuring magnetic contrast between an object or feature and its burial context. In soils where this contrast is low (e.g., those with high magnetic susceptibility), the detection of archaeologically interesting things is more difficult than where the contrast between the archaeological feature and the burial context is enhanced by a low soil susceptibility. In the latter case, the magnetism of the feature or its local field will produce little variation beyond that associated with orientation factors.

There are perturbations in the earth's field from both inside and outside the earth that manifest themselves in the total reading in our readouts. Variations in the earth's field are due to short- or long-term changes in the earth's dynamo, as well as to solar activity. Of the two, solar activity has the greatest impact on daily readings and can cause more than a 100 nT variation in readings over the course of a day's survey. In extreme cases, during major sunspot activity, surveys can be affected for days. This phenomenon is termed "diurnal variation" and can be compensated for by a variety of techniques, instrumental or numerical. Generally, the best appproach is instrumental, whereby one either uses two magnetometers in what is termed differential mode or uses one magnetometer in gradiometric mode. In the latter methods, a single instrument makes simultaneous measurements using two sensors in a vertical configuration such that the magnetic gradient between the two is measured. The objective of both approaches is to remove the effects of strong local gradients and diurnal variation.

Size, Shape, and Strength of Magnetic Features

Magnetic features that diverge from the earth's norm are called *anomalies* and are the phenomena observed with these devices. In earlier instruments, the precession signal was monitored as an audio signal and changes in the magnetic field were observed as alterations by

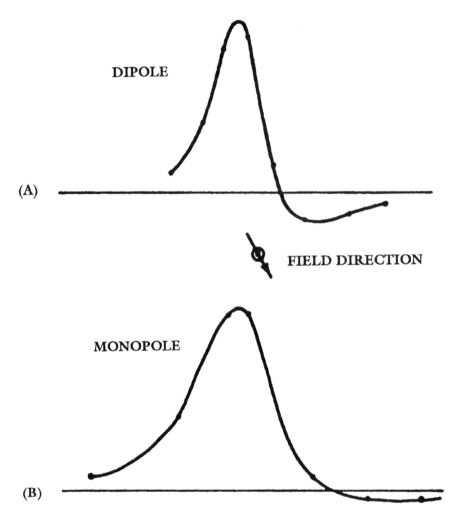

Figure 8.9 Magnetic anomaly shapes.

means of an easily heard sound frequency, or "beep". The strength of an anomaly was thus proportional to the change in the "beat frequency", resulting in the nickname for these devices of "beepers."

Although beepers were effective in locating everything from ore deposits to archaeological objects, one could not get more than a qualitative idea of the size and strength of the anomaly. To determine a quantitative value, the use of analog and, later, digital meters became common. The readings on the digital meters were directly in gammas or nanoteslas and could be recorded for later study and numerical manipulation. Much of the literature on magnetometers concerns the characterization of the size, shape, and depth of the buried anomaly.[39] The analytical procedures are legion and their aim is the same: to determine as nearly as possible the physical parameters of the magnetic feature. This is the conundrum in magnetic studies. A single reading or series of readings cannot tell us what is buried beneath the sensor or precisely where it is buried. We must deduce this through the use of models or algorithms for magnetic dipoles. Additionally we must understand that the anomaly value we record is the result of a composite of physical factors ranging from the magnitude of local and ambient field strengths, their direction, the material and shape of the object or feature, to the physical nature of the burial context. All these and other less obvious variables can intervene to make direct interpretation of the anomaly difficult.

The simplest and most direct method to resolve some of these difficulties is through the use of a simple heuristic device such as a nomogram. This graphical form of analysis relies on the typical response of buried metallic items to varying distance from the magnetometer sensor. It is undeniably a crude method and is only a way of obtaining a first approximation of the anomaly's nature. A complicating factor is the depth of an anomaly. For instance, a 50-nT anomaly could be the result of a deeply found ore body or a near-surface metallic object. Generally, the two can be readily differentiated, but it requires more than an intuitive evaluation of the anomaly. We must consider the variety of factors simultaneously at work before we can conceptualize what is being read with the magnetometer to any degree of certainty. Our interpretation of an anomaly is only as good as our initial assumptions about it; we can eliminate only so many possibilities for the anomaly's source but certainly not to the point that there is only one possible origin. There has to be an expectation or set of expectations for the range in the sources of magnetic phenomena in any magnetometer survey or the problem of interpretation becomes an iterative process similar to weather forecasting—and we all recognize how problematical that can be.

Still, we can make semiempirical attempts at deducing the form and kind of the anomaly—to be pleasantly (or unpleasantly) surprised upon excavation at the veracity (and prescience or lack of same) of our interpretation. Almost all of the so-called quantitative methods for anomaly characterization proceed with a recognition of the dipolar nature of magnetism and its orientation in a given context. The simple profile of an anomaly will change in symmetry according to the magnetic axis. For instance, in a relatively vertical ambient magnetic field, a discrete object, typically referred to as a "point source," will produce what is called a dipolar shape or signature that is more or less sinusoidal (figure 8.9). The symmetry, alone, relies on the uniform nature of the dipolar, local magnetic field of the anomaly in the context of the ambient field. If we simply change the direction of that ambient field by 180 degrees, we can, theoretically, shift the phase of the anomaly's symmetry by a like amount (figure 8.9). The amplitude of this wavelike projection of the anomaly will likewise vary in proportion to the distance of the feature from the sensor, hence the importance of burial depth. Again, depending in large part on the orientation, the magnetometer's response to a change in burial depth will vary by either the square or the cube of the distance of the anomaly from the sensor. This "falloff factor" takes into consideration the relationship of magnetic strength to distance. This recognition of the relationship between depth and orientation has led to empirical rules of thumb such as the "slope method" and "the half-width (error-curve) method." Simple nomograms or empirical techniques can assist the investigator in estimating the depth of an anomaly.

The slope method is a graphical procedure in which the slope of a line tangent to the point of maximum inflection of the plot of the magnetic anomaly's intensity is measured and drawn. Likewise, a tangent line to the point with half this slope is measured and drawn. The difference in horizontal separation, s, is measured, and this value is related to the depth of the anomaly by the formula $s = 1.6h$, where h is the depth.

The half-width method is the more common of the two methods. Again, a plot of the intensity of the anomaly is made. One then locates the point where the intensity is one-half the maximum value. The depth is then equal to the horizontal distance from this point to that of maximum intensity. It is valid for simple geometries such as spheres, cylinders, and edges of slablike features. The technique can be used for both vertical and horizontal fields. By treating an anomaly as a simple geometry, some analytical parsimony is possible. Almost all numerical considerations of magnetic anomalies proceed from a consideration of the anomaly as a sphere, a slab, or other uniform geometrical shape to allow some calculation efficiency. Here again, the difficulty in obtaining a straightforward determination of an anomaly is obvious just given the assumptions that have to be made before a numerical analysis can be done. Attempts to model magnetic data can be very helpful in assessing broad-scale representations of complex magnetic features. By bringing some assumptions to the analysis we can begin to obtain meaningful patterns in the data. Random anomalies can be seen to group into shapes which can be identified as features such as buildings, roads, or walls.

Ground-Penetrating Radar—An Active Electromagnetic Technique

In the last decade, ground-penetrating radar (GPR) has made significant contributions in engineering, geological, and archaeological applications.[40] It is basically a shallow seismic device which uses microwave-frequency radar pulses instead of sound. Its success depends upon two electrical properties of the soil: conductivity (σ) and the dielectric constant (E_r), which characterizes the relative electrical permittivity. Dielectrics allow the passage of electrical fields but do not conduct.

The quantity E_r determines the velocity of the radar signal through a material: $V_m = c/\sqrt{E_r}$ (in m/s), where c is the velocity of light. The strength of the reflected signal is determined by E_r and the impedance of the material: $r = (z_2 - z_1)/(z_2 + z_1)$, where z is the impedance of the respective soil layers and r is the coefficient of reflection. The importance of frequency, f, and the magnetic susceptibility to z, χ, are expressed in the following relationship: $z = \sqrt{j\acute{\omega}\chi/\sigma + j\acute{\omega}E_r}$, where j is imaginary, and $\acute{\omega} = 2\pi f$.

Conductivity and frequency largely determine the depth of penetration of the radar signal. High conductivity means poor radar performance. Table 8.3 gives the σ and E_r for a variety of materials. In the subsoil, the radar wave is reflected where there is a change in the dielectric properties of the materials. The similarity to acoustic seismic principles is readily apparent. Conductivity relates, as we have seen, to the resistivity and electrical impedance of a material. Highly conductive materials cause loss of energy in the electromagnetic pulse through heat absorption. In general, most dry soils (sands, silts, and clays) are reasonable conductors. The importance of water content cannot be understated, as the mobility of ions in a materials is significantly increased in saturated soils. Saline soils or any contexts with seawater present are highly conductive. Dry stone, asphalt, and concrete are relatively poor conductors.

GPR frequencies are in the megahertz range, typically from 100 to 1000. The relationship of frequency (f) to wavelength (λ) is roughly 10 mHz $\sim$1 cm wavelength. Velocities

Table 8.3 Approximate Values of σ and E_r for a Variety of Materials

Material	σ (mho/m)	E_r
Air	0	1
Pure water	10^{-4}–3×10^{-2}	81
Seawater	4	81
Freshwater ice	10^{-3}	4
Granite (dry)	10^{-8}	5
Limestone (dry)	10^{-9}	7
Clay (saturated)	10^{-1}–1	8–12
Snow (firm)	10^{-6}–10^{-5}	1.4
Sand (dry)	10^{-7}–10^{-3}	4–6
Sand (saturated)	10^{-4}–10^{-2}	30
Silt (saturated)	10^{-3}–10^{-2}	10
Seawater ice	10^{-2}–10^{-1}	4–8
Basalt (wet)	10^{-2}	8
Granite (wet)	10^{-3}	7
Shale (wet)	10^{-1}	7
Sandstone (wet)	4×10^{-2}	6
Limestone (wet)	2.5×10^{-2}	8
Copper	5.8×10^7	1
Iron	10^6	1
Soil		
Sandy dry	1.4×10^{-4}	2.6
Sandy wet	6.9×10^{-3}	25
Loamy dry	1.1×10^{-4}	2.5
Loamy wet	2.1×10^{-2}	19
Clayey dry	2.7×10^{-4}	2.4
Clayey wet	5.0×10^{-2}	15
Permafrost	10^{-5}–10^{-2}	4–8

are in the nanosecond range and are expressed in what is referred to as "two-way travel time" and measured in either ns/ft or ns/m. In practice, the radar pulse is propagated and focused by a waveguide antenna mounted on a tow sled. The tow sled also contains a receiver circuit, which converts the signal to a form suitable for display on a graphic recorder or for digital display. Figure 8.10 illustrates this relationship as well as salient components of the radar profile.

Figure 8.10 illustrates the introduction of analytical and modeling techniques more typically used in the seismic industry. The synthetic radargram was generated using a 1-D modeling program developed by D. Goodman and Y. Nishimura and incorporates thinking on the characterization of radar images by other investigators.[41] Forward modeling, as these techniques are generally known, attempt to corroborate geological (or archaeological) interpretations made from raw or processed data sets. A principal difference in the modeling of radar versus seismic data is where returning pulses are measured. In radar, the antenna or antennae are generally zero offset; that is, the radar pulse source is also the location of the receiver of the returning pulse. In seismic measurements, the sound source and the receivers (the geophones) are separated.

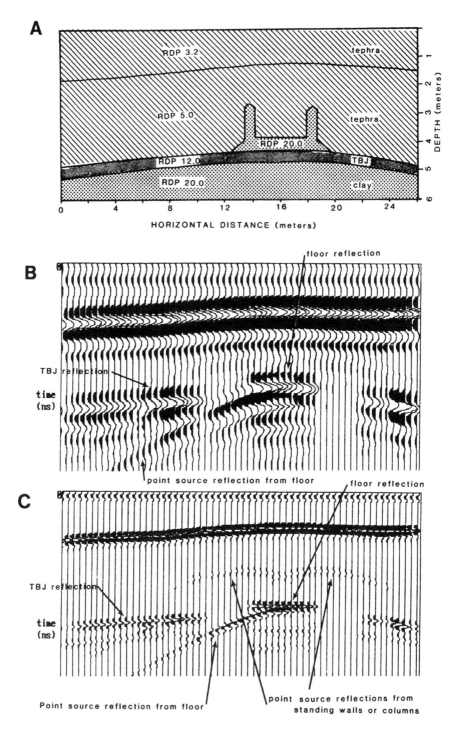

Figure 8.10 GPR images of a Classic Period Maya Structure at Cerén, El Salvador (cf. figure 7.7). Computer simulation using velocity-based seismic models. (Courtesy Payson Sheets.)

This difference can lead to problems as well as advantages in the modeling of data. In the analysis of seismic data, one can always use the direct-arrival portion of the wave on subsequent geophone locations to align the individual tracers. No such option pertains to radar data, as the zero offset between transmitter and receiver precludes the presence of a direct arrival. The GPR image is in a sense a straight-line profile in the near-field wave at or just below the antenna. Frequency dispersion of the wave (in the case of radar, a microwave) must also be considered. Dispersion in radar waves is most keenly related to moisture, particularly soil moisture content: more water, more dispersion.[42] Varied permittivity leads to greater or lesser dispersion. Goodman and others assume little dispersion in their modeling, although such an assumption might present problems in data with low permittivity contrasts.[43] Media with significant E_r values, such as sand ($E_r = 25$) and clay ($E_r = -1$), have frequency dispersion; the use of the same frequency through the synthetic model may obscure some detail, but iteration of frequency in multilayer cases can lead to realistic interpretations of the subsurface.

Another approach with important implications for archaeological interpretation of radar images is the use of "time slices" in the presentation of radar data. Goodman, Nishimura and others have successfully used this presentation of data in a survey of medieval Japanese burial mounds and kiln features.[44] A time slice represents a cut across a set of xyz data (here z = time values). It is then a horizontal slice through radar reflection data at a specific depth. Repeated such slices can examine varying depths of an archaeological site and reveal features as a function of time in the travel of the radar wave. Figures 8.11 and 8.12 illustrate both types of radar data manipulation: the first a synthetic radargram of a stone coffin and the second a time-slice representation of a medieval burial mound of the type in which such coffins are found.[45]

In archaeological contexts the effectiveness of electromagnetic techniques—resistivity, magnetic, and radar—has been conclusively demonstrated in a variety of settings in all parts of the world. The following example illustrates their complementary nature in locating and defining the complexity of a lost 16th-century outpost of Spain's New World empire: Mission Santa Catalina de Guale.

Geoarchaeological Prospection on St. Catherines Island,
Georgia: Mission Santa Catalina de Guale

In the late 16th century St. Catherines Island was the center of a Gaule-Tolomato Indian chiefdom. By 1587 both the town of the Guale chief, or *mico,* and a Spanish mission were to be found on the island.[46] By 1680 a military presence was established in a fortified *doctrina* in response to the establishment by the British of Charles Town, South Carolina, and the subsequent southward push by the British and their Indian allies. The mission was abandoned in 1684 after a British-led Yamasee attack. This frontier site of the Spanish colonial empire was never reoccupied and was quickly reclaimed by the pine and palmetto forests of the Georgia coast.

Beginning in the 1950s historical researchers and archaeologists began efforts to relocate the lost mission site.[47] Conclusive evidence of the exact location of the site was to resist the best efforts of archivists and archaeologists until a systematic magnetometer survey was performed in 1980. The enlistment of this geophysical method to find the lost mission came as a result, as most such things do, of a chance meeting at a presentation by David

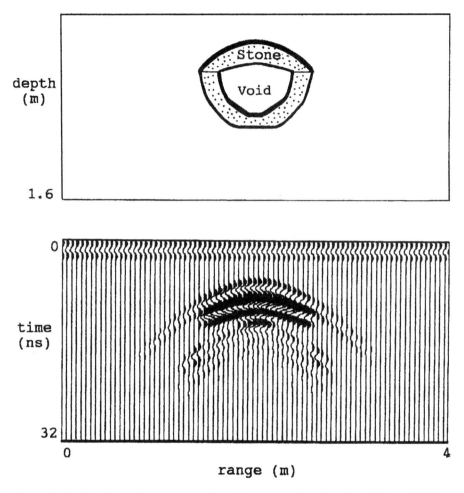

Figure 8.11 GPR simulation representation of a stone coffin from the Kofun period (AD 300–700) burial mounds, Japan. (Courtesy Dean Goodman.)

Hurst Thomas during a 1979 seminar at the University of Tennessee. In 1977 Thomas had begun a major archaeological survey and excavation project on the island under the sponsorship of the American Museum of Natural History (AMNH). As the project developed, the relocation of the Spanish mission became increasingly the principal focus. At the 1979 seminar a federal archaeologist, Rik A. Anuskiewicz, suggested the use of remote sensing in Thomas's search for Santa Catalina de Guale. As a result, a Texas A&M University team led by one of the coauthors of the present volume, Ervan Garrison, was on the island in early 1980.

Garrison and James S. Tribble, assisted by Deborah Mayer and Deborah Peters of the AMNH, followed the grid system established by Thomas on Saint Catherines (figure 8.13). Using a proton magnetometer the team found the heart of the mission in three days. This "heart" consisted of three anomaly areas originally called "right here," "over yonder," and

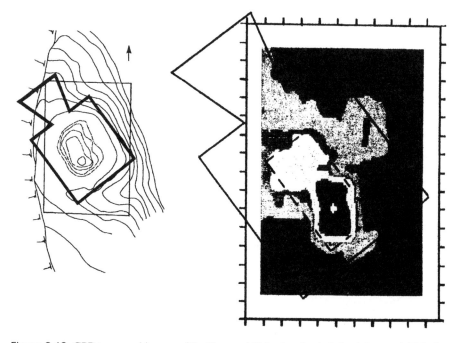

Figure 8.12 GPR topographic map of the Kanmachi Mandara keyhole burial mound, Nakajima Machi, Isinkawa Ken. (Courtesy Dean Goodman.)

"especially right here."[48] Upon excavation, Thomas's teams found at "right here" a Spanish, iron-hooped well a meter below the surface; at "over yonder," they found the collapsed burnt clay walls of the kitchen of the *convento,* the home of the Franciscan mission's friars; and at "especially right here" were the daub remains of the walls of the mission church, or *iglesia.* Thomas says it best: "Our search was over. We had discovered the *iglesia,* the house of worship at Santa Catalina de Guale, the oldest church known in Georgia. . . . the mag-netometer had detected the very heart and soul of Mission Santa Catalina."[49] Subsequent computer analyses of the magnetometer data yielded a first-order appreciation of the extent and complexity of the mission site.[50]

Thomas followed up this success with other magnetometer surveys. Realizing the po-tential of archaeogeophysical methods for not only discovery but also spatial and strati-graphic understanding of the large mission and Guale town complex, Thomas brought in other teams to use resistivity and GPR techniques. The resistivity surveys of 1982, directed by Mark Williams and Gary Shapiro, were highly successful in mapping the size, orienta-tion, and configuration of the unexcavated buildings at Mission Santa Catalina de Guale (figure 8.14). GPR surveys in 1984 by Red-R Services clearly produced detailed images of in situ features of the mission's stratigraphy to the level of individual graves and artifacts (figure 8.15).

The instrumental surveys at Saint Catherines Island conclusively demonstrated to American archaeologists what had been apparent for several years to their Old World col-leagues: instrumental prospection of archaeological sites is a robust, highly productive

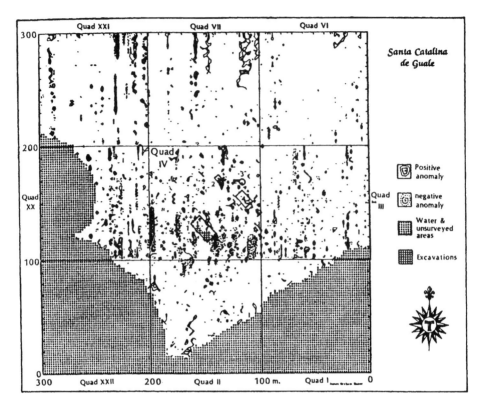

Figure 8.13 *left and right:* Magnetic contour map of Quads I–IV and XX–XXII at Santa Catalina de Guale.

methodology which represents the best of the marriage of technology and art in today's archaeology. For detailed and highly readable accounts of the history and research efforts at Saint Catherines the reader is directed to Thomas's accounts in his 1987 and 1988 publications.

Microgravity and Thermography

Gravity measurements have a long history in geological exploration.[51] Horizontal differences in gravity can be related to subsurface geological features. Microgravity meters measure local variation in gravity, i.e. anomalies, to a high precision such that cavities and fill features can be detected by reference to the normal value for gravity ($g = 9.80$ m/s^2, or 980 gals), typically expressed in microgals. At the site of the Roman sanctuary and town of Aqua Sextiae (modern Grand, France), archaeogeophysical studies have utilized microgravity to detect filled subterranean features found to be ancient channels or aqueducts under the modern town's church.[52] The contrast between the ambient gravity and that of the tunnels was on the order of 133 to -75 μgals. Also at Grand the French teams used a thermographic probe to measure small contrasts in the thermal or infrared environment.[53] The meter can

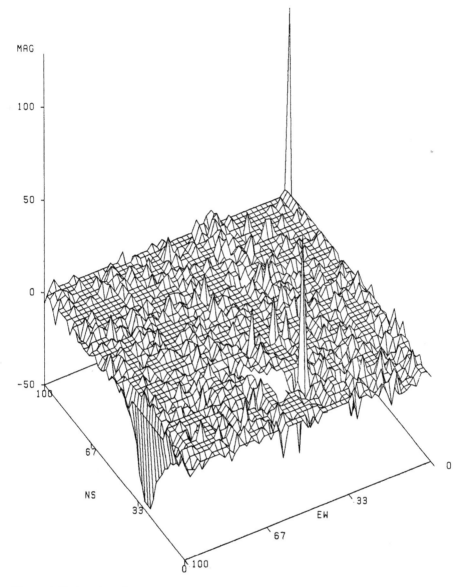

Figure 8.13 (*continued*)

read variations as small as 0.01 degree and at Grand confirmed the location and direction of ancient aquifers within structures like the village's church. Thermography units display temperature contrasts as a digital image which allows real-time visualization of the terrain and buried features.

Superconducting quantum interference devices (SQUIDs) are used in circuits of magnetometers and are the most sensitive. The 30-year-old technology has recently undergone

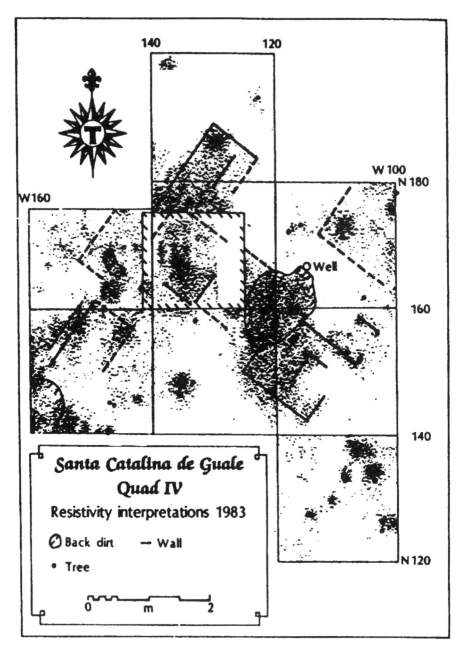

Figure 8.14 Dot-density map of soil resistivity data for a portion of Quad IV at Santa Catalina de Guale (from Thomas, 1987).

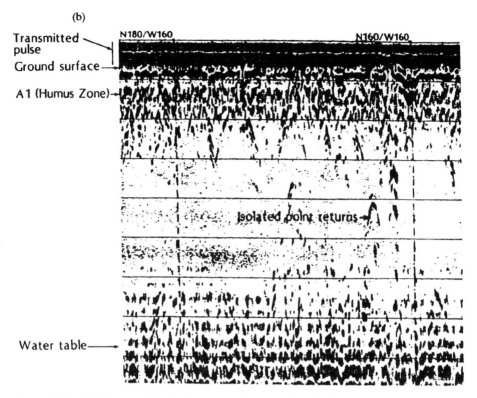

Figure 8.15 GPR images of in situ features at Santa Catalina de Guale (from Thomas, 1987).

a revolution in accessibility due to advances in high-temperature superconductors that operate at liquid nitrogen temperatures ($-196°C$).[54] In geophysics these devices have been used to map electrical resistivity in the deep earth. With low-energy requirements and enhanced sensitivity, the SQUID magnetometers can measure small changes in the magnetic properties of materials in the field or lab. To date, the SQUID devices have been used in the measurement of the remanent magnetic direction in burnt clays and rocks.

Magnetic Susceptibility

The magnetic property of soils is the most important factor in determining the detectability and geometry of magnetic features. We have pointed this out in our discussion of magnetic surveys. Susceptible soils and sediments tend to have a larger fraction of ferromagnetic minerals in their matrix.[55] Direct measurement of magnetic susceptibility (χ) has been recognized as a means of detecting human occupation or anthropogenic modification of past land surfaces.

In two recent studies, one in Switzerland and the other in Texas, the measurement of χ values in profiles correlate well with human activities such as occupation or fires.[56] χ was measured in situ and laboratory. In the Swiss study, a portable susceptibility meter was used, whereas in the Texas research, the soil/sediment samples were taken back to the laboratory and analyzed using a susceptibility bridge device. Both techniques are equally valid in the measurement of χ; the portable system is simply more convenient. Both studies were able to isolate, in the laboratory, the presence of very fine grained ferric oxides (magnetite/maghemite) in the archaeological layers. Aitken, Le Borgne, and others recognized early on the relative contributions of biological and human activities.[57] Soil bacteria tend to accelerate the transformation and fixation of ferric oxides while the effect of fire in oxidation/reduction and in "freezing" remanent moments is well understood. Where susceptibility is low in obvious archaeological strata, erosion or soil disturbance is blamed.[58]

The Future

These and other techniques are making their way into the battery of instrumentation available to archaeological geology. Combined with a burgeoning number of processing algorithms embedded in digital computer software, the ability to display and interpret the library of instrumental signatures is expanding daily. The correlation of instrumental signatures is not perfect, and many alternative interpretations exist for the same signatures, particularly for those of the electromagnetic methods. Nevertheless, the future is bright for geophysical methods in archaeological geology. The convergence of the results of more than one technique increases to our ability to discriminate between competing interpretations and supports the use of a multifaceted approach. Saint Catherines and Grand are eloquent examples of this trend.

9

Soil Phosphate in Archaeological Surveys

It has long been recognized that human activity chemically modifies the composition of the soil. This is especially true around ancient settlements that were occupied for relatively long periods of time. In areas that humans have inhabited, soil fertility is higher than in uninhabited areas because of an increase in plant nutrients derived from human and animal waste. Deep dark soils that contrast with neighboring lighter colored soils can define areas of intensive occupation with great precision.

Phosphate (PO_4^{-3}), an important plant nutrient, is highly concentrated at ancient sites and makes for an increased soil fertility. Arab farmers in the Near East have been known to use soils excavated from archaeological sites to fertilize their agricultural land. The soil phosphate has been derived from animal and human excreta and bones and dead bodies. Phosphate will be especially concentrated where animals have been enclosed.

Phosphate found in the soil can be bound chemically in a variety of ways. Since the soil is a dynamic system, its physical and chemical nature will constantly alter over time depending on local and temporal equilibria conditions.

The first studies of soil phosphate were by agronomists as a tool for agriculture. The observation that human occupation increased the phosphate concentration was noted at least by 1911 in Egypt as a result of agronomic studies.[1] O. Arrhenius, a Swedish agronomist, made the first attempt to apply phosphate studies to archaeology, in a series of papers beginning in 1929. He concluded that phosphate concentrations could be used to locate abandoned settlement sites, even where no visible evidence remained.[2] Thus, the initial application of soil phosphate analysis to archaeology was as a geochemical exploration tool to locate ancient settlements.

Human occupation should increase not only the phosphate found in the soil but also the nitrogen and carbon. These additions result from the decomposition of organic matter, principally human and animal remains and excreta. In desert or agricultural land, phosphorus in the soil ranges from 0.01% to 0.2% in the uppermost 10 cm and nitrogen ranges from

0.1% to 1%. A community of 100 people in an area of 1 hectare will add 125 kg of phosphorus and 850 kg of nitrogen to the soil annually.[3] Compared to the normal abundance of these elements in the soil, these are enormous amounts and will result in an annual increase of 0.7–7% of nitrogen and 0.5–1% of phosphorus in the uppermost parts of the soil. Although all three elements, P, N, and C, are added to the soil, only P, to a large extent, remains fixed and insoluble, especially in clays and in soils rich in Ca and Fe. The N and C are lost largely by biooxidation and in other ways.

The Nature of Phosphorus in the Soil

Phosphorus is most commonly found in nature as a phosphate ion, that is, as PO_4^{-3}. It is a principal mineral of bone largely as the carbonate apatite, dahllite, $Ca_5(PO_4,CO_3)_3(OH)$, or a fluoridated carbonate apatite, francolite, $Ca_5(PO_4,CO_3)_3F$. The phosphate must weather out of bone and go into soils to become available for plant growth. In soil solutions, two orthophosphate ions are found, $H_2PO_4^{-1}$ and $H_2PO_4^{-2}$. The relationship between these two ions depends on the soil pH, with $H_2PO_4^{-1}$ more common in acid soils (pH < 7) and $H_2PO_4^{-2}$ more common in alkaline soils (pH > 7). Phosphorus in the ionic state is "labile" and readily available to plants; that is, it is easily taken up by the plant roots. Labile phosphorus is often called *available phosphorus*. Phosphorus that is present in crystalline forms, as in the mineral apatite, is nonlabile and will become labile only slowly, adsorbed into the phosphorus pool as the mineral weathers.

In a recent study, it was found that the degree of bone preservation was related to the phosphate mineralogy. Although at the time of burial at the site, all the bones presumably had the same mineralogy, some had changed appreciably over time. Where bones were well preserved and intact, carbonate-apatite minerals were found, such as dahllite ($Ca_{10}(CO_3)_6(OH)_2$) e.g., a substitution of carbonate for phosphate in apatite, $Ca_{10}(PO_4)_6(OH)_2$. Where the bones had not been preserved, the apatite mineral was altered to the aluminum phosphate mineral variscite.[4] Thus, in soils there is an equilibrium between the labile or organic P in altered apatite and the largely inorganic P in fresher apatite. Good chemical methods to determine P in soils must be able to analyze and differentiate between the labile and nonlabile components.

In much archaeological surveying, only the available or labile P is determined. This is identical to analyses made for agriculture where available P is determined as a measure of the fertility of the soil[1] and to learn the kind and amount of nutrients that are available for plant growing conditions. The amount of labile P in a soil depends on many variables, none of which are related to archaeology, for example, soil water, soil textures and structure, and extraction of different amounts of P from the soil by different plants. The determination of available P in soil correlates only with the fertility of the soil and is controlled by modern conditions. The total amount of P actually present is not determined in an agricultural soil survey. Nonlabile P can be present as a phosphate mineral, as bone apatite, or also adsorbed onto the surfaces of oxides, hydroxides, and clay minerals.

The principal control over P distribution is pH, as shown in figure 9.1. In neutral-to-alkaline soils (pH > 6), calcium ions dominate and calcium phosphates are formed. The calcium phosphates are water soluble and are easily available for plants. The ions are either in solution or held loosely on mineral surfaces. Over time, the calcium phosphates will be

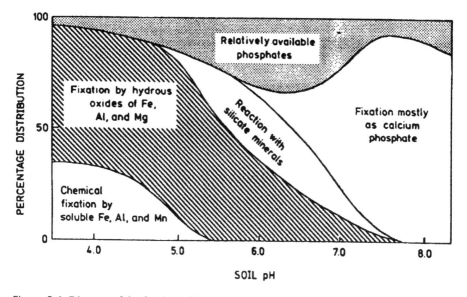

Figure 9.1 Diagram of the fixation of P vs. soil pH (modified from Brady, 1974).

gradually converted to relatively insoluble carbonate apatite. Fluorapatite, $3Ca_3(PO_4)_2 \cdot CaF_2$, is the most insoluble phosphate mineral of all and may be original, representing bone burial, even in deeply weathered soils.

In acid soils, the dominant ions are of iron and aluminum, which will also form phosphates. Commonly, variscite (the Al phosphate $AlPO_4 \cdot 2H_2O$), strengite (the Fe equivalent, $FePO_4 \cdot 2H_2O$), and barrandite (a combination of Al and Fe) will form.

In addition to its chemical nature, another important control on the availability of P is its physical distribution. As discrete phosphates precipitate on surfaces of calcium carbonate, clay, or iron oxides, P is readily soluble and available for plant intake. As discrete particles of Ca, Al, or Fe phosphates, it is less available, and as occluded phosphates (i.e., enclosed within calcium carbonate, iron oxides, silica, etc.), it is not available. The P taken up by a plant returns to the substrate when the plant, or a part of it, dies. The P then becomes available for soil organisms and microorganisms, but during this cycle, a part can also become mineralized and nonlabile.

Human activities can redistribute the soil phosphate. Animals and people eat and harvest plants which have taken up P from the soil. Animals enclosed in pens or in fields concentrate P in their manure, which can also be spread as fertilizer and returned to the soil. The P thus moves freely in the plant to animal cycle, but once it reenters a soil system, it remains until taken up again as a plant nutrient. This chemical behavior contrasts with that of N and C, which are also concentrated by human and animal activity. Whereas P is highly immobile and remains in the immediate environment, N and C are mobile and will be lost from the soil by many organic and inorganic processes. Both, for example, will form gases and enter the atmosphere and are soluble and will be carried away by groundwater.

Soil Phosphate Analysis

It should be kept in mind that soil phosphate analysis was first developed for agriculture and then, as a spin-off, found applications in archaeology. The aims of each type of survey are quite different. Agriculture is concerned with the amount of labile P in the soil—that is, the fraction available for plant growth—whereas archaeology is interested in the total P, both labile and nonlabile. This includes, for example, the very important P tied up as apatite in bone.

Many soil phosphate surveys for archaeology have followed techniques developed for agriculture and analyzed only for available P.[5] Because Arrhenius had pioneered the use of P analysis and was working as a soil scientist for agriculture, this is not surprising. The first survey to analyze for total P appears to be that of A. H. Johnson and D. Nicol (1949) in Galloway.[6]

In the history of both agricultural and archaeological analysis for P, the goals have been to make the analysis simpler to carry out and the results more discriminating. The basic analytical steps for both types of analysis involve extraction of P from the soil using an acid or alkali solution and the formation of a soluble orthophosphate. This is followed by reduction and complexing by an agent which yields a colored solution, the strength of the color being proportional to the amount of P present. A colorimeter that has been calibrated against solutions with known amounts of P is then used to determine the amount of P present in the soil sample. The intensity of the color in the sample is compared to the colors developed by solutions; the higher the P, the more intense the color. The amount and types of P extracted from the soil will depend on the strength of the acid or alkali solution. If total soil P is to be determined, then a strong digestion agent or ignition in a furnace is required to bring all the organic, inorganic, and compounds containing residual P into a form which can be taken into solution.

Comparison studies of analyses for available P versus total P have shown the importance of the latter in archaeology. A. H. Johnson studied a burial chamber where a body outline was still apparent. The available P in the chamber was only 2–11% of the total P, and the highest concentrations of the available P were under the skirts of the cairn, away from the body outline.[7] Other, later studies have shown available P as little as 1–3% of total P. Johnson's conclusion, echoed by the later workers, was that available P determinations were of questionable value to archaeology. If he had used only labile P in his cairn study, he would have missed proof that the body had been there altogether.

Sampling Strategies

If one keeps in mind the difference between labile and nonlabile P, geochemical surveys can provide a useful method to locate areas of past human activity. The method is exceptionally good where surface remains are sparse and geophysics is not diagnostic. In addition, even where sites have already been outlined, P surveys can help recognize different areas of human activity within a site, such as differentiating animal pens from residential or industrial activity.

Research on soil phosphate today is directed toward (1) a better understanding of soil processes and (2) improved strategies for collection and analysis of soil samples:

1. Research is concerned with an understanding of the processes by which phosphates form and are held in soil, the distinction between organic and nonorganic fractions, analysis of labile and nonlabile constituents, and the vertical distribution of P in soil horizons.
2. Sampling strategies include the need for care in collection and to collect a sufficient number of samples so that different populations can be distinguished. Control samples of the same soil type as found on a site should be collected outside the area to give an idea of the undisturbed P ranges to expect. D. A. Davidson found significant differences on and off the tell site of Sitagroi which allowed him to differentiate background P from P anomalies where human habitation was to be expected.[8]

Collection on site should be systematic. A collecting grid can be laid out with an initial 5 m² spacing to delineate areas of human activity. Then, where important, a secondary 1 m² grid can be used to help resolve the nature of the activities carried out in each small area. Another popular sampling strategy where the position of the site is known or expected is to lay out traverses radially from the site. Sampling is done at fixed intervals along the traverse. In any case, samples should be collected at the depth which represents the paleosoil developed at the time of occupation of the site.

When available P is to be measured, cheap, commercial kits are used, primarily to determine whether or not additional fertilizer is needed. This is a relatively rapid and simple field technique. To measure total P, which is necessary for archaeology, a 100-g sample and a lab to carry out the analyses are needed. Total P can be 100 times the available P.

Clearly, if there is a choice, the second field strategy should be followed. If resources are limited in the field, or a rapid reconnaissance is planned, then the first strategy may be implemented, with the understanding that any critical areas uncovered be sampled for total P. Care must be taken that samples are collected at the level of interest and not only modern topsoil. The P determined should be that which represents the ancient site and is not modern, as due to sheep husbandry. Care must be taken to avoid sampling around or too near tree roots or in fields where there has been recent or extensive manuring.

The Laconia Survey

W. C. Cavanagh, S. Hirst, and C. D. Litton carried out an extensive test of soil phosphate analysis in Laconia, the Peloponnesus, Greece (figure 9.2).[9] In general in Greece, where centuries of surface collecting has gone on, pottery sherds and other artifacts are not common. This study was designed to compare the usefulness of sherd concentrations with soil phosphate analysis in delineating boundaries between periods and sites. The sherds, which are subject to serious problems of migration and mixing of cultural remains, were collected by surface traverses.

Seven sites were sampled (figure 9.2):

LS31 is on a low terrace 300 m east of the Eurotas River. High amounts of building stone and glazed pottery were found. A Hellenistic villa was probably located on the site.
LS38 is a Late Classical or Hellenistic farmstead with glazed tile and other building material found. It lies close to the Eurotas River.
LS101: The pottery and glass rubble found suggest a small rural Roman dwelling.
LS108: The black glaze pottery and storage vessels confirm a Hellenistic rural settlement.
LS222 is a long-lived, terraced, Hellenistic-Roman site.

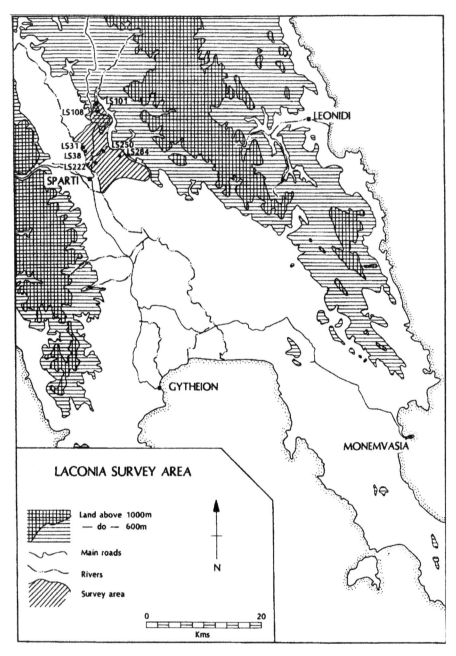

Figure 9.2 Soil phosphate survey, Laconia, Greece (modified from Cavanagh et al., 1988).

LS250 is smaller than LS222 and is Hellenistic or Roman.

LS284 is a complex site with a Byzantine church at the center. It overlaps Early Bronze Age, Early Helladic and medieval remains.

Sampling and Analysis

The center point for each site was selected by what appeared to be the density center of its sherd scatter. Four 30-m transects were laid out at right angles, following the cardinal compass directions in each direction from the center. Ten-g samples were collected at 2-m intervals along the length of each transect from 10–20-cm depth. Care was taken to avoid tree roots, manuring, and recent sources of phosphate deposition. Thus, 60 samples were taken from each site for phosphate analysis. Sherds within a 1-m area were counted every 2 m for comparison.

The method of chemical analysis used to determine total P was colorimetric.[10] In this method, the archaeologically significant total P is determined, not only the available fraction. The analytical technique was cheap and rapid and was carried out in the field. The soil sample was boiled in HCl, then 10 ml of Mo blue reagent was added. The intensity of the colors had previously been calibrated in the laboratory using a series of known concentrations of P. The soil sample color could then be compared to the standards. Reproducibility of the analyses was determined to be $\pm 3\%$; results were expressed in milligrams of P_2O_5 per 100 g of soil.

Survey Results

Forty-four background samples were collected off-site throughout the survey area, and an additional 24 samples collected from the neighborhood of specific sites were added as "off-site" samples. A total of 358 on-site samples were collected. As expected, the mean value in phosphate concentration of the inhabited sites was about double that of the background. The on-site samples also show a stronger normal distribution.

Comparing the sherd densities to the phosphate concentrations (figure 9.3) shows that P is maintained over a much larger area. This is probably due to the fact that the sherds reflect the area covered by roofed buildings and perhaps rubbish pits, whereas P delineates animal pens, compost areas, and other additional features. As a crude calculation, the phosphate might cover 0.25 ha, considerably more than the area covered by roofed buildings. Sherd surveys thus give only a minimal indication of a site's complexity and extent.

The area suggested for each site is clearly much greater using information obtained from P than from sherds. In addition, sherds are subject to degradation and movement by cultivation. Phosphate could be moved by geomorphic processes such as soil erosion and slope wash, but this was not found to have taken place in this study. Phosphates are less subject to degradation and movement out of site than sherds. For this study, the sampling and analysis for soil phosphate were clearly worth the effort.

The Ecuador Survey

At Nambillo, on the western flank of the Andes in Ecuador's Pichincha Province, fieldwork began in 1984.[11] The first two years uncovered 230 prehistoric sites in this region of rugged

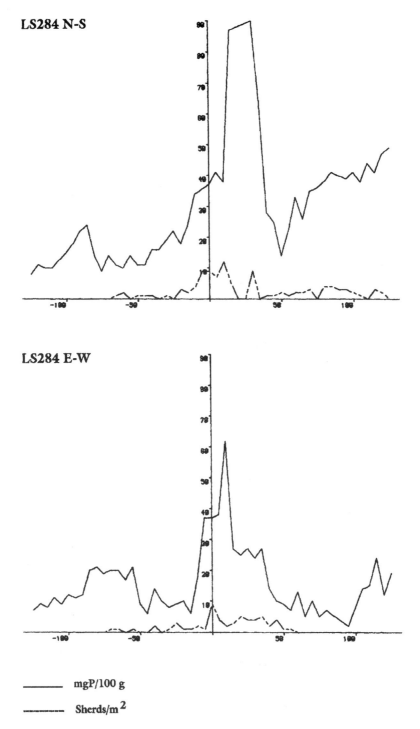

Figure 9.3 Plot of phosphate concentration vs. sherd densities at Laconia, Greece (modified from Cavanagh et al., 1988).

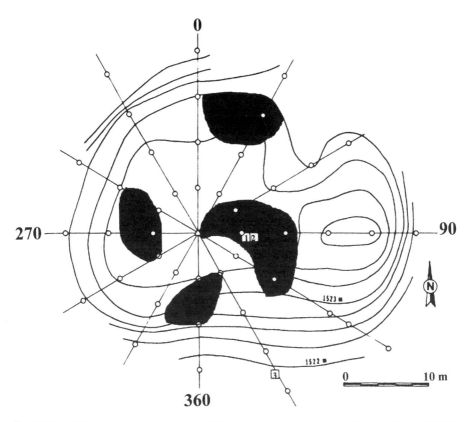

Figure 9.4 Paleotopography and areas of highest phosphate concentrations at Paleo-Hill B1 (modified from Lippi, 1988).

sub-Andean slopes and subtropical and tropical rainforests that separate the highland basin of Quito from the Pacific lowlands.

The second stage of the study involved the subsurface exploration of selected sites, including this study at Nambillo, which was carried out using soil phosphate analysis. The distribution of artifacts, coupled with historic ethnographic data, suggested a dispersed population in numerous, isolated farmsteads rather than nucleated towns.

The site of Nambillo occupies a long, narrow ridge top, about 300 m above the nearby valley floor and 1500 m above sea level. Three well-developed buried paleosols were uncovered, each yielding cultural remains, with ages ranging from 1500 BC to the Spanish conquest. Soil coring and phosphate analysis were chosen as the primary testing strategies when other, more traditional methods were ruled out by the nature of the site: large, deeply buried, heavily forested.

The paleotopography was figured out from a study of cores and by identifying paleosols. Figure 9.4 shows the topography of Paleo-Hill B1, as well as the sampling strategy used in this study. The upper boundary of each paleosol was abrupt and well defined. Its depth could be determined with a precision of ±3 cm. The base of each was transitional. The stratigraphic data were used to make maps of paleosol. Paleosol I was found to be 10 cm below the surface of the modern hill. It developed in a volcanic ash (stratum IV) and represented the ground surface before 1660, when an eruption took place.

Sampling was carried out carefully at each paleosol level. The samples were initially analyzed in the field using a colorimetric commercial kit that determined only the P available for agriculture. Since these analyses could not give the precision needed to outline areas of human activity, additional phosphate tests were run in a chemical laboratory in Quito to determine the total phosphate. As a reconnaissance tool to be used in the field, the first method often suffices, but complete phosphate analyses will eventually be needed in an archaeological study.

Figure 9.4 is a representation of the phosphate concentration in Paleosol I, collected 10 cm below the surface of Paleo-Hill B1. Four zones with high phosphate concentrations were found. The high P concentrations and the paleotopography suggested two possibilities. (1) Each high P area might represent a separate household. (2) Given the fact that the zones are of different sizes and in different relationships to the hilltop, the entire site represents one agricultural group. This is the more probable interpretation. The zone of high P on the hilltop itself, just west of datum, may be the residence. To the south of the residence on a steep slope, the high P concentration could be the principal refuse area. The light zone just south of datum, with a very low P concentration, was free of organic residues. The lower-than-normal P for the area suggests activity that did not promote organic residues, such as stone flaking. Northeast of the datum point and the extreme southeast are P anomalies whose meaning is not apparent and which deserve further study. R. D. Lippi concluded that the soil phosphate study yielded considerable information about the nature of the site with relatively little time and effort.

Phosphate Fractionation

R. C. Eidt suggested, along the lines of Lippi's analysis, that soil phosphate fractions could be used (1) to identify land-use areas and (2) as a potential relative dating technique.[12] In Eidt's scheme he used relative quantities of three inorganic phosphate fractions. Fraction I comprised the loosely bound aluminum and iron phosphates (variscite and strengite); fraction II comprised occluded aluminum and iron phosphates; and fraction III comprised the calcium phosphates. Eidt speculated that fraction I occluded over time to become fraction II. The ratio of II/I could be used to measure the time since the soil's last enrichment.

K. T. Lillios evaluated Eidt's hypotheses in a study of medieval and Bronze Age soils at the site of Agroal, Portugal.[13] This site had six occupation and abandonment episodes, beginning in the Early Bronze Age (2000 BC). It was abandoned and reoccupied in the Late Bronze Age (1500–1000 BC). It was abandoned again and not reoccupied until the Middle Ages. Occupation lasted until the Renaissance. Control soils reflect the local lithology—karstic limestone of Jurassic Age. The Postglacial soils reflect the inorganic mineral phosphate content of sedimentary rocks (ca. 200 ppm).[14]

In general, Eidt's proposition of a significant relation between land use and phosphate fractions was supported by the Agroal study. Eidt's hypothesis concerning the use of the ratio of II/I produced mixed results. The ratio appears to be less related to a linear buildup of phosphate over time than to a more complex and varied use of agricultural practices over history.

PART IV

ARTIFACT
ANALYSIS

10

Archaeological Materials

Rocks and Minerals

This chapter is only a brief introduction to lithic archaeological materials. Archaeologists with but little knowledge of rocks and rock-forming minerals are urged to learn about them in greater detail than that presented here. Lithic resources are abundant in almost every archaeological site, and lithic artifacts are invariably the best preserved of any remains. Early societies learned how to exploit these resources, and the use and production of lithics go back to the earliest known sites, at least 1.5 million years. In fact, the earliest cultures are distinguished on the basis of their lithic industries and lithic artifacts.

Horror stories in misidentification of lithics abound. Not only have misidentified artifacts proven embarrassing to the archaeologist, but also they have made it difficult to make meaningful comparisons of different societies using published descriptions. In addition, conservation strategies for historical monuments cannot be developed without an understanding of the nature of the material used in their construction. Some egregious examples of ignorance of the rocks and minerals from our personal experience include the following:

1. An archaeologist asked if a quartzite scraper was either flint or chert. When told that it was neither, he asked, "Well then, which is it more like?" (answer, still neither).
2. Egyptian basalt statues have been called limestone in publications (and several other rock types).
3. Sources for alabaster were searched to explain a trading link between a site and elsewhere when the geological map showed the site was adjacent to a mountain of gypsum, the mineral component of alabaster (the gypsum may have merely rolled down the hillside to the workshops, where it became the more salable alabaster).
4. Conservators searched for methods to preserve an allegedly granitic historic monument, or so it had been identified. Chemical analysis revealed only abundant Ca, Mg, and carbonate. Fossils were also abundant in the "granite," which dissolved easily in hydrochloric acid (the "granite" was clearly limestone).

Petrology is the branch of geology that deals with the occurrence, origin, and history of rocks. Petrography is concerned with descriptions of rocks, their mineralogy, structures, and textures. A rock can be defined as a naturally formed, solid material composed of one or more minerals and having some degree of chemical, textural, and mineralogical constancy[1]. Rocks are divided into three groups:

1. *Igneous:* formed from a liquid magma at high temperatures and including granite, basalt, and rhyolite
2. *Sedimentary:* deposited by the action of water or wind and including limestone, shale, and sandstone
3. *Metamorphic:* igneous, sedimentary, or previously metamorphosed rocks altered and recrystallized by the action of heat and pressure, commonly in a tectonic event, and including marble (metamorphosed from limestone), quartzite (from sandstone), and schist (from shale).

Minerals are naturally occurring inorganic substances having a definite chemical composition and a characteristic crystal structure. Mineraloids are also naturally occurring, but they can also be organic substances and lack a uniform composition and crystal structure. They include resins (such as amber and jet), pearl, amorphous glasses, and opals.

Matter and Materials

Archaeological materials are either organic or inorganic in origin and (1) occur naturally in nature and are used in the form in which they are found, with no or minimal processing needed, or (2) require manufacture before their final use. Manufacturing processes include extraction of valuable metals, such as copper and gold, from ores by smelting; baking as pottery and glass; or working as sculpture (table 10.1).

In early primitive societies, the first type of utilization was more important, and materials were collected and used as is, including wood, hides, and stones. Manufacturing became increasingly more important as societies developed sophisticated technologies to produce pottery and metal artifacts. Modern analogies exist for the first type and include mining and production of industrial rocks and minerals, for example, salt, agricultural lime, and sand and gravel.

Materials that are considered homogeneous show similar properties throughout, such as marble, quartz crystals, water, and metals. Heterogeneous materials are nonuniform mixtures of several materials with disparate properties, such as granite, pottery, and metal alloys. Any number of new materials can be produced by mechanically combining two or more homogeneous materials, as is done in the manufacture of pottery.

At the molecular level many apparently homogeneous materials are seen to consist of mechanical mixtures or solid solutions of different materials, called end members. The proportions of each material can be varied infinitely and the substance itself may exist in any form, as a liquid, gas, or solid. Metal alloys and many minerals are good examples of materials in solid solution. Plagioclase feldspars exhibit uniform physical properties and appear homogeneous at a macrolevel, including by optical examination. X-ray diffraction analysis, however, reveals plagioclase to consist of a solid solution of different end members: the molecules albite ($NaAlSi_3O_8$) and anorthite ($CaAl_2Si_2O_8$), with variable amounts of a third molecule, orthoclase ($KAlSi_3O_8$).

Table 10.1 Examples of Archaeological Materials

Material	Naturally occurring	Manufactured
Inorganic		
Metals	Gold, silver, copper (native elements)	All metals found as compounds (sulfides, oxides, carbonates, etc.)
Building	Limestone, marble	Brick, tile
Ceramics	Clay, obsidian	Clay body ceramics, glass
Pigments	Ochre, chalk	Egyptian blue, lead white
Organic		
Vegetable	Wood, resin	Dyes, fibers, cosmetics
Animal	Skin, bone, ivory, feathers	Glue, leather, fibers

Substances can be completely described by their chemical composition and physical properties. Chemical properties are involved with chemical changes and, as such, control rusting (oxidation) and solubility. Physical properties include hardness, density, luster, crystal structure, ductility, malleability, melting and boiling points, and electrical and thermal conductivity.

Elements and Compounds

There are more than 100 elements, of which 94 are found in the earth's crust. The others are radioactive products of man-made nuclear reactions. However, only 12 elements account for 99% by weight of the continental crust (table 10.2). These are the major elements and are reported in most complete chemical analyses of materials as oxides, as they are shown in table 10.2. In the earth's crust, 9 out of 10 ions are either O^{2-}, Si^{4+}, or Al^{3+}, and 94% of the volume of the crust actually consists of O^{2-}.

All other elements are minor or trace elements (table 10.3). In a chemical analysis, elements present over 1% are considered major, from 1% to 0.01% are minor, and less than 0.01% are trace elements. Traditionally, major and minor elements are reported as oxides, in percentage by weight. Trace elements are reported as elements in parts per million (ppm, 0.01% = 1,000 ppm). Clearly, most of the archaeologically significant metals exist only as trace elements in the earth's crust (table 10.3).

Although on average, these metals are rare in the earth's crust, they also occur in anomalously high concentrations as ores and can be exploited by humans. An ore is a geochemical anomaly in which the abundance of the element is concentrated many times over its average crustal abundance. Ores must also be defined in terms of economics; that is, they must yield their metal at a profit. With market and technology conditions constantly changing, definitions of ores (economically exploitable under the conditions of the times) versus reserves (potentially exploitable with improving economic conditions) will change as well. In classical Greece, because of the availability of slave labor, much lower grade deposits of gold were worked on the island of Thasos than can be worked today. Some ancient workings on Thasos apparently had a tenor (average amount of metal or mineral in an ore) of 2 g/ton of gold.[2] The lowest economic grade gold deposits today have at least 2.5–4 g/ton.

Table 10.2 Major Components of the Earth's Crust

Compound	Percentage by weight	Common name
SiO_2	59.26	Silica
Al_2O_3	15.35	Alumina
Fe_2O_3	3.14	Ferric iron
FeO	3.74	Ferrous iron
MgO	3.46	Magnesia
CaO	5.08	Lime
Na_2O	3.81	Soda
K_2O	3.12	Potash
TiO_2	0.73	Titania
H_2O	1.26	Water
P_2O_5	0.28	Phosphorus pentoxide
Total	99.23	

However, lower grade copper deposits can be worked today than in antiquity because of improved extraction and smelting technology. With gold, the ores of antiquity are the reserves of today, whereas the reverse is true of copper ores.

In ore deposits, the valuable metal is found either in the uncombined state as native metal (such as gold and some copper) or as discrete minerals forming chemical complexes with sulfur and sulfate (such as lead and some copper), oxygen (such as manganese and tin), or carbonate (such as zinc and copper). The Bronze Age marked the initiation of rapid progress in the art and technology of discovery and exploitation of the earth's relatively rare ore de-

Table 10.3 Abundances of Selected Trace Elements in the Earth's Crust

Element	Symbol	Abundance (ppm)
Antimony	Sb	1
Arsenic	As	5
Cobalt	Co	23
Copper	Cu	70
Gold	Au	0.005
Lead	Pb	16
Manganese	Mn	1,000
Mercury	Hg	0.5
Nickel	Ni	80
Platinum	Pt	0.005
Silver	Ag	0.1
Strontium	Sr	300
Sulfur	S	520
Tin	Sn	40
Tungsten	W	69
Uranium	U	4
Zinc	Zn	132
Zirconium	Zr	220

posits when humans learned how to find and mine ores and extract their metals from chemical complexes by metallurgical processes. In chapter 12 we shall expand on this topic.

Physical Properties of Archaeological Materials

The physical properties, as well as the chemical composition, of archaeological materials are important in determining their ease of manufacture and their usefulness. As already noted, important physical properties include hardness, density, luster, crystal structure, ductility, malleability, melting and boiling points, and electrical and thermal conductivity.

Density and specific gravity are measures of the mass of a substance per unit volume; that is, the higher the density, the heavier the object. Density is defined as the ratio of grams per cubic centimeter, and specific gravity is the ratio of the density of a substance compared to the density of water. The density of various archaeological materials varies from pure gold, which is 19.0; to Egyptian alabaster, limestone, and marble, all about 2.7; to amber, about 1.1; and wood, pine at 0.43 and oak at 0.79.

Thermal conductivity is the ability of a material to transmit heat energy. Materials considered to be good thermal conductors include most metals. Poor conductors act more as insulators, such as brick and wood.

The workability of a material can be described by its ductility and malleability. Highly workable materials undergo deformation without failure—that is, without rupturing. Deformation occurs in raw material as it is worked in the manufacture of artifacts. Many materials that can be worked easily have a high ductility; for example, they can be drawn into wires. Most easily worked materials have a high malleability as well; for example, they can be hammered and rolled into thin sheets. Highly ductile and malleable materials include the metals gold, silver, and copper; metals with lower ductility and malleability include iron, tin, and lead.

Hardness of materials is measured by their ability to resist scratching (Mohs Scale) or penetration (Knoop Scale). This property should not be confused with toughness or resistance to breaking or shattering, although all these properties are controlled by chemical bonds and crystal structure. Mineral cleavage is a visible manifestation of these controls: quartz and diamonds have no obvious cleavage and resist breakage until very high stress is applied, at which point they shatter. Calcite and feldspar have well-developed cleavage and will start to break up under very minor stress.

Hardness measured by the Mohs Scale is based on the increasing difficulty of scratching of 10 minerals; those higher in the scale have greater hardness and will scratch those lower (table 10.4). Intervals on the hardness scale are roughly equal on a logarithmic scale, except for the gap between corundum and diamond, which is considerably greater. On a true logarithmic scale, with talc as 1 and corundum 9, diamond should have a hardness of 42.5. Some useful hardness standards for the scale are your fingernail (2+), a copper coin (3), a steel knife blade (5+), window glass (5.5), and a steel file (5.5).

Archaeological materials that are monomineralic have a hardness roughly equal to their mineral. Agate, flint, and chert, which are composed largely of cryptocrystalline quartz (silica), have a hardness close to 7. Marble composed of calcite and dolomite (hardness 3.5–4) will have a hardness between 3 and 4. Most alabaster is a form of gypsum and has a hardness of 1.5–2. However, so-called Egyptian alabaster is actually calcite and has a hardness of 3.

Table 10.4 Mohs Scale of Hardness

Hardness	Mineral
1	Talc
2	Gypsum
3	Calcite
4	Fluorite
5	Apatite
6	Orthoclase feldspar
7	Quartz
8	Topaz
9	Corundum
10	Diamond

Common Rock-Forming Minerals

Rocks are described primarily on the basis of their mineral composition and textures. Although there are about 2,400 described minerals, fewer than 40 account for the great bulk of the earth's crust. These are the major mineral components of common rocks and are called the *rock-forming minerals.* Considering the abundance of silicon and oxygen in the earth's crust, the most abundant mineral groups are, not surprisingly, the silicates. Comprising less than 5% of most rocks are the *accessory minerals.* The final group, rarer but widespread, is made up of the *ore minerals,* most of them oxides or sulfides, which are locally concentrated into ore deposits.

The chemical composition of minerals is written so that the cations are shown first (they yield electrons, such as Pb^{2+}) and then the complex ions or anions (they receive electrons, such as S^{2-}). Thus, dolomite, $CaMg(CO_3)_2$, is immediately obvious as a carbonate and related to calcite, $CaCO_3$. This fact would not be apparent if the formula were written C_2CaMgO_6.

Many minerals form complex solid solutions. This can be represented, as in the case of olivine, by the formula $(Mg,Fe)_2SiO_4$, which shows that Mg and Fe occupy the same cation sites in the mineral structure. If chemical analysis has determined the exact amounts of Mg and Fe that are present, then the formula can be written $(Mg_{0.3}Fe_{0.7})_2SiO_4$. In this case, 30% of the cation sites are occupied by Mg and 70% by Fe. Another way to show this is by using the names of the end members. Pure Mg olivine is called forsterite, abbreviated Fo, and pure Fe olivine is fayalite, Fa. Such a mineral formula could be written $Fo_{30}Fa_{70}$.

Mineral groups, other than the native elements, are classified on the basis of their anions. This system is convenient because most chemical properties are controlled by the anions present rather than the cations. The largest groups of minerals are the oxides, silicates, sulfides, and carbonates. The value of this system is seen in the case of calcium: it can form calcite, $CaCO_3$, a carbonate with properties similar to the Mg carbonates dolomite (CaMg) and magnesite (Mg) but completely different from Ca silicates such as anorthite $(CaAl_2Si_2O_8)$ and Ca fluorides such as fluorite (CaF_2).

Minerals crystallize in one of six symmetry groups, from highest to lowest symmetry: isometric, tetragonal, hexagonal, orthorhombic, monoclinic, and triclinic. In most rocks, except for extrusive or near-surface volcanic rocks, the mineral grains do not show a clear crystal habit (shape), so their crystal system is not immediately obvious. Since detailed information on crystallography, mineral properties and identification, and chemistry is far beyond the scope of this book, the reader is referred to any good text on mineralogy, such as that by Zoltai and Stout (1984), for more information.[3]

The Silicate Minerals

Since oxygen and silicon are the two most abundant elements in the earth's crust, it is natural that the silica and silicate minerals are also the most abundant minerals on earth, making up over 90% of the crust and mantle. The basic building block for silicate structures is the silicate anion $(SiO_4)^{4-}$, with four unsatisfied negative charges (figure 10.1). To form Mg silicate, two Mg cations, Mg^{2+}, will tie on to a silicate anion and make forsterite, $Mg_2(SiO_4)$. The Al^{3+} ion can substitute in some minerals for Si^{4+}, which will then require different cations to satisfy the different charge requirements. Other cation substitutions allow for solid solutions to exist in mineral series. Such cations have similar ionic radii and generally similar charge to facilitate the substitution. Thus, Mg^{2+}, Fe^{2+}, and Mn^{2+} readily form substitutions in olivine and pyroxene, and Ca^{2+} and Na^{1+} in plagioclase feldspars.

The eight common rock-forming silicates of the igneous rocks and their characteristic properties are the following:

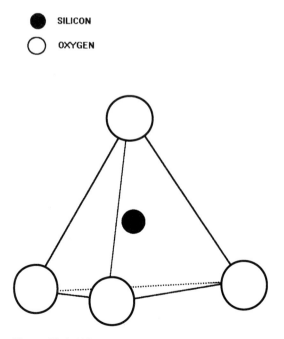

Figure 10.1 SiO_4 tetrahedral unit.

Quartz: poorly defined interstitial grains, no cleavage, colorless glassy appearance

Plagioclase feldspar: commonly lathlike, white or gray or slightly greenish, right-angle cleavage, parallel markings may be visible on cleavage (twins)

Alkali feldspar (orthoclase): commonly equidimensional grains, white or pink, right-angle cleavage

Amphibole (hornblende): lathlike grains, dark green to black, cleavage about 60°

Pyroxene (augite): equidimensional or lathlike grains, dark green to black right angle cleavage

Olivine: rounded grains, small, pale green, no cleavage

Biotite (mica): perfect platy cleavage, sheetlike, black to dark brown

Muscovite (mica): perfect platy cleavage, sheetlike, silver-white to light green

Nonsilicate Minerals

Although there are many compositional families in addition to the silicates, only eight are important as rock-forming minerals: oxides, hydroxides, sulfides, sulfates, carbonates, phosphates, halides, and native elements.

Oxides: include magnetite (Fe_3O_4), hematite, (Fe_2O_3), and ilmenite ($FeTiO_3$)

Hydroxides: principally of aluminum, e.g. diaspore ($Al_2O_3 \cdot H_2O$); iron, e.g. goethite ($Fe_2O_3 \cdot H_2O$); and manganese, e.g. manganite ($Mn_2O_3 \cdot H_2O$)

Sulfides: iron sulfides—pyrite, marcasite, and pyrrhotite—are common rock-forming minerals; sulfide minerals—copper, lead, zinc, etc.—generally occur as ores

Sulfates: gypsum ($CaSO_4 \cdot 2H_2O$) is the most common rock-forming sulfate, a result of evaporation of seawater

Carbonates: the two most important are calcite, $CaCO_3$, hardness $= 3$; and dolomite, $(Ca,Mg)(CO_3)_2$, hardness $= 3.5$–4; both with good cleavage; they are most abundant as extensive sedimentary limestone and metamorphic marble

Phosphates: apatite important as an accessory mineral in all rock types and as a principal constituent of bone

Halides: form by evaporation of seawater, such as halite ($NaCl$), an important rock-former

Native elements: occur as accessories or as ores, including native copper, gold, platinum, silver, and carbon as graphite

Petrography

Rocks are composed of an aggregate of minerals; their classification and names depend on both their mineral composition and their texture. They are divided into three large families depending on their origin: igneous, sedimentary, and metamorphic. Igneous rocks (from the Latin word for *fire*) form by the consolidation of molten or partially molten rock material. Sedimentary rocks are formed by the cementation of materials derived by the physical and chemical breakdown of preexisting rocks or by diverse chemical or biochemical processes. The materials are transported and sedimented by water or wind action. Metamorphic rocks form as a result of solid-state transformation of preexisting rocks under conditions of high temperature, high pressure, and/or a changed chemical environment.

Pyroclastic rocks are transitional between igneous and sedimentary. They are made up of fragments of material ejected from volcanoes and then cemented or welded to a coherent aggregate. Another important transitional rock type exists between sedimentary and metamorphic. These are rocks in which lithification has taken place at low temperature and has resulted in minor changes in composition and texture.

Igneous Rocks

Igneous rocks are formed by the solidification of liquid magma generated by the melting of preexisting rocks in the earth's crust or mantle. They include large granite masses (batholiths), basaltic flows, and glassy rhyolite. Table 10.5 shows a simplified classification of the principal igneous rock types. The table is arranged vertically by texture (glassy in extrusive rocks to coarse grained in plutonic) and horizontally by chemistry and mineralogy. The horizontal divisions represent a change from "acid" rocks (originally thought to have a high silicic acid content), also called "felsic" (with high feldspar and quartz—i.e., high SiO_2, Na_2O + K_2O), through a decrease in silica and alkalis to the "basic" (low silicic acid) or mafic (high Ca-Fe-Mg) rocks. The mafic rocks have a concomitant increase in dark-colored (Fe-Mg) minerals.

The igneous rocks are also classified by grain size. At the top of the table are the rocks that formed near or on the surface and thus cooled very rapidly. They commonly are glassy or very fine grained and may contain scattered small crystals called phenocrysts. The acid volcanic rock obsidian and the mafic basalt are good examples of this rock type. Medium-grained, hypabyssal rocks are intruded as dikes and small masses at shallow depths. They cooled more slowly than the extrusive lavas and commonly develop a porphyritic texture in which large phenocrysts contrast with a finer-grained groundmass. The plutonic, abyssal rocks cool slowly at great depths and develop a coarser-grained texture than the other rock types. They form huge batholithic granitic intrusions that often underlie mountain ranges.

Sedimentary Rocks

Sedimentary rocks are derived from the physical and chemical breakdown of preexisting rocks and by diverse organic processes. They are transported and deposited by water or wind

Table 10.5 Common Igneous Rocks

	SiO_2, Na_2O, K_2O; felsic minerals $\rightarrow$ CaO, MgO, FeO; mafic minerals				
Chemical type	Acid	Intermediate		Basic	Ultrabasic
Fine-grained, extrusive, volcanic lavas	Rhyolite Obsidian Pitchstone	Trachyte	Andesite	Basalt	
Medium-grained, hypabyssal, dikes, sills	Porphyritic granite Felsite	Porphyritic syenite	Porphyritic andesite	Diabase dolerite	Lamprophyre, etc.
Coarse-grained Plutonic batholiths	Granite	syenite	Diorite Tonalite	Gabbro	Peridotite Pyroxenite[a]

[a]Metamorphosed equivalents are soapstone and serpentine.

Table 10.6 Classification of Siliceous Clastic Sedimentary Rocks

Name	Grain size	Mineral composition and characteristics
Conglomerate	>2 mm	Rounded rock or mineral fragments
Breccia		Angular rock or mineral fragments
Sandstone	0.062–2 mm	Mostly quartz, grains generally round
Arkose		Mostly quartz and feldspar, angular grains
Graywacke		Clay, quartz, rock fragments; angular grains
Siltstone	0.005–0.062 mm	Generally massive, quartz and clay, gritty feel
Shale	<0.005 mm	Generally laminated, quartz and clay, smooth feel

action as rock fragments or organic debris, or they are precipitated by chemical or bio-chemical action or by flocculation of a colloidal suspension. Some sediments are precipitated directly to a solid aggregate, but most must be consolidated—lithified by diagenetic processes—to form sedimentary rocks. Typical structures of sediments are stratification (i.e., layering or bedding showing their plane of deposition) and fossils (the former life-forms preserved in the rock).

Wind-laid sediments are lithified sandstones and siltstones; water-laid sediments are mainly marine but also include lacustrine, fluvial, or glacial deposits. Water-laid deposits include limestone, sandstone, siltstone, shale, mudstone, and conglomerate. Limestone and sandstone are the two most important groups archaeologically, forming the majority of building and sculptural stones.

Limestone composition varies from almost pure $CaCO_3$ to dolomitic and, with increasing impurities, siliceous or clay rich (marl). Textures are highly variable, from fine-grained homogeneous chalk to a layered fossil-rich hash. The carbonate can form as a chemical precipitate or from organic processes. Travertine and tufa are carbonates that have been chemically deposited in carbonate-rich hot springs, as at Tivoli, near Rome.

Sandstones consist largely of quartz grains but are highly variable in color, grain size, structures, and mineral content. Eolian, or wind-blown, types form red beds, commonly with a high amount of feldspars, and show cross-bedding. Marine sandstones generally form in relatively shallow water, are white to gray in color, and have varying textures and mineral content depending on their source. Graywacke, with a high percentage of impurities, including clay, feldspar, and mica, was a popular honestone. "Brownstone," a highly impure sandstone, was a popular building stone in the northeastern United States but unfortunately is not resistant to weathering in modern urban environments. A classification of these clastic sedimentary rocks based on mineral content and grain size is given in table 10.6.

Other important archaeological siliceous lithic materials, found largely in sedimentary rocks, include flint, chert, and jasper, popular terms which are often used interchangeably. Some sedimentary-appearing materials used in construction are actually weathering products: laterite, bauxite, calcrete, silcrete, and ferricrete. They are near or at the surface, can be easily extracted, and are moderately resistant to weathering as building materials, especially in an arid or semiarid environment.

Metamorphic Rocks

Metamorphic rocks are rocks that have been altered and recrystallized due to the application of pressure and heat (P/T). The original parent rock could have been sedimentary, metamorphic, or igneous. The most extensive metamorphic rock formations generally form in a tectonic event when tectonic plates collide (regional metamorphism), or by the thermal effects of igneous intrusions on the country rock (contact metamorphism). The transformation processes of metamorphism take place in solid rock at P/T conditions higher than those required for diagenesis but lower than those under which the rocks would melt.

Metamorphic changes are textural and mineralogical: the rock becomes coarser grained, and minerals form that are stable under the new P/T conditions of the metamorphic environment. The only common chemical change is the loss of the volatiles, water and carbon dioxide. With rapid cooling, as takes place in the contact zones of plutons emplaced in shallow crustal levels, new minerals form in response to thermal metamorphism. In relatively deeply buried zones involved in mountain-building activities, new minerals will form as a result of applied heat and pressure. Cooling is relatively slow so that new minerals, called porphyroblasts, can grow to relatively large size. Metamorphic rocks formed in the near-surface environment are produced by contact metamorphism; in deeper tectonic zones they are products of regional metamorphism. The second process is by far the more important.

Many metamorphic rocks possess a good foliation—that is, they have a preferred orientation of their constituent tabular and platy minerals. This is especially pronounced in rocks where micas and amphiboles are well developed. Rocks that do not develop foliation are called massive. Some important examples of each include the following:

Contact Metamorphic Rocks
  Hornfels: hard, dense, fine grained, with or without scattered large porphyroblasts, massive

Regional Metamorphic Rocks
  Massive, nonfoliated
    Marble: recrystallized limestone and dolostone, largely fine- to coarse-grained calcite and dolomite
    Quartzite: recrystallized sandstone, largely quartz
    Granulite: formed from many rocks, produced by extremely high P/T conditions
    Metaconglomerate: formed from very coarse detrital sediments, conglomerates
    Serpentinite and associated rocks: form from altered mafic igneous rocks; called *verde antique* or green marble in the building trade but is not a true marble
  Foliated
    Phyllite, slate, argillite: transition stages in metamorphism of shale; grain size increases with increased P/T from argillite to phyllite but too fine grained to identify individual minerals
    Schists: form largely from clay-rich sediments, are generally biotite, muscovite, quartz, and are plagioclase rich, also with some amphibole, garnet, and aluminum silicate minerals.
    Amphibolite: forms from a variety of igneous and sedimentary rocks, medium- to coarse-grained, rich in amphibole and plagioclase; better foliated varieties may have abundant biotite
    Gneiss: high-grade metamorphic formed from a variety of rocks; quartz- and

feldspar-rich, with varying amounts of mica, amphibole, garnet and aluminum silicate minerals.

Archaeological Use of Stone

Worked-stone artifacts are found in the oldest hominid sites, at least 1.5 million years old. Stone tool production and artifacts are commonly the most important keys to technology and culture found in early sites. In fact, cultural phases in the Stone Age are defined largely on the basis of stone tools and the type of stone used in their manufacture. Early African Stone Age sites reveal a wide variety of rocks, including limestone, chert, and quartzite. At first, all the stone used was local, but in later African Stone Age sites, imported raw material appeared, such as chalcedony and obsidian. Clearly, the artisans of the stone implements became more sophisticated and geologically knowledgeable about the properties of the raw material. A hunting community would need easily worked yet high-quality material, depending on the end use for the implement. Lithics are invariably the best preserved of artifacts in archaeological sites. The evidence they offer for technology and trade is most often not otherwise available.

In many Neolithic sites of Great Britain, the earliest implements were overwhelmingly made of chert, which is widespread in Cretaceous limestone formations. Import of exotic igneous and metamorphic rocks, such as rhyolite and quartzite, became popular because of their improved properties and ease of working over chert for the manufacture of stone axes, querns, etc. With the further evolution of societies, stones were selected not only for their utilitarian properties but also for their beauty and were used for other purposes, such as ornamentation. In England, chalk was popular because it is easy to cut and was made into cups, figurines, and symbolic tools such as axes. In the late Neolithic in the Franchthi Cave of mainland Greece, obsidian and marble were imported from the Aegean islands of Melos and Naxos. Obsidian was used for arrowheads and many other tools. It could be worked with greater ease than the chert found on the site and developed much sharper cutting edges. Marble was used for jewelry and for symbolic grave goods, such as cups and figurines.

From Neolithic times onward, knowledge of the properties of rocks accelerated. Quarrying of stone and underground mining for metals started. Artisans knew about jointing, layering, and foliation of rock, which helped in its extraction and fabrication.[4] Knowledge of the durability, strength, hardness, and resistance to weathering of local and imported stone led to the increasing use of stone as a material for building and monuments. With this knowledge and sophistication, the demand for imported stone raw material increased, and early trade patterns were developed.

Building Stones

From earliest times, the most important stones used in building were limestone, sandstone, and granite. According to figures of the U.S. Geological Survey, that has not changed even in modern times. In 1995, 42% of the dimension stone used in the United States was granite, 31% was limestone, and 13% was sandstone. Of the rest, 3% was marble, 3% was slate, and 8% was other rock types. Ease of extraction, beauty, strength, and resistance to weathering have been the most important factors in the choice of these stones.

Sandstone

Sandstone is resistant to weathering if it is cemented with silica and diagenesis has produced an interlocking texture of sand grains. The brownstone that was popular at the turn of the century for houses in Philadelphia and New York is an attractive Triassic sandstone quarried in the nearby Connecticut River valley. However, its cement is carbonate and over the years has weathered badly. Carboniferous sandstones, cemented with silica, are popular in England and resistant enough to be used as paving stones. In Egypt, sandstone was used in building and sculpture but took second place to limestone in important monuments, except for the Colossi of Memnon at Thebes. They were carved from a coarse-grained monolithic block of quartzite weighing 720 tons. The Incas of Peru could handle sandstone blocks weighing up to 100 tons. The Moghul emperors of India prized the red sandstone of the Precambrian Vindhyan series, which had been used from the end of the 1st century BC to construct the Red Fort at Delhi and the palace at Fatehpur Sikri. Sandstone was supplied from quarries 240 km south of New Delhi.

Limestone

The carbonate rocks limestone, marble, and dolomite probably account for more than half of all building stone used over time. However, because of reactivity with sulfuric acid and sulfate particles in the atmosphere, many historic limestone monuments have weathered badly in modern times. The most popular building stone of ancient Egypt was the nummulitic limestone used for construction of the Pyramids. The ancient Greeks used limestone as foundations and structural elements that were faced with marble for their most important temples. Many temples were also completely constructed of limestone where marble was not available or the limestone was esthetically pleasing, as at Corinth, Epidaurus, and Aegina. At Epidaurus, the local drab gray limestone was faced with the golden limestone from Corinth. In Great Britain, much limestone, considered especially attractive, was transported over great distances. Between 1175 and 1400, Jurassic oolitic limestone from Somerset was taken to southeastern Ireland to construct monasteries and castles. Jurassic Portland limestone, one of the world's most famous, was introduced in London in 1619 by Inigo Jones and has been used for public buildings ever since. Another classic limestone is the Jurassic Caen stone, imported into England first by the Romans and later by the Normans to construct Canterbury Cathedral. In the United States, oolitic Indiana limestone has been a popular stone in construction for many years.

Marble

The ancient Greeks and Romans used a variety of marbles for sculpture and in building construction. The Mediterranean basin has abundant attractive white marbles, with quarries exploited since classical times in the Pyrenees, Alps, Apennines, Greece, Turkey, and North Africa. The Mediterranean marbles are largely Mesozoic limestones in origin, metamorphosed in later Alpine orogenies. Much work has gone into the study of ancient marble quarries to characterize each petrographically and geochemically. Since the time of operation of each major quarry is known, developing "fingerprints" can help source, date, and authenticate marble artifacts. Marble was initially used as jewelry, figurines, and grave goods as early as the Neolithic in Greece. It is found as artifacts in sites where no local marble occurs, so that determining the provenance of marble artifacts has helped decipher ancient trade patterns.

Granite and Granitic Rocks

Commercial granite and archaeological granite are not always geological granite. One of the most famous granitic rocks is the rose syenite of Egypt; outcroppings occur between Aswan and the first cataract. It has been used since the 1st Dynasty for pavements in tombs and later more extensively in buildings, sarcophagi, and statues. Although a very hard stone, granite has regular jointing patterns which make it relatively easy to extract and use as a building material. Its use is most widespread where it occurs locally, as in the Acropolis of Zimbabwe and for lower storeys of Hindu temples. The Romans prized very highly the granite of Mons Claudianus in the Eastern Desert of Egypt, 500 km south of Cairo. Extraction of the granite and transport to Rome took a mighty effort: first blocks were carried to the Nile, some 120 km away, where they were loaded onto barges and floated downstream to Alexandria, 800 km north, and then transferred to special ships designed for carrying stone to Rome across the Mediterranean.[5]

The Geology of the Sphinx

The Great Sphinx of Giza lies 10 km west of Cairo in the Libyan Desert. It is believed to be part of Pharaoh Khufu's pyramid complex (4th Dynasty, ca. 2613–2498 BC). The Great Sphinx was carved directly from the natural limestone of the Giza Plateau. It is 20 m high and 70 m in length (figure 10.2). The Giza Plateau is part of the Motattam Formation, which formed from marine sediments deposited when the waters of the sea covered northeast Africa during the Eocene period. Ancient Egyptian builders built the Sphinx from the lowest layers of limestone. Member I is a hard and brittle reef rock. The Member II rock forms most of the lion body and south wall. Member II is the rock which is the most deteriorated on the monument. It is a poorer quality limestone; its deterioration pattern allegedly indicates it was built between 5000 and 7000 BC according to Boston University geologist Robert Schoch and writer John A. West.[6] Throughout Member II, there are laminated layers which contain different percentages of salt, sand, silt, and clay. The recessed layers, where the percentages are highest, are the areas where erosion, which is now considered normal and not indicating a longer period of weathering than pharaonic, is occurring most rapidly. Member III is a weather-resistant rock whose durability has left the mysterious face of the Sphinx in relatively excellent condition.[7] In 1988 a 3-ton chunk of limestone fell from the Member II rock from the shoulder of the Sphinx. Member II rock is rapidly flaking off because of salt expansion in the surface layers of the stone. Other erosional forces assault the Sphinx: sun, wind-blown sand, modern acidic pollution in the air, and natural seismic activity. The troubled Member II stratum contains significant amounts of sodium chloride and calcium sulfate; these mixed with water, salts, and changing temperatures cause breakage in the limestone. Erosion would have most likely destroyed the Sphinx long ago if sand had not covered the monument for most of its existence. Yet another fear, perhaps the most important, is that the neck of the Sphinx is becoming too weak for the heavy, well-preserved head.[8]

The Sphinx was not completely perfect when it was built. For example, the back and

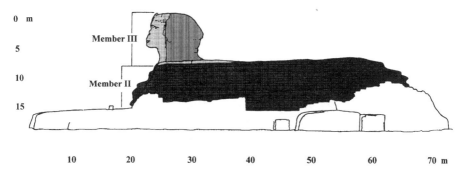

Figure 10.2 The Sphinx and its component geological members (I–III). Members I–III are from the Motattam Formation, composed of limestones. Member I is encased in masonry; the head-dress, formed from Member III, has significant portions made of cement.

sides of the Sphinx were never meant to be seen and were surrounded by mud-brick walls. When these walls eroded, large natural crevices and fissures in the Sphinx were left exposed to the desert elements. This is one of the reasons the sides and back of the monument are in such poor condition today.

Stonehenge: Stones and Sources

At Stonehenge, the most impressive features are its five great trilithons (figure 10.3). Structures exactly like them, only smaller, occur inside round and long megalithic barrows as gateways to underground burial chambers. Some barrows are surrounded by a circle of upright stones, the inner peristaliths. Comparable to these inner peristaliths are the uprights of the bluestone ring encircling the trilithons at Stonehenge.

Stonehenge I consisted only of the circular bank and ditch, the 56 so-called Aubrey Holes, and the Heel Stone (figure 10.4). Postholes in the causeway probably supported a gate but there is no indication that any structure ever stood at the center of the enclosure. Stonehenge I is not unique. It resembles a number of other henge monuments in the area, Avebury for one. A henge is a circular bank-and-ditch structure with avenues leading into it. In Old English the verb *hengen* means *to hang,* in this case lintels resting on pillars. The word *Stonehenge* may be a corruption of Old English *stan-hrycg,* meaning *stone-ridge,* or stones on the ridge. Avebury has two causeways although most have four, and all the henge monuments point in different directions. Some of the later ones seem to have a sunward orientation, though only Stonehenge is significantly oriented to the rising sun of midsummer.[9]

There is nothing in the manual labor required for the building of any phase of Stonehenge that could not have been accomplished by a Neolithic technology. The bluestones (dolorite—an igneous rock) of Stonehenge II were set up as a circle of unshaped boulders. Later, after the sarsens (a sandstone-conglomerate) of Stonehenge IIIa were put up, the bluestones were reerected and shaped by bashing and abrading with stone mauls. The most unusual feature of Stonehenge is the presence of stones that have been shaped. No other megalithic site in prehistoric Europe has this feature.

The technique of bashing and abrading applied to the bluestones was used only for spot

Figure 10.3 Stonehenge trilithon. (Photograph by Ervan G. Garrison.)

Figure 10.4 The Stonehenge Heel Stone viewed through the sarsen ring. (Photograph by Ervan G. Garrison.)

finishing of the sarsen stones; they are too hard to hammer into a form. Sarsen stone ranks 7 out of 10 on the Mohs Scale of hardness. Sarsen being a sandstone is as hard as quartz and harder than steel. It is not easily pulverized under the blows of a stone hammer which can remove only 6 cubic inches per hour. This was clearly not the way the huge upright of the largest trilithon, 50 tons in weight, 29 feet 8 inches (8.4 m) high and 2 feet 8 inches thick (82 cm), was shaped. The method of working sarsen follows from certain characteristics of its original formation.[10]

The sarsen stones were formed during the Eocene epoch of the Tertiary period, about 60 million years ago. Sarsens are the remains, in the form of boulders, of a bed that overlay a bed of chalk. The chalk formation, a marine deposit, toward the end of the Cretaceous period was uplifted and covered over with a shallow marine sand deposit. Uplift during the Eocene epoch exposed the chalk with its superimposed sand beds as dry land. Some of the lithified sand deposit was carried away by weathering, leaving the sarsen boulders lying on the chalk. Continued uplift raised the chalk layer hundreds of feet above the sea, resulting in the present chalk areas of Salisbury Plain and Marlborough Downs with their boulder-sized relics of the ancient sand beds. The sarsen structure of Stonehenge III was built from these boulders, which were likely once scattered around the site.

The bluestones of Stonehenge II are a spotted dolerite, which are local as well. These are petrologically related to formations on Mount Preseli in Pembrokeshire, southern Wales. The bluestones were once thought to have come from that source. Additionally, a long history of trade existed with Lancashire and Wales for dolerite axes. It is now understood that the bluestones were found on Salisbury Plain itself. They were dumped there at the end of a newly discovered glacial moraine during the last major advance of the Pleistocene ice sheet.

The sarsen boulders retain the stratification of the sandstone from which they formed, and split readily and evenly in planes parallel to the bedding. They formed either at the top or the bottom of the original sandstone beds. When fractured at right angles to the bedding, a well-shaped rectangular block is separated from the parent body, one set of faces flat, the other set rough. If the boulder occupied the entire thickness of the stratum, then its two original faces would be flat and parallel. These are the two types of sarsen boulders that once lay on Salisbury Plain, unlike the irregular, shapeless boulders, formed wholly within the middle of the sand bed—the so-called greywethers of Marlborough Downs.[11]

O. Williams-Thorpe and R. S. Thorpe, in their recent geochemical analysis of the Stonehenge stones, used x-ray fluorescence to determine major elements and 12 trace elements. Petrographic thin sections were prepared of the archaeological samples, as well as 40 samples of outcrops of dolerite and rhyolite from the Preseli Hills, 6 samples of glacial erratics from Preseli and nearby Lampeter Velfrey, and 6 samples of igneous erratics from near Cardiff[12]. While the Preseli outcrops show some regional variation a clear geochemical correlation can be made between them and the Stonehenge samples, glacial erratics and dolerite of southwest Wales. Petrography corroborates these findings. No chemical studies of the sarsens has yet been done, but heavy-mineral studies of five excavated fragments and two specimens from Marlborough Downs showed mineralogical differences.[13]

The archaeological controversy over the origin and mode of transport of stones in megalithic monuments such as Stonehenge appears to have a geological solution. While it is more dramatic to suppose a herculean labor in moving great stones over large distances from their sources to locations such as the Salisbury Plain, such a hypothesis requires a more sober assessment and appreciation of our Neolithic ancestors' skills and achievements.[14] As Michael J. O'Kelly has commented on the building of another megalithic monument, Ireland's magnificent Newgrange, "We are not dealing with questions of brute force and mere strength of numbers so much as with intelligent and well-organized method."[15]

Petrographic and Isotopic Studies of the Gallo-Roman
Mosaics, Vallon, Switzerland

The magnificent mosaics found in 1985 in the burned ruins of a Gallo-Roman villa at Vallon, Fribourg, in Switzerland were studied petrographically and isotopically. Together with other mosaics located near Orbe, in the Swiss canton of Vaud, some 70 km to the northwest of Vallon, they constitute some of the finest examples of Roman mosaic art. The villa of Vallon was clearly associated with the Roman presence at Aventicum (modern Avenches), some 5 km to the south. The seat of Roman authority in the late Republic to the 3rd century AD, Aventicum was likely surrounded by villas, such as those of Vallon, Orbe, and Colombier.

The analysis of the tesserae (mosaic chips) from Vallon by R. Bollin and M. Maggetti of the University of Fribourg followed a similiar study done on those from Orbe.[16] The tesserae were classified into seven major color groups (white, black, yellow, brown, red, green, and "mixed"). They come from seven separate lithologies: clear and dark marly limestones, oolitic and sparry limestones, sandstone, radiolarites, marbles, metamorphic rocks, and glasses. Ninety-eight percent of the Vallon tesserae are limestone, 1% are sandstone, and the remainder come from the other lithic types.

Microfossils identified the limestones as Jurassic to Cretaceous in age. These are com-

Figure 10.5 Roman mosaic from Vallon, Fribourg, Switzerland (modified from Perzynska, 1990).

mon to the Swiss Jura. The green tesserae were found to be Tertiary molasse typical of the Swiss Plateau lithology. Vallon lies on the plateau proper, roughly 20 km south of the Jura fold. Radiolarites, metamorphic and igneous rocks most likely originated in glacial erratics from morainic deposits of the Rhône glacier. The black tesserae were determined to be micritic limestones originating, in most cases, in southern Germany. Marble tesserae were believed to be imports from northern Italy (Carrara, for example). Isotopic comparison of the Vallon and Orbe white, yellow, and black tesserae indicated quite distinct signatures (interspecific) resulting from various local limestones. Figure 10.5 shows the Bacchus/Ariadne mosaic (26 m^2) from Vallon.

Instrumental Analytical Techniques

Instrumental analysis has become a mainstay in the study of provenance of artifacts and their materials. A veritable "alphabet soup" of acronyms shorten the often ponderous names of the large number of techniques available today: XRF (x-ray fluorescence), XRD (x-ray diffraction), NAA (neutron activation analysis), AAS (atomic absorption spectroscopy), PIXE (proton-induced x-ray emission), ICP (inductively coupled plasma spectroscopy), FTIR (Fourier transform infrared spectroscopy), EMP (electron microprobe), RIS (resonance ionization spectroscopy), ESR (electron spin resonance), CL (cathodoluminescence spectroscopy), STM (scanning tunneling microscopy), AFM (atomic force microscopy), NSOM (near-field scanning optical microscopy), and SEM/TEM (scanning/transmission electron microscopy).

Neutron Activation Analysis

In 1934 Enrico Fermi irradiated 63 elements with 2×10^7 neutrons/s, using a radon source of 800 millicuries. He confirmed 3 types of neutron ($_0^1n$) reactions:

1. $_Z^A X + _0^1 n = _{Z-2}^{A-3} X + _2^4 He$: (n, α)

2. $_Z^A X + _0^1 n = _{Z-1}^A X + _1^1 H$: (n,p)

3. $_Z^A X + _0^1 n = _Z^{A+1} X$: (n,$\gamma$)

Examples of these are

1. $^{27}Al + _0^1 n$ (n,α) ^{24}Ni

2. $^{27}Al + {}_0^1n$ (n,p) ^{27}Mn

3. $^{27}Al + {}_0^1n$ (n,γ) ^{28}Al

The reactions of types 1 (n,α) and 2 (n,p) occur chiefly among the lighter elements, while type 3 occurs among the heavy elements. Nuclear reactions caused by neutrons are of interest in archaeological geology and geochemistry because such reactions are used for analytical purposes to measure the concentrations of trace elements in both archaeological and geological materials. Neutrons produced by fission of uranium 235 atoms are emitted with high velocities and are called "fast" neutrons. For a controlled reaction, the fast neutrons must be slowed down because the fission reaction requires other than "fast" neutrons. Slowing down neutrons requires a "moderator" such as water or graphite with which the fast neutrons can collide without being absorbed. Ordinary water typically serves this purpose in the so-called swimming pool reactors.

Slow neutrons are readily absorbed by the nuclei of most of the stable isotopes of the elements. The neutron number of the product nucleus is increased by 1 compared with the target nucleus, but Z remains unchanged, so the product is an isotope of the same element as the target. The product nucleus is left in an excited state and de-excites by emission of gamma rays. The absorption of a slow neutron by the nucleus of an atom can be represented by the following equation:

$$\ _Z^AX + {}_0^1n \rightarrow {}_{\ Z}^{A+1}Y + \gamma, \qquad \text{for example, } ^{39}K(n,\gamma)^{40}Ar \qquad (11.1)$$

or

$$\ _Z^AX(n,\gamma)\ {}_{\ Z}^{A+1}Y, \qquad \text{for example, } ^{35}Cl\ (n,\gamma)^{36}Cl \qquad (11.2)$$

The products of (n,γ) reactions may be either stable or unstable. Unstable nuclides produced by neutron irradiation are radioactive and the product nuclides decay with characteristic half-lives. As a result of slow-neutron irradiation, a sample composed of the stable atoms of a variety of elements produces several radioactive isotopes of these "activated" elements. This induced activity of a radioactive isotope of an element in the irradiated sample depends on many factors, including the concentration of that element in the sample. This is the basis for using neutron activation as an analytical tool.

For most elements the ratios of stable isotope abundances are invariant. The relative abundances of elements found in a particular sample is a function of where the sample is found. This makes the identification of prehistoric trade goods by element composition possible. For example, when iron ores from one mine are compared to iron ores from other mines, certain elements may be found in greater abundance in one of the ore samples. The metal products of one mine can be traced through the trace elements characteristic of that mine. Element abundances which might change through processing of the ore should not be used. The pottery of a particular group of people can be traced through the location of the source of clay used for the pottery and determination of the trace elements in that clay. However, inclusion of temper in the clay before firing might vary the trace elements characteristic of the clay deposit itself. Even when inclusion of temper creates a problem in variation of trace elements of the clay deposits, the temper might itself be characteristic. The elemental composition of the temper, if sufficiently specific to a particular group, might then serve as a marker of that group regardless of where the clay was obtained. Neutron activation has the advantage of being nondestructive. The sample does become radioactive, but it is not so radioactive as to prevent handling a few weeks after irradiation. Only a very small number of the atoms are affected, and the fundamental composition of the sample is unchanged.

Neutron activation can be used to determine either the absolute or the relative quantities of specific isotopes present in a sample. Absolute determinations are needed for chemical analysis, to find (1) the total amount of each parent element, generally present in minor or trace amounts, or (2) for some age determinations techniques, like potassium-argon. Relative determinations are used (3) as isotopic fingerprints of artifactual material to determine provenance, or (4) for other age determination techniques such as $^{40}Ar/^{39}Ar$.

Neutron activation involves (1) the irradiation of some sample with neutrons to produce radioactive isotopes; (2) the measuring of the radioactive emissions from the irradiated sample; and (3) the identification, through the energy, type, and half-life of the emissions, of the radioactive isotopes produced.

Irradiation of a sample with neutrons is carried out with one of three types of neutron sources. Neutrons can be obtained from nuclear reactors. In this case, the sample is directly inserted within the reactor. This type of source represents the strongest source in the sense of producing the most radioactivity. Very often, the radioactivity induced within a sample may be too weak to measure when produced by neutron bombardment in ways other than nuclear reactors. Large particle accelerators such as Cockroft-Waltons, Van der Graffs, and cyclotrons can be used where they are available. The least desirable technique for neutron irradiation is through the use of radioactive sources. Radioactive sources do not emit neutrons but alpha particles which can be absorbed by materials wrapped around the source to produce neutrons.

The stronger the neutron source, the more isotopes will reach a detectable level of radioactivity. The measuring of radioactive emissions from the irradiated sources almost always involves the detection of the gamma radiation from the sources. There are two reasons why gamma rays are used. First, of all the radioactive emissions, gamma rays are the least influenced by the structure of a material. A sizable fraction of the gamma rays which escape the radioactive material usually do so with their energy unchanged. Second, since gamma rays are emitted in monoenergetic groups the groups of gamma energies can be used to determine the isotope which has emitted them. The half-life of the emitted gamma rays serves as an additional check on the radioactive isotope created by neutron bombardment. The gamma rays are detected and displayed using devices called gamma-ray spectrometers, which exhibit the energies of the detected gamma rays as a series of peaks. The spectrum is calibrated by an energy-versus-distance plot made with reference to gamma rays of known energy. These spectra have a tendency to have relatively sharp peaks, which correspond to gamma rays absorbed in the detector through the photoelectric effect. These sharp peaks are usually superimposed on and/or interspersed with relatively broad peaks, which result from absorption of the gamma rays through the Compton effect. Few investigators use the Compton peaks for interpretive work. Most work is done with the relatively sharp photoelectric peaks, which will be spread along the energy spectrum.

Generally, more than one gamma-ray energy spectrum is taken at successive periods of time, which are on the order of the half-life of the radioactive isotopes expected. In successive spectra, some series of peaks diminish in height faster than others and in a manner predictable from the half-life of the isotope producing the particular gamma rays. The height of the photoelectric peak, if it is not superposed on a broad Compton peak (also called the

"Compton tail"), is directly proportional to the intensity of the radioactivity of the isotope that produces the peak.

The energy of the gamma rays present and the half-life of the isotopes associated with those gamma rays is determined. If an isotope emits, for example, three gammas, these gammas will generally be of measurably different energies. Inspection of successive gamma-ray spectra will show all three gammas diminishing with the same half-life. Successive spectra then should tell which gamma rays are associated with the same isotope. All such gamma rays will diminish at the same rate with time. From these gammas, hopefully, the isotope can be identified. The half-life can be used to confirm the guess.

Analytical Procedures

Materials irradiated in neutron activation analysis (NAA) have a rate of production that is effectively constant. The rate of production is

$$P = \Phi \sigma N_i \tag{11.3}$$

where N_i is the number of inactive atoms present, Φ is the flux of bombarding particles (in particles/cm^2/s), and σ is the activation cross section for the reaction (in cm^2 or barns). The rate of disintegration or activity of the radioisotopes produced is given by the relation

$$A = \Phi \sigma N_i (1 - e^{-\lambda t}) \tag{11.4}$$

The quantity f is the fractional abundance of the isotope involved. The number of atoms can be obtained from the weight, W, of the element present with atomic weight, M, and Avogadro's number. The quantity λ is the decay constant for the element. Hence,

$$A = \phi \sigma \frac{Wf}{M} (1 - e^{-\lambda t}) \times 6.02 \times 10^{23} \tag{11.5}$$

In general, the activity will not be determined until after a time t after the bombardment has passed, so that the activated atoms will have decayed by a factor $e^{-\lambda}$. Therefore,

$$A = \phi \sigma \frac{Wf}{M} (1 - e^{-\lambda t})(e^{-\lambda t}) \times 6.02 \times 10^{23} \tag{11.6}$$

This equation can be rewritten as

$$W = \frac{MA}{f\sigma\phi} \frac{e^{-\lambda t}}{(1 - e^{-\lambda t}) \times 6.02 \times 10^{23}} \tag{11.7}$$

In principle, all the factors on the right side of the equation are known or can be measured and be used to calculate the weight of the element present. In practice, the amount of activity from the sample is compared with that from a standard amount of the element being determined. It is a simple matter to calculate the weight of the element in the sample from the relationship

$$\text{weight of element in sample} \frac{\text{ug}}{\text{g}} = \text{weight of element in standard} \times \frac{C_r}{C_s} \tag{11.8}$$

where C_r is the observed counting rate of the sample, and C_s that of the standard, measured under comparable conditions.

Sensitivity

If irradiation is carried out for a sufficient number of half-lives so $(1 - e^{-\lambda t})$, known as the saturation factor, becomes effectively unity, and if the activity is determined within a small fraction of a half-life after irradiation ceases, the weight of the element in the samples is given by 11.9.

$$W = \frac{MA}{f\sigma\phi \times 6.02 \times 10^{23}} \tag{11.9}$$

The sensitivity of the method will be greater the higher the efficiency of the detection equipment, the lower the atomic weight of the element, the greater its activation cross section, the greater its isotopic abundance, and the greater the flux of bombarding particles, ϕ.

The efficiency of detection where efficiencies of 10–40%, can be achieved without much difficulty. Total (100%) efficiency is only possible in favorable cases and then only with special apparatus and technique. The atomic weight is fixed for the isotopes of any particular element, but other things being equal, activation analysis is more sensitive for the lighter-Z elements. The activation cross section is also fixed for any given activation process and is really the main factor affecting the sensitivity of the method since it can vary from less than a thousandth of a barn to several barns.

In actual practice, it is difficult to separate peaks in gamma spectra differing by less than 5%. The 0.83 MeV gamma from nickel would overlap the 0.84 MeV gamma from manganese. Such overlapping is characteristic of neutron activation studies. Overlapping peaks can frequently be separated by their half-lives since the shorter half-life will die off first, leaving a peak which is decaying at a longer half-life.

Neutron activation is the most sensitive of analytical techniques, although it has been challenged recently by techniques such as ablation laser inductively coupled plasma—mass spectroscopy (AL ICP-MS).[1] With detection levels of 0.1 ppm and accuracies of ±5% for most elements, INAA is a routine, high-precision tool in archaeometric analyses today.[2] The following example demonstrates its use in the evaluation of ancient coinage and coin manufacture.

The Coins and Coin Molds of the Celtic Hill Fort at Titelberg: Instrumental Neutron Activation Analysis

The Iron Age hill fort called the Titelberg sits on a hill in southwestern Luxembourg near where the present-day borders of Luxembourg, Germany, and France join (figure 11.1). It was built by Celtic peoples ca. 200 BC and was in the heartland of the Treveri tribe at the time of the Roman conquest (58–51 BC). Stratified above the Iron Age occupation period was a later Gallo-Roman settlement that continued until around AD 400.[3] Excavations by the Luxembourg State Museums (1950 to present) and the University of Missouri—Columbia (1972–82) have revealed a high level of manufacturing at the site, including iron forges, a glass factory, potteries, and mints. It is the mints that concern the use of NAA in the study of Titelberg materials.

Most of the structures excavated by the University of Missouri teams were used as mints or mint foundries. The evidence for bronze, silver and gold coin manufacture is abundant: unstruck coins ("flans"), coin dies, and ceramic molds.[4] The coins and molds, shown in figure 11.2, were studied by NAA.

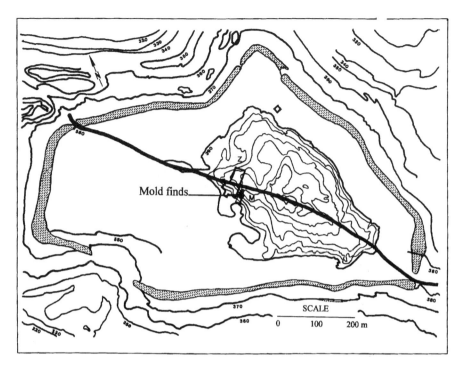

Figure 11.1 The Titelberg, an Iron Age hill fort in southwestern Luxembourg. Location of the coin molds shown. (Courtesy Ralph M. Rowlett.)

The presence and study of coin manufacture has a relatively long history in modern Celtic archaeology.[5] At the Titelberg more than 1,000 coin mold fragments were recovered. As can be seen in figure 11.2, they take the form of flat bricks, although their generally fragmentary condition makes it difficult to determine their exact size. They are about 2 cm thick and 18 cm on a side. Cylindrical depressions made by dowels of appropriate size provided the forms for the coin blanks. Similar forms have been found at the French site of Villeneuve-St. Germain.[6]

The molds from the excavation level "light brown I" (table 11.1), a late Augustan/Tiberian period structure, showed persuasive evidence (elevated amounts of Au and Ag) of noble coinage manufacture, as did molds from stratum Light Brown II, a floor dated to between 30 BC and AD 1, or early Augustan period. It was in this latter level that the largest number of molds and coins were found.[7] High Au/Ag ratios in the coins appeared to be concordant with the presumed ages of deposition—late Republic/early Empire—golds coins having ~0.7% silver and silver coins a like amount of gold. In Roman silver coinage after Vespasian (ca. AD 80), debasement begins, with copper percentages increasing to 12–30%. Given the amount of copper observed in some Titelberg coins, it is plausible that such a process had begun even earlier.

Similar, but less quantitative, results, have been obtained by P. Chevallier et al. using x-ray fluorescence (XRF) in the assay of coin molds from the late La Tène sites of Villeneuve-St. Germain and Mont Beuvray.[8] Although XRF was capable of detecting a suite of metal-

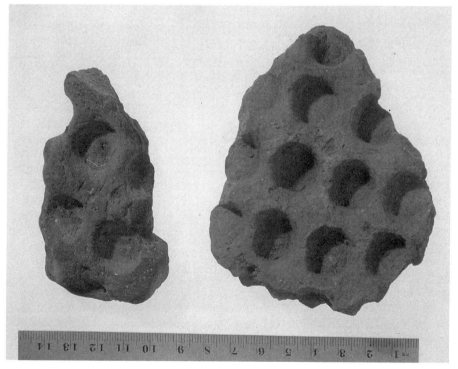

Figure 11.2 Coin mold fragments from the Titelberg. (Photograph courtesy Ralph M. Rowlett.)

Table 11.1 Titelberg Provenances and Ages

Mold number	Location in excavation	Age (estimated)
10-76	25/I light brown I exterior	Late Augustan-Tiberian
275-76	24/J south Dalles floor	Early Augustan
342-76	21/K + L Dalles cellar IVD	Early Augustan
547-76	221/K + L Dalles cellar IVD₃	Early Augustan
425-76	23/K dark brown posthole in pale brown III floor	This posthole seems to belong to the Dalles floor, so this is probably an early Auustan coin mold
617-76	21/J south spotted orange IIG	Ca. 75–50 BC
790-76	22/J light brown I tile Bankette	End of late Augustan-Tiberian
648-76	25/J compact brown IIIb above rubble	End of late Augustan-Tiberian

lic elements, the study was unable to quantify their exact amounts due to the limitations of this particular instrumental technique. The analytical robustness of NAA is further underscored by its increased usage in the chemical analysis of all forms of materials from coins to ceramics.

NAA and Other Archaeomaterials

Since its first uses in archaeologically related studies, NAA has become the "radiocarbon" of analytical techniques. This is to say that the access and reliability of NAA have allowed it to reach a similar level of usage and recognition within the archaeological community. Early workers such as Garmen Harbottle, Edward Sayre, Frank Asaro, and Isaac Perlman were veritable proselytizers for NAA, particularly in its application to ceramic studies.[9] The long list of archaeologically interesting materials analyzed by NAA has led to a large number of databases as well as interlaboratory standardization of methods and calibrations. The two drawbacks of NAA which are insurmountable are (1) the "invisibility" of the elements H, B, C, N, P, and Pb and (2) the radioactivity of the sample after exposure to a reactor-level neutron flux. Neither have proven detrimental, as the technique's analytical popularity shows. Cost of analysis was an early limiting factor but subvention by agencies such as the U.S. National Science Foundation, France's CNRS, and others have made the procedure available to large numbers of investigators. Beyond NAA are a host of capable modern instrumental techniques which have developed into standard practice in archaeological material analyses. We shall next examine the suite of techniques that, in one form or another, rely on the stimulation and emission of x rays.

X-Ray Diffraction

In XRD analyses, collimated x rays are allowed to fall on the specimen and a proportion will be diffracted at angles depending on the crystal structure of the specimen. An essential feature for diffraction is that the distance between scattering points is the same magnitude of wavelength of the incident radiation. XRD works only on crystalline minerals; there is no diffraction in any glass. The orderly arrangement of atoms in repeating structures in minerals is based on the unit cell. 100 unit cells are roughly 1 μm (1000 Å). Every mineral has its own unique unit cell. In practice, the XRD device scans a sample, generally a powder, with a fixed x-ray wavelength of 1.54059 Å, the $K\alpha_1$ (L → K shell) transition for copper. As a comparison, the unit cell spacing in SiO_2 (α-quartz) is 1.6 Å.

The Unit Cell and Wavelength

Constructive interference occurs only when the angles $\Theta = \Theta'$, or the path length difference − DC + CE (pathlengths) = λ (wavelength)

So $DC + CE = \lambda; \theta = \theta'$
        Path length, $DC = AC \sin \theta$
        + Path length, $\underline{CE = AC \sin \theta}$
        $DC + CE = 2AC \sin \theta$

or

$$\lambda = 2d \sin \theta \qquad \text{(Bragg's Law)}$$

where DC + CE = λ and AC = d (d is the unit cell)

In quantitative analyses, one can use (1) absolute intensity or (2) reference intensity.[10] Whichever the analytical approach, one obtains a diffractogram (figure 11.3), which is a binary plot of I (in cps) versus the angle 2Θ, which is the angle swept during analysis. One can evaluate XRD versus XRF easily using Bragg's law. For instance, in XRF one obtains E through determining $n\lambda$ where the relation is $E = h\lambda$. The unit cell (d) is known and we hold Θ constant. In XRD, the wavelength, λ, is held constant and Θ is varied to obtain d.

XRD has been very useful in archaeological geology. Linked with the geologist's traditional friend, thin-section petrography, together they make a formidable conjoint analytical procedure.[11] This is particularly true in ceramic studies since clays undergo phase changes when fired. Above 550°C, they may become amorphous, and above 1000°C, they may vitrify or recrystallize as mullite and cristobalite. Associated with clays are accessory minerals such as plagioclases, feldspars, quartz, chalcedony, amphiboles, pyroxenes, and micas which form an excellent matrix for study by XRD and petrography.

X-Ray Fluorescence Spectrometry

In x-ray fluorescence spectrometry (XRF), the specimen is bombarded with x rays from a conventional x-ray tube and, in turn, emits secondary (fluorescent) x rays of wavelengths characteristic of the elements present in the specimen. These secondary x rays are then examined with a spectrometer. Each element present in the specimen is identified from its Bragg angle of refraction and the intensity of x rays received by the detector. An important variant of XRF is the microprobe or focusing x-ray fluorescence spectrometer, which is specifically designed for the microanalysis of rocks and minerals (see the following section).

One way to raise an atom to a higher state is with absorption of energy. If Na is exposed to incident radiant energy of 5890 or 5896 Å, the transition is to the higher-energy excited

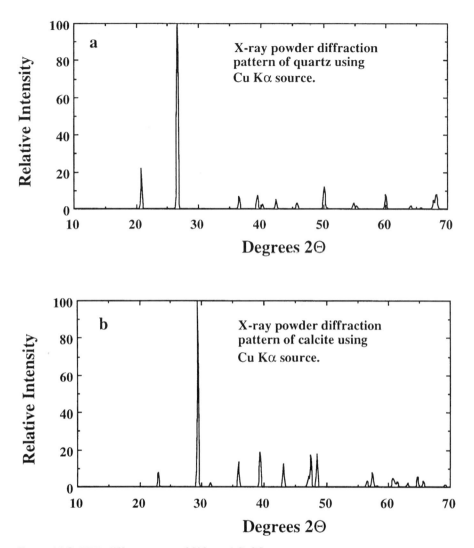

Figure 11.3 XRD diffractograms of SiO_2 and $CaCO_3$.

state. After a short time, called the lifetime of the excited state, which is 1.6×10^{-8} s for Na, the atom returns to the normal state via emission of x-rays and yellow light.

The Na atom produces x-ray lines at 2594 Å, 2544 Å, 3302 Å, and 2853 Å as well. Energy considerations require $E = h\nu$ where h is Planck's constant and ν is the frequency. In the normal absorption cell the ratio of n_o (normal atoms) to n_e (excited ones) is small and given by Boltzmann's equation:

$$n_e/n_o = \exp(-E/kT)$$

where k = Boltzmann's constant (1.38×10^{-23}), T = temperature, and E = energy in electron volts (eV).

There are two processes for x-ray emission:

1. Some electrons are stopped by a target and their kinetic energy is converted to x rays.
2. Some electrons "excite" the target's atoms, which decay, or "de-excite," and emit x rays.

The second process is characteristic of a target's atoms; the first process is not.

Again, recalling the "shell" structure of atoms, the electrons fill these levels according to quantum rules; for example:

$$K = 2, L = 8, M = 18, N = 32$$

where K, L, M, N are electronic shells, from nearest the nucleus. Na contains 11 electrons in the configuration of 2, 8, 1 for K, L, and M shells, while molybdenum (Mo) has 42 arranged 2, 8, 18, 14. The outer-shell electrons are responsible for the atom's optical spectra because little ΔE is required to remove them and raise the atom to an excited state. More energy is required to remove the inner electrons (e.g., K shell). Their displacement and the decay to normal levels give rise to higher-energy x rays. When an electron or x ray displaces a K-shell electron, it is filled by an L, M, or N electron with an x ray emitted. A series of three such transitions is called the K series; for example,

$K_{\alpha} \Rightarrow$ L electron to K level
$K_{\beta} \Rightarrow$ M electron to K level
$K_{\gamma} \Rightarrow$ N electron to K level

Mo, Tn, and Cu all have K series spectra. Other elements with more electrons have more complex spectra with L series, and so on. "Primary" x rays of sufficient energy can produce "secondary," or fluorescent, x rays. These secondary x rays are diffracted by a crystal and the intensities of element lines are measured by a photon detector, Si(Li).

XRF has little or no sensitivity for elements below $Z = 7$. The method is best suited for estimating the abundance of most elements in the $>0.1\%$ range, and under favorable circumstances, trace elements in the ppm range. It requires the preparation of small powdered samples or is nondestructive in the examination of surfaces. Many kinds of artifactual material can be analyzed by XRF, including lithics, pottery, glass, and metals.

Each chemical element produces a characteristic x-ray spectrum whose line spacings and heights are a function of the element itself and its concentration. XRF tells which elements are present and their amounts; XRD (x-ray diffraction) tells which minerals are present and their relative amounts. Since minerals have known chemical compositions, XRD also gives information on the chemical compounds, telling us, for example, that calcium detected by XRF is present as fluorite, CaF_2 rather than wollastonite, $CaSiO_3$.

Electron Microprobe Analysis and Proton-Induced X-Ray Emission

Analysis by electron microprobe and by proton-induced x-ray emission (PIXE) have become widely used in elemental analysis of archaeomaterials. In electron microprobe analysis, electrons from a filament are accelerated by about 30 KeV toward the specimen and focused by two magnetic lenses into a minute spot 0.001 mm in diameter. When the beam strikes the specimen, x rays are produced that are characteristic of the elements present in the specimen. These x rays are analyzed by a spectrometer as in the XRF technique. In PIXE

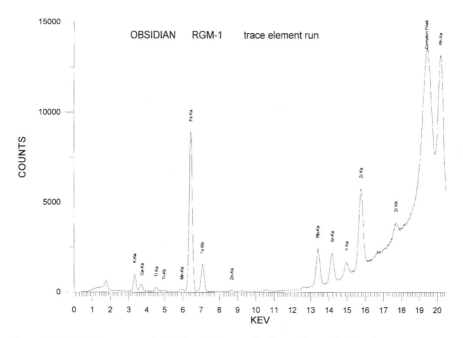

Figure 11.4 XRF spectrum of obsidian. (Courtesy Professor Steve Shackley.)

analysis, the electron is replaced with the ionized hydrogen nucleus (proton) as the incident particle. The technique is based on the excitation of surface atoms by use of an energetic proton beam of a few (≈ 2.0) MeV/amu (amu = atomic mass unit).[12] In theory, both techniques are nondestructive methods, the restriction of sample size often necessitates the removal of a specimen from a large object. Also, cross sections may have to be prepared because the penetration of the electron beam is very small. A limitation with the microprobe is the fact that the electron path must be in a vacuum. This was a drawback with PIXE as well until development of an external beam method allowed the analysis to be performed in air.[13] An advantage of these methods is that objects can be examined in minute detail— that is, a succession of layers can be examined when each is only a few microns thick in a very small area (1 μm^2 for the microprobe and 1 mm^2 for PIXE). Very small quantities of an element can be detected, but since the total amount examined is also very small, the lower limit of percentage composition detected is only about 0.1% with the microprobe; PIXE can attain ppm levels. Using the electron microprobe, researchers have the advantage of (1) quantitative, elemental determination, (2) use of the dispersive nature of x rays for a semi-quantitative characterization of large suites of elements, and (3) production of secondary electron images of the sample's surface (figure 11.5).

Atomic Absorption Spectroscopy

Atomic absorption spectroscopy (AAS, or just AA) has a long record of use in modern analytical chemistry. Elemental concentrations can be determined at the same level as the bet-

Figure 11.5 X-ray dispersive spectrum/secondary electron image of garnet. (Courtesy Chris Fleischer.)

ter x-ray techniques (1 ppm to 100%), with an accuracy ranging from $\pm 1\%$ for major elements to $\pm 15\%$ for trace elements.[14] AAS is simply the measurement of the absorption of optical radiation by atoms in the gaseous state. In a physical sense it is the reverse of emission spectroscopy. Absorbance can be regarded as that energy required for transition between atomic levels which are exactly the same as those of emission. Free atoms in the ground state can absorb radiant energy at precisely defined frequencies ($h\nu$), so the light absorbed measures the number of these atoms per volume.

To quantify this absorption, AAS uses the Beer-Lambert law in the following form:

$$\Phi_{tr} = \Phi_o \cdot e^{-xNl}$$

where Φ_o and Φ_{tr} are the radiant fluxes prior to and after light passes through a layer of thickness l, X is the atomic absorption coefficient, and N is the number of free atoms. Generally, this reduces to

$$\alpha = \frac{\phi_o l - \phi_{tr}}{\phi_o} = 1 - T$$

where α is the absorbance and T is the reciprocal transmittance. In practice, AAS is a relative method where a linear relationship exists between N and their absorbance.[15] The output of these devices are spectra with precise data on the amount of individual elements in the vaporized material. Using the graphite furnace technique, solids may be analyzed. Using the flame method, 39 elements may be evaluated; with the graphite furnace, 33. Light metals such as lithium (Li), hard to detect using x-ray emission, can be evaluated using AAS.

As powerful an analytical tool as AAS has proven to be it is interesting to note the shift in interest from AAS to atomic emission spectroscopy (AES) using inductively coupled plasma (ICP) as the excitation source.[16] It is difficult to point to one specific reason for this. Old-time users of AAS are familiar with the sometimes perplexing need to change the "lamp" (radiation source) to measure a given element. These hollow cathode lamps can be either single or multielement. For a broad-scale analysis of an artifact or sediment, several lamps must be used. Today the hollow cathode lamps have been almost universally replaced by electrodeless discharge lamps (EDL).[17] Nonetheless, two ICP techniques have come to be routinely used in the analysis of archaeomaterials.[18]

Atomic Emission Spectroscopy and ICP Mass Spectroscopy

With the coming of ICP in the 1970s many dire predictions were heard for the future of AAS. Because of the efficacy and sensitivity of ICP, it was heralded as the natural successor to AAS. While ICP-AES has many appealing features such as simultaneous multielement determination, relatively high precision, and low detection limits, in terms of these latter two aspects, the graphite furnace AAS is superior by at least two orders of magnitude.[19] AAS has great sensitivity and specificity but lacks speed in multielement studies. ICP-AES is weak in the detection of trace elements but excels in the characterization of B, Ta, Ti, U, and so on. As a replacement technique for AAS, ICP-AES is more a complementary than an alternative method. It is perhaps with ICP-MS that the discrepancy between AAS and ICP is most manifest.

ICP, in brief, is an emission spectroscopy technique based on the vaporization of a sample in a plasma. This ionized gas mixture burns at 6000–8000° K. The plasma is ignited by induced microwave heating, hence the name "inductively-coupled plasma." The carrier gas is inert argon (Ar). The emission spectra are taken as different materials are injected into the carrier gas stream. At this point ICP-AES becomes like any spectrophotometric method in analyzing the appropriate emission frequencies or wavelengths.

Along with quadrapole charge/mass spectroscopy, the analytically robust quantitative method of ICP mass spectroscopy (ICP-MS) has developed since the 1980s.[20] The system uses either a standard ICP system coupled with a quadrapole mass spectrometer or, in its most esoteric form, both ICP and MS are joined to a Nd-YAG laser, which performs the ablation (vaporization) of a localized portion of a material's surface. In either mode, ICP-MS or AL (ablation laser)-ICP-MS, 65 elements can be determined at ppm levels. This clearly makes the technique a true challenger to NAA's dominance in instrumental archaeometric analyses. Still, as Porat et al. point out, Harbottle documented an NAA database estimated at 40,000–50,000 entries in 1982, with an annual production of thousands.[21] Neither is the sensitivity of the two methods identical: standardization of ICP to NAA remains an outstanding problem.

Cathodoluminescence Miscroscopy

Cathodoluminescence (CL) is the emission of photons in the visible range of the electromagnetic spectrum after excitation by high-energy electrons in an electron microscope. It is an effect coincident with several other forms of stimulated emission such as primary and secondary x-ray emission, and secondary, and backscattered electrons. The excited mineral, calcite, dolomite, etc., produces different colors of luminescence that originate from various impurities in the crystal lattice. The lattice defect in carbonates, most used for CL, is that of Mn^{2+}, zircons with dysprosium (Dy^{3+}) and europium (Eu^{2+}), apatites with Mn^{2+}, Dy^{3+}, and samarium (Sm^{3+}), and fluorite Eu^{2+} and Eu^{3+} centers. CL microscopy is practiced in two differing procedures: hot-cathode CL and low-temperature CL. To date, the hot-cathode method has been most used in archaeological geology.

Beginning in 1965, CL became of interest to geologists.[22] The hot-cathode CL microscope consists of a hot tungsten cathode under a vacuum, the accelerating voltage ranges between 2.5 and 50 keV, and current densities are between 5 nA/mm^2 and 10 nA/mm^2. The sample is viewed by an optical microscope fitted with a still-frame camera. A monochromator with a spectral range of 350–850 nm is fitted between the cathode and viewing optics.

In archaeological geology studies, CL has made its biggest impact in the provenance studies of white marbles.[23] CL studies have divided white marbles into three major families based on luminescent color. Calcitic white marbles have a dominant orange or blue luminescence, while dolomitic marbles show a red luminescence (figure 11.6). By combining CL with grain size, texture, and, most important, stable isotope analysis, finer distinctions can be obtained between marbles.

The orange luminescence family contains most of the calcitic marbles from quarries such as Mount Pentelikon, and Thasos in Greece; Naxos, Paros, and Pteleos in the Cyclades; Carrara and Lasa in Italy, and Dokimeion in Turkey. Blue luminescing marbles are less easy to characterize because the cause of the CL coloration is not yet established.[24] The blue emission is observed in marbles with manganese present in the calcite at levels below 5 ppm. Red luminescing marbles are exclusively dolomitic (greater than 50% dolomite). Representative marbles include those from Crevola, Italy; Villette, France; and Naxos, in the Cyclades.

Resonance Ionization Spectroscopy

Resonance ionization spectroscopy (RIS) is a laser-based spectroscopic technique developed in the last thirty years.[25] The basic science is straightforward and involves atomic, molecular, and optical physics. A laser induces an electron to excite (move from its ground state to a higher state). At this point other photons, collisions, or electric fields cause ionization of the excited atom. A proportional counter detects the photoelectron or photoion produced. Any kind of mass spectrometer (ion, quadrapole, etc.) can then detect the Z value (atomic number). Selection of the Z value beforehand and the use of a tunable laser allow great detection efficiency. The "resonance" absorption only occurs at a set or fixed wavelength (λ) as determined by the laser pulse. Sensitivity and selectivity require great precision in the pulse length of the laser; that is, pulse duration must be shorter than or commensurate with the lifetimes of the excited states to avoid "optical pumping," where the excited atoms spontaneously drop back to an intermediate energy state and avoid ionization. In devices such as cesium magnetometers, optical pumping is the process that allows the proxy measurement of the local magnetic field. RIS has a sensitivity high enough to detect a single atom during a single pulse. Deeply bound or ground-state atoms can be brought

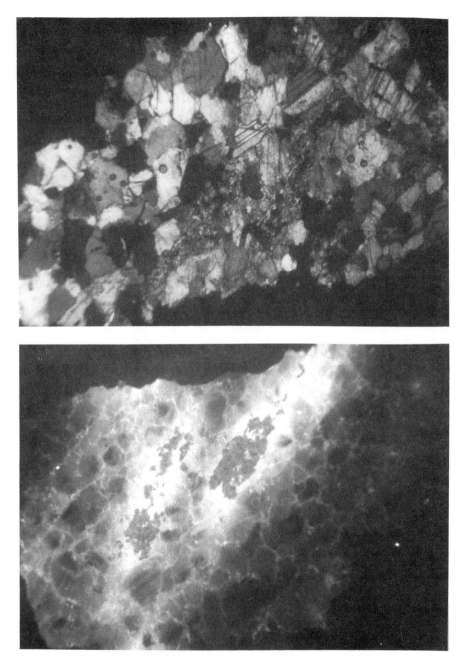

Figure 11.6 Cathodoluminescence of white marble: (*top*): petrographic image; (*bottom*): CL image. (Courtesy Professor V. Barbin.)

to ionization levels by multistep resonance (e.g., multipulse) procedures. The versatility of RIS is such that *any* atom or molecule can be excited and identified. Continuous-wave lasers coupled with simple mass spectrometers can detect small numbers of atoms ($\sim$3,000) such as ^{210}Pb or radon in environmental samples. In earth science the RIS has been applied to the K/T Boundary Anomaly, which was first explicated by Luis Alvarez and coworkers. These investigators raised the extraterrestrial iridium present to postulate an extinction event caused by meteorite impact.[26] While extraterrestrial iridium is present in parts-per-thousand levels, other corroborative extraterrestrial elements such as rhodium occur at ppm levels or lower. NAA can easily detect the iridium, as it did for the Alvarez team. Bekov measured iridium and rhodium in K/T sedimentary rocks and noted a sharp rise in both elements at the K/T boundary event. In turn, the RIS measurement of the Rh/Ir ratio in these levels coincides nicely with those typical of meteorites.[27]

Scanning Tunneling (Electron) Microscopy and Atomic Force Microscopy

Since the awarding of the Nobel Prize to Heinrich Rohrer, Ernest Ruska and Gerd Binnig in 1986 for the development of the scanning tunneling electron microscope (STM), its use as a nanoscale surface imaging technique has steadily increased.[28] Based on the quantum mechanical property of electron tunneling, a sharp needle is scanned across a surface between which a small bias voltage exists. A quantum mechanical tunneling current between the tip of the needle and the surface is sensed by a feedback loop which, in turn, drives piezoelectric actuators that control the tip height above the surface within an angstrom (Å).[29] Spatial changes in the tunneling current or tip height are used to develop an image of electron density on the surface. STM resolution depends on the square root of the product of the tip radius (a few Å) and the exponential decay distance for the tunneling current (again on the order of a few Å).

Atomic force microscopy (AFM) is the first offspring of STM.[30] The heart of the AFM, like the STM, is a tip that interacts with a sample surface. The tip attaches to a spring, usually a cantilevered wire. The force between the tip and the sample surface is monitored by tunneling, optical interferometry, or the deflection of a laser beam. Imaging of a graphite surface indicated resolution of 1.5 Å in the horizontal and fractions of an angstrom in the vertical.

Both STM and AFM, in their far-field modes, 75–90 μm, show great promise in many areas of archaeomaterial analysis. Morphology of rocks and minerals; artifact surfaces, residues, striations; and metallurgy—corrosion, defects, cracks, and phases—are only a few of the possibilities.

Near-Field Scanning Optical Microscopy

In 1993 Eric Betag and Robert J. Chichester, physicists at Bell Laboratories, announced a significant advance in optical microscopy that will enable archaeologists and geologists to visualize surfaces at a level of resolution approaching that of electron and atomic microscopes.[31] The instrument, called a near-field scanning microscope (NSOM), has a resolving power of 500 Å. Compared to the STM or the AFM, which have imaged individual atoms, this is not as impressive, in itself, as it would have been 10 years ago. The NSOM system, however, combines the direct optical viewing of nanometer surfaces with a spectroscopic capability these other devices do not have. Moreover, the NSOM is a nondestructive system which does not require a conducting surface of the material being studied.

The breakthrough that allows the NSOM to optically resolve previously unapproachable levels with light photons is a subwavelength light source.[32] In traditional optical microscopes, resolution is defined by the light diffraction limit. The light source is several wavelengths above the surface of a sample in these instruments, placing a bound on their capacity to "see" below this level. To resolve at subwavelengths, the light source must be moved to that distance from the surface. Betag and Chichester did this while scanning the source across the surface as well. Using the optical luminescence of the sample, the NSOM can spectroscopically identify what it is observing. This is something that STM and AFM devices cannot do at present. In a sense, much like the high-sensitivity mass spectroscopy of atom counting (e.g., AMS), the NSOM can focus on individual molecules or small regions of atoms and spectroscopically analyze them. In this method of "atom counting," one does not have to dissociate the specimen in high magnetic fields and rely on individual charge states to separate the atomic species.

As yet untested in archaeologically interesting materials, one can see the analytical potential the NSOM promises to provide to researchers of minerals and artifacts. Indeed, physical phase seems little hindrance to the new instrument, as it appears capable of imaging from condensed matter to solutions.

Transmission and Scanning Electron Microscopy

We close this chapter with a brief note on the "STMs" and "AFMs" of the past decades: transmission and scanning varieties of the electron microscope. These instruments revolutionized the ability of science to image geological, artifactural, and/or biological specimens at microlevels unattainable by the optical microscope. The importance lies in the difference between microstructure and nanostructure. Then we could see crystals; now we view the atoms of those crystals. Nevertheless, much like the case for the AAS and ICP systems, we should not consider these older electron systems as outmoded. As M. S. Tite observed recently, the SEM, in combination with quantitative elemental analysis, is a powerful technique for the study of ancient ceramics.[33] In their many varieties, such as HV-SEM (high-voltage SEM), which is combined with a backscatter electron (BSE) detector capable of imaging successive layers of a material,[34] they will remain useful to the archaeological geologist for years to come.

12

Metallic Minerals and
Archaeological Geology

Archaeological Economic Geology

Economic geology had its inception in the ancient utilization of rocks and minerals. The first economic materials were nonmetallic and include flint, quartz, diabase, rhyolite, obsidian, jade, and other stones, which were sought for weapons, implements, adornment, and even art. Beginning with the Upper Paleolithic Aurignacian period, clay began to be widely used for simple figurines, then brick and finally pottery.[1] S. H. Ball identifies 13 varieties of minerals—chalcedony, quartz, rock crystal, serpentine, obsidian, pyrite, jasper, steatite, amber, jadite, calcite, amethyst, and fluorspar—as economic within the Paleolithic.[2] Add to this list the use of ochres and mineral paints together with nephrite, sillimanite, and turquoise. In the standard reference on the nonmetallic deposits, "Industrial Minerals and Rocks", 6th edition published in 1994, deposits are classified by use and the minerals and rocks described as commodities.[3] The fourteen use groups include such items as abrasives, constructions materials, and gem materials; the 48 commodities include clay, diamonds, feldspar, etc.

Metalliferous minerals as ore deposits are unevenly distributed throughout the world.[4] The formation of a mineral deposit is an episode or series of episodes in the geological history of a region and reflects three broad categories: (1) igneous activity, (2) sedimentary processes, and (3) metamorphism. Table 12.1 summarizes general features of the three categories of mineral deposits.

Admixtures of metals are by far the most common form of mineral deposits. Gold, silver, and copper occur either as native metals or admixed with other metals and compounds. Most ore deposits are actually mixtures of metals: silver commonly with lead, zinc with cadmium, iron with copper (table 12.2).

Many metallic ore deposits are products of igneous activity (table 12.1). Conditions change in the magma chamber as the principal rock-forming minerals crystallize, temper-

Table 12.1 Major Mineral Deposits by Origin. Ref. Carr and Herz, 1989

Genesis	Process	Representative Deposits
1. Igneous Processes		
Intrusive rocks	Cooling from Magma	Construction materials
Magmatic processes	Crystallization or concentration by differentiation of mafic magma at great depth	Diamonds, nickel, platinum-group, chromite, titanium-iron
Hydrothermal	Form from felsic, water-rich near surface magma	Copper, lead complexes and sulfides, tin, pegmatites-feldspar, beryl
Hydrothermal alteration	End stage of magmatic activity, water-rich solutions	Clays, porphyry copper, sulfides
Replacement and vein deposits	Final stage, fills fractures and replaces original minerals	Lead and zinc sulfides, silver-lead, gold, tin
Porphyry deposits	Disseminations into country rock and borders of intrusive	Copper, tin, molybdenum
Extrusive rocks	Low temperature in volcanic rocks	Silver, gold, sulfur
Exhalative	Submarine volcanogenic, "black smokers"	Iron-copper-zinc sulfides
2. Sedimentary Processes		
Solution transport	Precipitates by changing Eh-pH	Iron ore, limestone, copper, manganese, phosphorus
Bacteriogenic	Anaerobic sulfur-fixing bacteria organic reduction	Sulfur, gypsum, uranium, iron, manganese, sulfides
Evaporation	Evaporation of sea water in enclosed basins, lagoons, etc.	Gypsum, salt
Clastic	Weathering and transport of heavy residual minerals	Placer deposits, gold, gems, cassiterite
Residual	Weathering in place, residual "lag" minerals	Clay, iron
3. Metamorphic Processes		
Contact metamorphism	Around or near igneous intrusions	Graphite, Al-silicates, tin, gold-silver, lead
Regional metamorphism	Widespread, associated with orogenesis	Slate, marble, serpentine, asbestos, graphite, talc, emery, garnet

ature falls as the magma cools, pressure is lowered as the magma rises in the crust, and volatiles increase in the magma chamber. The increase in volatiles, especially of water and halides—chlorine and fluorine—is due to two factors: (1) minerals crystallizing—the early minerals to form, such as quartz, feldspar and pyroxene, are all anhydrous allowing a relative increase of volatiles in the liquid magma, and (2) an influx of surface waters. The relative amounts of ore elements also increase because they do not enter into the crystallizing silicates to any great amount. The increase in the concentration of metals and of volatiles needed to transport the metals, the falling temperatures and pressure and the temperature gradients set up between the hotter core of the chamber and the cooler rims are the processes primarily responsible for the formation of metallic ore deposits.

Table 12.2 Principal Metallic Ores

Group and name	Formula
Sulfides	
Galena	PbS
Sphalerite	ZnS
Covellite	CuS
Chalcocite	Cu_2S
Argentite/acanthite	Ag_2S
Stibnite	Sb_2S_3
Pyrite	FeS_2
Chalcopyrite	$CuFeS_2$
Bornite	Cu_5FeS_4
Stannite	Cu_2FeSnS_4
Oxides	
Cuprite	Cu_2O
Magnetite	Fe_3O_4
Hematite	Fe_2O_3
Limonite	$Fe_2O_3 \cdot 3H_2O$
Cassiterite	SnO_2
Carbonates	
Malachite	$CuCO_3 \cdot Cu(OH)_2$
Azurite	$2CuCO_3 \cdot Cu(OH)_2$
Siderite	$FeCO_3$
Smithsonite	$ZnCO_3$
Cerrusite	$PbCO_3$
Chlorides and silicates	
Cerargyrite	$AgCl$
Atacamite	$CuCl_2 \cdot 3Cu(OH)_2$
Chrysocolla	$CuSiO_3 \cdot 2H_2O$
Hemimorphite	$2ZnSiO_3 \cdot H_2O$
Arsenates	
Enargite	Cu_3AsS_4
Olivenite	$Cu_2(AsO_4)(OH)$
Domeykite	Cu_3As
Tennantite	$Cu_{12}As_4S_{13}$

Nernst's law is representative of the processes that control mineral precipitation in solutions. When two intermingling liquids contain salts, precipitation of minerals of differing solubility occurs by the decrease in one salt's solubility in the presence of another salt that has a common ion. Siderite ($FeCO_3$) or smithsonite ($ZnCO_3$) is precipitated in the presence of calcite ($CaCO_3$), in solution. Metal sulfides are formed when solutions of H_2S oxidize near the surface to H_2SO_4 (sulfuric acid), percolate downward, acidify alkaline sulfides, and cause precipitation. When a concentrated mineral solution mixes with groundwater, precipitation ensues. The ores and gangue minerals precipitated from hypogene to supergene solutions can replace earlier formed minerals. A replacement mineral or metasome such as copper sulfate can replace limestone to yield copper carbonate; pyrite to yield chalcocite. Replacement is the most important process in the formation of epigenetic ores (ores formed later than the rocks in which they are found)—by either hot gases and vapors or by hot and cold water solutions. Epigenetic copper ore minerals are malachite and azurite—$CuCO_3 \cdot Cu(OH)_2$ and $2CuCO_3 \cdot Cu(OH)_2$, respectively. Where carbonate is lacking, copper can form chrysocolla ($CuSiO_3 \cdot 2H_2O$).[5] Silver does not precipitate as a carbonate but does form epigenetic deposits of native metal or cerargyrite ($AgCl$), such as at the famous Laurion mines near Athens. At Laurion, the geology—alternating schists and limestones—provided zones of metasomatism for hydrothermal solutions where their cooling at the impervious mica schist contact precipitated argentiferous galena (PbS).[6]

Weathering and redeposition are another sequence of events in the formation of economic mineral deposits. Weathering proceeds as either a mechanical or a chemical action; at times, both operate in concert. The end result of chemical and then mechanical weathering is freed-up minerals that can be transported and redeposited in concentrations such as placers. Placers are syngenetic; that is, the ore deposits are the age of the enclosing rocks. The most famous placer deposits are those containing gold, such as in the California Sierra Nevada. Tin, such as cassiterite, is another important ancient placer mineral.

Metallic Minerals: Techniques of Identification

In chapter 11 we discussed many instrumental methods for the characterization of minerals including ore minerals: XRF, ICP, ICP-MS, AAS, PIXE, and EMP, INAA. Heavy-ion mass spectroscopy remains the most reliable means of determining the isotopic provenance of metals, slags, and ores (see chapter 14). More mundane, less "high-tech" means of studying metallic minerals and their products remain important. These are the old standbys of optical petrography and its lineal relative—optical metallography.[7]

Metallography is the basic technique for revealing the internal microstructure of metals. Important ore minerals are opaque, so reflected polarized microscopy is best for examining their optical properties. Under reflected light, two properties, color and isotropy/anisotropy, help identify the unknown mineral. While the colors of opaque minerals are not the vibrant rainbow hues of the nonmetallics, the creams are generally sulfides and the grays are oxides.

Native Metals: Monometallurgy

In following the elegantly straightforward 1819 sequence of Christian Jurgensen Thomsen—stone, copper, bronze, and iron ages—it is interesting to see how ancient metallurgy evolved from using native copper alone to monometallic compounds and then polymetallic mineral exploitation. The condition for economic exploitation of metals is volition and the development of the skills to extract and manipulate these minerals.

In moving to the discussion of metals and their sources in the context of archaeological geology, we begin with a definition of *ore*. An ore is a naturally occurring mineral aggregate from which one or more minerals (metals and nonmetals) can be extracted economically.[8] Mixed with the ore is the host or waste rock gangue. Ores are extracted from the earth by mining, a process more ancient than metallurgy by hundreds of millennia.[9] It began with the search for geological materials for fabrication into tools or accessories by ancient humans.

The intentional search and recovery of first stone and then pigments, such as ochre, are well established in human history. By the beginning of the Pleistocene, early humans were selecting specific types of stone—generally flints or volcanics—for use as core/flake tools. Certainly by the mid-Pleistocene, in what is termed the Middle Paleolithic, the Neanderthals were burying their dead with large amounts of iron oxide (red ochre).[10] Other early "economic" minerals include cinnabar (HgS); wad, a manganese oxide; and galena (PbS).[11] In terms of metals, the natural, or native, varieties of copper, iron, silver, and gold were the first to be used.[12] These native "ores" occur on or near the earth's surface and require little effort to recover.

Of the native metals gold must have been the first utilized. By 1000 BC it had attained wide usage as a metal for currency.[13] Gold is the most obvious in its occurrence, mainly as flakes and nuggets in placer deposits, but also in polymetallic veins. Native gold can be almost pure gold, but substantial amounts of silver and copper may also occur with gold. Native copper can be found as irregular masses and disseminations filling voids in lavas and sedimentary rocks.[14] It is relatively easy to smelt and so was used for fabricating tools. Native iron is exceptionally rare and is of either terrestrial or meteorite origin. Terrestrial iron is associated with basalts or is in sedimentary contexts and is also called *telluric*.

Much of our knowledge of the early use of native metals, particularly iron and copper, comes from the New World. The "Old Copper culture," an Archaic period (ca. 3000 BC) cultural horizon that arose in the midwestern United States and southern Canada, exploited vast copper deposits of the Great Lakes region. G. Rapp, Jr., and others identified over 50 geologic sources, using NAA, in the United States and Canada.[15] The principal source is the Keweenaw deposits of northern Michigan. Native iron use was widespread among the Inuit (also called Eskimo) in the circum-Arctic, with meteoritic and telluric sources in the Cape York area of northwest Greenland and telluric sources in west Greenland on Disko Island.[16] This is a late prehistoric culture dating from ca. AD 1000 in contact with the European Norse.[17]

In the Old World the first archaeologically demonstrated use of copper dates to the period 7250–6750 BC at the Turkish site of Cayönü. Studies of these Turkish copper finds (metallic and malachite beads) indicate the material came from the great Ergani Maden mines, 20 km north of the site.[18] It is not surprising that copper and gold are the most prevalent in early metal-using cultures. In other ancient cultures, such as those of Mesoamerica and western South America, these metals, first in monometallic and then polymetallic forms, also characterize the increasing richness and complexity of these evolving societies.

The demand for metal reflects the emphasis that societies place on its role. In the copper-using American Archaic groups, the number of artifacts is impressive, yet their individual size parallels the size of the communal structures—small, mobile hunter—gatherer bands, the New World equivalent of the Old World's Mesolithic. The native copper tools were both functionally utilitarian and items of status worthy of accompanying their owners to the grave.

Native gold, while not as easily found, nonetheless was ubiquitous enough to provide early "smiths" the material for easy melting and forming into the simple or complex items we admire in the examples surviving from the graves of elites from ancient Peru to Neolithic Bulgaria. Still, the size of the native gold artifacts mirrors the technological immaturity of their makers. The mastery of the metal lay outside the material itself. Whether the metal is gold, silver, copper, or iron, the exploitation of monometallic ores now seems more a variation of ancient metallurgy than a necessary stage to the use of polymetallic ores. In his treatment of the use of monometallic copper and iron in the Arctic, A. P. McCartney terms this "epi-metallurgy"—a dependency on, but not a production of, metals.[19]

Copper to Bronze: Polymetallic Metallurgy

The production of copper from polymetallic ores in antiquity required a pyrotechnology that could smelt the copper-bearing ores such as malachite, chalcocite, and chalcopyrite—$CuCO_3 \cdot Cu(OH)_2$, Cu_2S, and $CuFeS_2$, respectively (table 12.2). As these ores are generally low grade (less than 2% copper), the copper must be concentrated by the smelting procedures. In some instances native copper or high-grade metallic ores such as those found in the Great Lakes area were "cold-worked" into tools and weapons.[20] This epimetallurgy occurred in both the New World and the Old World. Mining for native copper in the Lake Superior region dates to the Middle Archaic period (5000–3000 BC) and that in the Anatolian region of Cayönü to as early as the 9th millennium BC. Sites such as Çatal Hüyük in Anatolia have copper implements in the 7th millennium BC.[21] A general model for Old World metallurgy is given in table 12.3.

By the early 4th millennium, copper smelting was used to produce tools such as knives, sickles, and woodworking adzes. Copper metal in pure form is quite malleable and has decidedly less tensile strength ($\leq$40,000 psi) than stone or bronze, an alloy of copper. This tensile strength was enhanced by hammer working of the tool after casting. To produce the copper, the smiths roasted the sulfide ores to oxidize (calcination) the sulfides of impurities such as arsenic (As) and antimony (Sb). Cu_2S remains as a *calcine*. The carbonate ores pre-

Table 12.3 Evolution of Metallurgy in the Old World

1800–1200 BC	In Anatolia, iron asserted itself as a dominant metal and spread gradually to Assyrian, Ugaritic, Iranian, Egyptian, Palestinian, and Minoan peoples.
2300–1800 BC	The earliest manifestations of iron capped the first period of industrial metallurgy. This was marked by the extensive production of copper, silver, and lead in Anatolia and trade by Assyrian and Babylonian merchants in copper and tin. Swords from Alaca Hüyük in Anatolia were of native iron. Iron slag was present in a copper furnace at Alaca Hüyük.
Late fourth and third millennia	The invasion of the sulfide zone of copper ores accompanied the birth of the Bronze Age and marred the full advent of polymetallism.
Fifth and fourth millennia	Smelting of a variety of metals from their ores and casting of copper introduced smiths to arsenic, antimony, bismuth, and tin as impurities. Iron, silver, and zinc were major by-products of smelting.
Sixth millennium	Annealing of copper lead to attainment of 1083°C, temperatures at which copper can be cast. Experiments with smelting other ores proceeded apace.
Ninth to seventh millennia	Copper was first hammered as a stone, then annealed and hot-worked, drawing on a tradition of annealing obsidian and other stones as a way of working them.

Source: Modified from T. A. Wertime, 1973, Pyrotechnology: man's first industrial uses of fire, *American Scientist,* 61(6):670–682, Table 1.

sented less of a problem to the smelters and were processed in the simple bowl-shaped hearths to form "puddles" of 98–99% pure metal. Because of the primitiveness of the early smelt furnaces and the generally large amounts of ore needed, the metal produced was utilized for exclusive items such as jewelry or for small tools.

As smelting techniques improved in intensity and efficiency, the demand for ore increased. It was probably due to experimentation with various ores that led to the discovery of bronze in either of two forms: arsenical and tin-bronze.

Arsenical copper ore such as enargite, Cu_3AsS_4, produces a natural bronze alloy.[22] This metal improves the tensile strength properties of the alloy to roughly double that of the unworked pure metal. It is possible to form tin-bronze from stannite, Cu_2FeSnS_4, by direct smelting.[23] As either an accidental alloy or, by 3000 BC, an intentional metal process, bronze was the predominant metal until the development of iron as a useful metal by 1000 BC. The transition from arsenical copper to tin-bronze took place relatively rapidly as copper-based artifacts containing more than 5% tin for Europe and the Near East are found for the periods 2700–2200 BC and 2200–1800 BC.

Major tin sources in Europe and the Near East include Cornwall (Britain), Spain, and Turkey (Anatolian Plateau). Sources for Egyptian tin have been identified in the Sudan during the reign of Pepi II (22nd century BC). Outside Europe and the Near East, major Bronze Age cultures are known in Thailand (Ban Chang) (ca. 2000 BC) and northern China (Hunan, the Shang dynasty, ca. 1400 BC).[24]

The development of metallurgy comes in the Neolithic period and seems to have actually preceded the production of pottery in many areas. Early metal was used primarily for weaponry and tool manufacture. The smelting and casting procedures (and cost) in antiquity precluded its use as a structural material.

Iron Metallurgy

The first smelting of iron may have taken place as early as 5000 BC in Mesopotamia.[25] Throughout the Bronze Age, iron was produced sporadically as iron droplets, a by-product of copper smelting, formed in lumps on top of the slag from the smelting process. This was due to the increased efficiency in copper smelting by the use of fluxes that contained iron oxides mixed in with the copper ores, particularly carbonate ores. The first reaction in the process is $(CuCO_3)CuOH_2 \rightarrow 2\ CuO + CO_2\uparrow + H_2O\uparrow$. Then the CuO is reduced by carbon monoxide: $CuO + CO \rightarrow Cu + CO_2$. Silicate impurities in the ore react with the CuO to form the slag: $2CuO + SiO_2 \rightarrow (CuO)_2 \cdot SiO_2$. The iron oxides in the copper ores increased the copper yield thus: $(CuO)_2 \cdot SiO_2 + Fe_2O_3 + 3CO \rightarrow 2CU + (FeO)_2 \cdot SiO_2 + 3CO_2$.

To reduce iron in large amounts required a CO/CO_2 ratio of 3:1, whereas only 1:5,000 CO/CO_2 was required for the reduction of CuO. Forging these lumps of early iron was difficult because the iron contained copper and sulfur. By 1200 BC, the smiths had begun to roast Fe_3O_4 in smelting furnaces. In the process, at 1200°C, the iron did not melt as copper did, but only became pasty. Iron requires 1540°C to melt (figure 12.1). The production of iron is not a melting operation; it is a reducing operation. The object is to reduce Fe_3O_4 to metallic FeO. Today, we accomplish this with blast furnaces, but in the early days the smiths roasted the iron ore in a forge called a *pit forge* or later a *bloomery forge*. To achieve the required temperature, a device was needed to create a draft. The first smiths utilized tuyeres

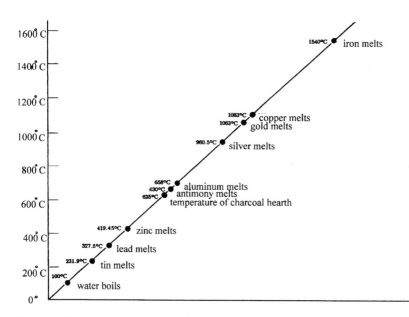

Figure 12.1 Melting temperatures of various metals.

(blow pipes). The result of this crude smelting was a pasty mass of iron, slag, and charcoal called a *bloom.* With further heating and hammering at 1250°C, the bloom could be consolidated into a wrought iron implement. When hammered, the fluid slag and oxides squirted out and the iron particles welded together. This silicate slag gangue was composed largely of *fayalite,* the iron-rich olivine.

Bloomery iron is a poor substitute for bronze. Its tensile strength is about 40,000 psi. The strength of pure copper is 32,000 psi. Hammering brings the strength of iron to almost 100,000 psi. A bronze with 11% tin, however, has a tensile strength after casting of 60,000 psi and a strength after cold-working of 120,000 psi. Further, bronze could be melted and cast at temperatures reached by early furnaces. Iron could be cast only after adding as much as 4% carbon to the bloom, which in turn caused the cast metal to be extremely brittle. Bronze corroded little; iron rusted easily. Why, then, did iron replace bronze? One theory suggests that iron's popularity was due to the fact that, properly carburized (the addition of carbon), it produces an alloy with desirable cutting and durability properties. This alloy is steel.

A carbon content of 0.2–0.3% gives steeled iron a strength equal to that of unworked bronze; raised to 1.2%, steeled iron has a tensile strength of 140,000 psi. If the blacksmith then cold-hammers the steeled iron, a tensile strength of up to 245,000 psi can be obtained— more than double that of cold-worked bronze!

In the bloomery furnace, the fuel and ore charge pass down the stack, where at 800°C, CO reduces the Fe_3O_4 to FeO flakes. With rising temperature, the metal agglomerates and forms wüstite FeO, some of which reacts with SiO_2 to form an iron sulfate-fayalite-slag. This seals off the rest of the batch from furnace gases, the oxidizing zone above the tuyeres. With high-grade ores, SiO_2 must be added to form the slag.

Below the tuyeres, the furnace atmosphere is reducing. Silicate-rich slag flows to the bottom of the furnace, and the bloom forms a layer above the bowl. The bloom is extracted and is ready for the forge, where it can be steeled by the introduction of carbon. The carbon is picked up by the iron from the white-hot charcoal at 1200°C. The diffusion of carbon into the iron is temperature dependent. After 9 hours at 1150°C, the concentration of carbon is 2% at 1.5 mm of depth. In modern metallurgical terms, the carburized iron has a microstructure known as *austenite.* As the temperature falls to 727°C, the austenite breaks down into ferrite (pure iron) and cementite (iron carbide) into a new microstructure called *pearlite.* Pearlite is comprised of alternating layers of ferrite and cementite. If the iron contains 0.8% carbon, the entire (100%) microstructure will be pearlite.

The ancient smiths developed a method to harden the steeled iron even further. This method was quenching, or rapid cooling in air or water. Rapid cooling less than 1 second produces the microstructure called *martensite*—hard but brittle. Slower cooling, up to 3 seconds produced fine pearlite, or more than 10 seconds coarse pearlite. A third technique for manipulating the end result of forging steeled iron was tempering. After the ancient smiths realized that quenching made hard but brittle steel, they discovered the process of tempering, whereby the steel is heated to the temperature of transformation, 727°C. The iron carbide (in the cementite) precipitates and coalesces, increasing the hardness and ductility of the metal by creating pearlite. The temperature attained and the time held at that temperature determines the amount of iron carbide formed and thus the final hardness and ductility of the metal.

Beyond the Artifact: Slags and Tailings

Slags of metal smelting are excellent indicators of the particular minerals being utilized and, to a large degree, the sophistication of the smiths. Artifacts are not in themselves evidence of metallurgy, at least not of their production in their locus of discovery.[26] Slags, themselves the remains of actual metal production, are not always that easy to find.[27] Their prevalence is related to the specific metallurgy and its intensity. Copper smelting is unlikely to produce large quantities of slag, whereas industrial-level bronze and iron production produced small hills.[28] The early phases of metallurgy are ephemeral: a few hearths or furnaces; little in the way of slags; and only fragments of crucibles, molds, or other ceramic accouterments. In Eneolithic-Chalcolithic sites of Europe and the Middle East, sometimes only tiny droplets of copper are found. For geochemical analyses only a droplet is often all that is necessary.

The copper content of Cypriote slags produced by melting is higher than that produced by smelting.[29] Crucible (melting) slags from Cyprus contain 20–28% copper, while the highest content of early smelting slags was 5.8%.[30] In the early southeastern European metal culture of Vinča, copper-smelting slags from Selevac and Gornja Tuzla have been analyzed by energy-dispersive spectroscopy (EDS) and EMP.[31]

The Selevac slags were in actuality small glasslike droplets. The slag matrix was principally an iron-copper-silicate ($FeO-CuO•SiO_2$) and copper-calcium-iron silicate ($CuO•CaO•FeO•SiO_2$) and some copper metal. The composition of the slag is consistent with the local ores of malachite ($CuCO_3•Cu(OH)_2$) and azurite ($2CuCO_3•Cu(OH)_2$). The Gornja Tuzla slags had bands of pure metal (Cu), cuprite (Cu_2O), and copper silicate. It is interesting to compare these slags, compositionally, with those of the Chalcolithic site of

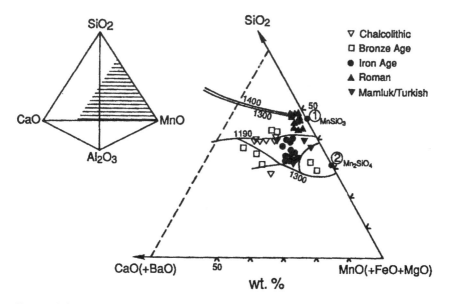

Figure 12.2 Ternary plot of Jordanian copper slags.

Timna in the Negev Desert of Israel. After the 4th millennium, all the copper slags at Timna are ferrous silicates.[32]

A. Hauptmann, G. Weisgerber, and H.-G. Bachmann, in their studies of ancient copper production at Chalcolithic and Early Bronze Age sites in the Feinan region of Jordan, demonstrate the strength of a combined petrographic, metallographic, and geochemical methodology in the characterization of ores and slags. Compositional plots of the Feinan slags from early to late periods reflect a rough chronological framework for at least that area (figure 12.2,3). Their results show a narrow range of variation in the chemical and miner-alogical composition of the slags which reflects a relatively homogeneous mineralized ore body.[33] This contrasts with Hauptmann's work on slags from Oman, where vertical primary and secondary zoning of ore deposits resulted in significant differences in slags.[34] Hauptmann describes slags produced in two periods: the Early Bronze Age (third millen-nium BC) and the Sumerian Umm an Nar period (third to second millennium BC). Their bulk compositions were similar (SiO_2, FeO, MgO, and CaO), with hematite and limonite added as fluxes.[35] The slags differ markedly in their copper content. The later period contains mostly below 2% copper, while the Early Bronze Age slags have up to 31% copper. This incomplete smelting is attributed to low temperatures. The main component of the silicate slag is fayalite-rich olivine, along with ferrosilite and hedenbergite ($Fe_2Si_2O_6$–$CaFeSi_2O_6$), which are varieties of clinopyroxene. The slags indicate both strong reducing and highly oxidizing conditions in the furnaces. The reverberatory furnace used in Oman produces ox-idizing atmospheres, which agrees with F. Czedik-Eysenberg's studies of the smelting of copper in the Alps.[36]

Studies of the later European Iron Age reveal the small size of slag heaps compared with

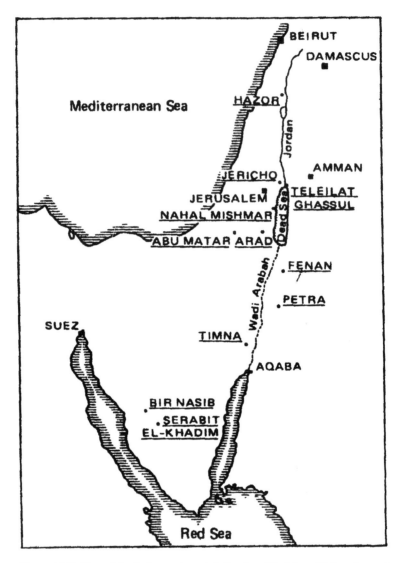

Figure 12.3 Map of Jordanian early metal sites (modified from Weisgerber and Hauptmann, 1986).

the Roman period. Because of more efficient production, early British slag heaps are on the order of 8 kg (20 lb) rather than the tons of slag and tailings of Roman industry.[37] In comparison to Early Iron Age slags, the Roman examples are lower in phosphorus.[38] In Roman lead-silver slags, the efficiency of extraction is seen in the negligible amounts of the noble metal found.[39] Ancient slags from Hubei Province (China) indicate a bronze technology using oxide ores with as little copper as 0.7%.[40]

Chemical and mineralogical analysis is a powerful tool for reconstructing ancient metallurgical processes. Analysis of the slag gives evidence on the nature of the ores, fluxes, and physical-chemical conditions of the metallurgy. A greater problem is the linkage of ore-slag-artifact using chemical analysis. Variations are caused by the metallurgical process: some volatile elements may be lost, all elements will partition themselves differently from ore to slag to metal, and other metals may be added in alloying. The most promising method to trace an artifact to its ore source is by use of isotopic ratios, especially those of lead (see chapter 14). Lead ratios will not be affected by any metallurgical process and remain invariant from ore to metal to artifact, although they may be modified by different ratios of an alloying metal.

Mines and Metallic Minerals

Extraction of mineral resources is carried out by mining. Mines are either surface open-cut, which includes quarrying and dredging, and underground mining. The earliest mines world-wide were simple shallow open pits to recover ores found at the surface. With the advance in ingenuity and technology, early miners used picks of durable antler horns and mauls of diabase to ease the extraction of ores. An ochre ore was exploited in the Paleolithic using antlers of a now-extinct elk on the Aegean island of Thasos. With lowered sea level of the time, early humans could walk over from present-day Macedonia to work the ore. The ochre deposit was followed from the surface into a cave, making this site one of the oldest examples of underground mining in the world.[41] With tools, miners could advance beyond the surface exposures to follow ore-bearing veins. Our admiration for their accomplishments with this simple technology is continually reinforced by our discoveries of their ancient works across the globe.

The Sumerian word for mining was *ba-al,* "to dig." In Akkadian the term for "mine" was *hurru,* literally "a hole." The copper from such a mine was *si-it hurru,* "the product of the mine." Before the mine was the recognition of the mineral. Generally, this was by color. Table 12.4 lists names of variously colored minerals known to the ancients. The following description is ancient Chinese in origin: "Where there is cinnabar above, yellow gold will be found below. Where there is magnetite above, copper and gold will be found below. Where there is *ling shih* (a plant) above, lead, tin and red copper will be found below. Thus it can be seen that the mountains are full of riches."[43]

Ancient mines often are simply open pits, such as those of the Great Lakes. These open workings could easily reach depths of 6 m.[44] To do this the aboriginal miners used wedges (sometimes of copper), stone mauls, and wooden levers as found on Isle Royle, Michigan.[45] Neolithic miners at Grimes Graves, England, dug bell-shaped pits with antler picks, then galleries were run outward from them along the veins of flint, some up to 100 m in length.[46]

Z. Baoquan, H. Youyan, and L. Benshan describe similar impressive works by ancient miners in China. Hundreds of galleries and shafts, with wooden pit props, were found, exploiting stone, copper, and iron. The earliest shafts were begun in the eleventh century BC and were as deep as 60 m.[47] By contrast, Chalcolithic Timna mines in Israel had relatively short (2–4-m) shafts into the Nubian sandstone, not a particularly challenging material, with wide chambers excavated at their ends. In the Bronze Age, with better implements, the shafts at Timna reached depths of 15–20 m.[48] Table 12.5 presents a general outline of the development of mining and its associated metallurgy. Mining and metallurgy had its own independent development in the New World.[49]

Table 12.4 Ancient Names for Various Minerals

Color and English word	Sumerian	Akkadian	Egyptian	Greek	Latin
Black					
Hematite	$^{na}_4$KA-GÍG	šadanu	b'i3	aematite	hematite
Sulfur	KI-A-ÍD	kibritu	—	—	—
White					
Mercury	IM-KAL-GUG	—	—	hydragyros	argentium vuvim
Yellow					
Ochre (yellow)	IM-MAC-LI	kalû	sty	ochra	sil
Limonite	$^{na}_4$KA-SIG$_7$-IM-Guškin	—	—	—	schistos
Green					
Malachite	^{a}NINI-MUŜ-GIR	^{a}mušgarra	w3ḏ	chrysocolla	chrysocolla
Blue					
Azurite	^{a}ZA-GIN-KUR-RA	tâkiltu	tfrr	kyanos	cyanus
Copper carbonate	il-lu A-KAL-URUDU	hilueri	—	armenion	armenium
Red					
Ochre (red)	^{a}KA, IM-SA$_3$	šaršerru	tmś	miltos	sinopis
Lead (red)	^{a}ZA-TU-BE	sâsu	—	—	sandaraca
Cinnabar	GUG	sându	—	kinnabari	minium

Source: R. J. Forbes, 1956, Studies in Ancient Technology, Vol. 7 (Leiden: Brill), Table VI, p. 116.

Tin revolutionized early copper metallurgy by becoming the alloy of choice and replacing its arsenical variety.[50] Rather than mined in stratified deposits, it was first extracted from placers as cassiterite (SnO_2).[51] Major Old World sources were in the wadis of the Eastern Desert in Egypt, Cornwall, the Hartz Mountains, Turkey, and Spain. In the New World the west Mesoamerican and Bolivian sources are famous. Placer mining is ephemeral and leaves few remains, unlike underground works.

The copper-mining center at the Mitterberg, Austria, located 45 km south of Salzburg, flourished from the 2nd millennium B.C. until the beginning of the European Iron Age (ca. 800 BC), named for that other famous Austrian mining center, Hallstatt. Mitterberg copper is most closely associated with the *pfahlbau,* lake dwellings, of the Mondsee. Copper mining in this part of Europe has taken on new importance with the 1991 discovery of the "Iceman" in the Tyrolean Alps. Speculation surrounding the mummy's copper ax, together with a radiocarbon age of 5000 BP, revised thinking on the beginning of polymetallic metallurgy in central Europe. The jury is still out but the possibility of the ax having been made from Mitterberg ores is an attractive one.

The Mitterberg works are truly impressive: a 1000-m gallery with shafts to depths along the lode to 100 m. Some stopes were 5 m wide and 30 m in length. Timbering was used and ventilation shafts of 1–2 m in diameter were driven from a single level. Fire was used to detach the ore. Tools of sandstone and metal were used. Great heaps of slag containing 0.25% copper were found. Resmelting at later dates is a possible explanation for this low amount.[52]

The "Mount Everest" of ancient mining was clearly the great silver mines of Laurion. The number of detailed and scholarly studies of the technical, social, and economic aspects of these mines is reflective of their historical importance in the rise of Athens and its influ-

Table 12.5 Ancient Mining Techniques, Minerals, and Metallurgical Stages (Adapted from Forbes, 1956)

Cultural period	Methods	Tools	Ores/stone	Metallurgical stage
Paleolithic	Scavenging; open pits	Digging sticks; horn/bone; stone; torches	Quartzite; flint/chalcedony; obsidian; steatite; amber; jadite; calcite; quartz ochre	
Neolithic/ Eneolithic	Shallow quarries; open pits; shallow shafts to deeper galleries	Antler picks; stone hammers and mauls; fire; lamps	Pigments (various); obsidian; jade; lapis, marble; salt native metals; copper arsenates and oxides	Epimetallurgy; annealing(?); early smelting and melting of polymetallic copper ores
Chalcolithic	Shaft galleries; bell-pits; placers	Picks; mauls; hammers; fire; simple shoring; early metal tools (copper)	Copper ores; native metals; pigments; building stone-alabaster, marble	Annealing; casting and hammering; reduction of oxide carbonate and arsenical ores
Early Bronze to Middle Bronze Age	Systematic workings; outcrop mining; placers; shaft/galleries	Stone and metal tools; shoring and bracing; crude ventilation; fire	Copper; tin ores; iron ores in small amounts; galena and argentiferous ores	Hammering and casting; alloying with tin, lead, and antimony replaces the arsenicals; sulfide ores oxidized silver from galena.
Late Bronze Age	Shaft works; outcrop, placer mining; open works	Stone and metal tools; shoring; ventilation; simple drainage	Copper ores; tin ores; increased use of iron deposits; noble metal ores	Elaborate smelting; shaft furnaces casting; hot and cold hammering; full polymetallic metallurgy
Iron Age (to Roman times; "Warring States" period in China)	Open works; placers; outcrops; drained workings in shaft mines	Copper, bronze, and iron tools; shored works; ventilation; drainage; pumps	Magnetites, hematites, ilmenites, pyrites, chromite; oolithic and limontic ores; galena deposits at Laurion at peak	Ore roasting; charcoal furnaces; casting in Asia; bloomery forges in Mideast and Europe; wrought and steeled iron; New World copper alloys and gilts; brass mercury.

Source: R. J. Forbes, Studies in Ancient Technology, vol. 7 (Leiden: Brill), Table VII, pp. 116–17.

ence on "Western" civilization. Indeed, the fuel for the ascendancy of Athens in classical Greece was its great wealth generated by the miners of Laurion. For the most recent detailed work on the mining activities, we direct you to Shepherd's 1993 book.[53]

Exploitation of the silver-rich galena deposits began in the Mycenean period (ca. 1500 BC). The mines didn't become important to the Athenian treasury, however, until the 6th century BC, and were in decline by the 4th century BC. The deep shafts (110 m) required multiple ventilation shafts. Older pillar and stall workings left support pillars called *mesokrineis*. Mining slaves lived in the mines.

Geologically, Laurion consists of three beds of limestone and marble separated by mica-schist.[54] As we noted earlier the silver-rich PbS deposits are found at the contacts of the beds. The PbS mineral, galena, is rich in silver: 1200–4000 g silver per ton of ore. Other ore minerals include zinc-sphalerite (ZnS), cerrusite or calemine (carbonate). Iron is found as pyrite (FeS) and as oxides including hematite which was worked for iron tool manufacture.

Over two thousand shafts were used in the mining operations. The shafts were turned every 10 m by 8–10°, giving the shafts a corkscrew aspect when viewed from above. Ladders and notches were used for access by the miners. The deepest shaft reached 117.6 m at the watertable.

Iron Replaces Bronze

The coming of iron to Old World metallurgy swept away copper-bronze metallurgy. In the New World it never took hold. The very ubiquity of the varieties of iron ores contributed to the triumph of iron. Developed techniques for wrought iron in the Middle East and Europe and for cast technology in Asia created "metasomes" for every aspect of bronze technology.[55] Add to the mix the relative scarcity of tin for the bronze alloy and the stage was set for iron.

Development of an iron metallurgy produced weapons and tools eventually superior in strength to copper alloys. Economics tells us that scarcity or abundance drives demand and price. Iron is not only abundant and cheap but the utensils and tools from it were sharper than copper alloys as well. A copper ax was a rarity in the 4th millennium; iron became commonplace. Iron differs from copper in that the composition of the copper alloy determined its metallurgy, while the working of iron determined its final properties, for example, the control of carbon content and the formation of metal phases in the microstructure of iron from wrought to steeled. With steady improvements in iron extraction and manufacture, it became the metal of choice in the Old World. Abundant iron artifacts found in archaeological sites clearly reflects this.

A Metal for the Common Man: African Iron

Iron metallurgy in Africa—smelting and mining—has enjoyed a well-deserved scrutiny in recent years.[56] Sub-Saharan iron metallurgy represents a contrast in the evolution of the acquisition and use of metals. Here the copper-bronze metallurgy common elsewhere (Northern Africa, Asia, Europe, and the New World) was not dominant, where bloomery iron rose to prominence. This iron metallurgy was small scale and is characterized by a simple, relatively uniform technology.[57] The demand for iron in southern Africa after AD 1000 grew along with the growth of towns and widespread trade in other metals as well. Major sites such as Great Zimbabwe and Bosutive show significant change in fabrication technology, diversification in the metals mined to include tin and gold, and the trade in cast ingots of copper and tin, particularly at Zimbabwe.

In this section we will use a study of African metal tradition to illustrate the role geology has in the archaeology of metallurgy. This study was carried out by David Wenner and Nicholas van der Merwe on iron mining and smelting at two well-preserved Iron Age sites on the Nyika Plateau of Northern Malawi.[58] The first mine, Ncherenje, had around 1600 m^3 of ore mined over a linear distance of 3 km from shallow pits, large excavations, and a number of underground shafts and tunnels. The second mine, Mbiri, is a single excavation where only about 15.5 m^3 of magnetite-rich gneiss was taken. Wenner and van der Merwe's study concentrated on determining why the largest quantity was mined at the site of poorer ore, Ncherenje.

The Ncherenje ore was low grade, ranging from only 6.6 to 21.3 wt. % Fe_2O_3. The Mbiri ore averaged 58% Fe_2O_3. Why would the metallurgists focus on the extraction and use of such poor ore? Wenner and van der Merwe determined that the answer lay in the indigenous smelting technology available. Ncherenje and Mbiri both lie within a heterogeneous association of Precambrian igneous and metamorphic rock, termed the Malawi Basement Complex. Much of the bedrock is covered by an extensive regolith that varies, below the soil, from 1 to 8 m in thickness. Ncherenje contains three basic rock types: gneiss, cataclastite, and granite. The Early Iron Age miners obtained their ores from soft, highly weathered material, abandoning these as they encountered unweathered rock at depth. Table 12.6 shows the four weathering types defined at Ncherenje. Analyses of these types indicated that illite replaced feldspar in the quartz feldspathic zones, with kaolinite and fine-grained iron oxides forming from the mafic phases. It is in the highly weathered type 4 rock that the iron oxides (and clay) are most abundant (5–20%).

Table 12.6 Weathering Index of Samples from the Ncherenje Mine Site

Weathering type	Texture	
	Megascopic	Microscopic
1	Gneissic texture well preserved	Quartz > 25%, coarse grained > fine grained; trace feldspar sometimes preserved; quartz largely confined to 1–2 mm wide bands
2	Gneissic foliation partially preserved	Quartz > 15%, coarse grained ≅ fine grained; Quartz invariably disseminated in 1–2 mm wide bands
3	Gneissic foliation may or may not be preserved	Quartz < 5%; coarse grained < fine grained; Quartz only slightly concentrated in 1–2 mm wide bands
4	Fine grained and homogeneous; no vestige of original gneissic foliation	Quartz ranges from <5% to trace amounts; Quartz largely disseminated throughout matrix

Note: Coarse-grained quartz is 1–0.5 mm, whereas fine-grained quartz is <0.1 mm.

Table 12.7 Sample Descriptions from the Mbiri Mine

Sample number	Location	Modal analyses	Additional comments
14	Bottom of excavation	55-Mt; 5-FO; 30-il; 10-kaol	
14V	Same	Somewhat enriched in Mt	Hand-selected ore concentrate
15	Same	Not observed in this study	Hand-selected by Phoka smelters in 1982

Note: Mt = magnetite; FO = fine grained amorphous ferric oxides; il = illite; kaol = kaolinite.

The Mbiri mine occurs within a single homogeneous rock unit mapped as a cordierite gneiss, although most of the mining activity took place in the regolithic portion. Magnetite veinlets (1–2 cm wide) occur within this partially weathered bedrock. Table 12.7 lists the sample descriptions for the Mbiri mine. To corroborate their initial suspicions as to the variety in grade of the two mines' ores, Wenner and van der Merwe carried out XRF studies. The difference in the Fe_2O_3 content between the two is obvious (table 12.8).

Surely the ancient smiths had a practical understanding of ore quality. This was apparent after further study of the slags produced by the respective ores in earlier ethnographic studies done in 1982–83.[59] The modern smiths of Malawi, using a two-step natural-draft and forced-draft smelting procedure, maximized the iron-poor Ncherenje ore's potential in one critical aspect of iron manufacture: production of slag (SiO_2). While rich in iron, the Mbiri magnetite could not produce fayalitic slag (Fe_2SiO_4) due to a lack of silica. Additionally, the low-temperature (1200°C), natural-draft furnaces could not melt the high-

Table 12.8 Whole rock XRF chemical analyses of samples from the Ncherenje and Mbirii mines

	Ncherenje Mine				Mbiri Mine		
	1	2	11	13	14	14V	15
SiO_2	56.95	37.37	68.06	40.94	15.66	11.37	40.70
TiO_2	1.38	1.16	0.66	1.10	4.94	5.07	5.07
Al_2O_3	19.73	31.5	15.27	28.26	13.07	10.77	14.92
Fe_2O_3	11.57	15.22	6.55	12.87	58.71	64.82	47.48
MnO	0.00	0.08	0.08	0.12	0.28	0.85	0.26
MgO	0.56	0.56	1.18	3.04	0.40	0.38	0.42
CaO	0.00	0.00	0.02	0.00	0.00	0.00	0.01
Na_2O	0.12	0.92	0.26	1.10	0.16	0.18	<0.1
K_2O	3.30	5.96	1.76	7.00	2.97	2.62	2.68
P_2O_5	0.10	0.06	0.14	0.08	0.14	0.20	0.13
H_2O^-	0.38	0.3	0.58	0.18	0.56	0.50	0.57
LOI	5.84	6.38	5.74	5.24	3.52	2.66	4.97
Total	99.93	99.51	100.30	99.93	100.41	99.42	99.59

Note: LOI = loss on ignition

grade Mbiri ore. To circumvent this problem, the ancient smiths chose the "better" ore for their metallurgy. That ore was the silica-rich, iron-poor Ncherenje stock. Ncherenje is also more deeply weathered and richer in water-content as shown by its higher "loss or ignition." This facilitates fluxing of the components during smelting. The first smelt was followed by a resmelting of the fayalite in the smaller, forced-draft furnace. The product was a bloom sufficient to produce the utensils and tools required by the consumers—farmers and herders of ancient Africa. The abundance of the weathered ore was largely the reason iron metallurgy spread throughout Africa and elsewhere—a common metal for the common man.

The emphases, or at least the breadth, of archaeometallurgy are indeed shifting from a detailed description of the form and fashion of metal items to a wider discussion of the role of metals in human social groups.[60] Far from restricting its methodological frame of reference, archaeometallurgy has widened its scope to consider the relationship between metals and humans. That relationship is a continuing dynamic in the modern world system that metallurgy helped create.

13

Ceramics

Archaeological ceramics refers to products made primarily of clay and containing variable amounts of lithic and other materials as well. The term *ceramic* is derived from the Greek *keramos,* which has been translated as "earthenware" or "burned stuff." Ceramics include products that have been fired, primarily pottery but also brick, tile, glass, plaster, and cement as well. Since pottery is by far the most important archaeologically, and the methods of sampling and study are largely applicable to the others, this chapter is devoted primarily to pottery.

Pottery then is the general term used here for artifacts made entirely or largely of clay and hardened by heat. Today, a distinction is sometimes made between *pottery,* applied to lower-quality ceramic wares, and the higher-grade product *porcelain.* No such distinction will be made here, so the term *pottery* alone will be used. Raw material that goes into the making of a pot includes primarily clay, but also varying amounts of *temper,* which is added to make the material more manageable and to help preserve the worked shape of the pot during firing. Of primary interest in ceramic studies are

1. the nature and the source of the raw materials—clays, temper, and slip (applied surface pigment)—and a reconstruction of the working methods of ancient potters;
2. the physical properties of the raw materials, from their preparation as a clay-temper body through their transformations during manufacture into a final ceramic product;
3. the nature of the chemical and mineral reactions that take place during firing as a clue to the technology available to the potter; and
4. the uses, provenance, and trade of the wares produced.

Much of the information needed to answer these questions is available through standard geochemical and petrographic analysis of ceramic artifacts. Insight into the working methods of ancient potters also has been obtained through ethnographic studies of cultures

Table 13.1 Characteristics of Ceramic Bodies

Type	Porosity	Firing Temperature	Applications	Notes
Terra-cotta	High: ≥30%	<<1000°C	Roof tiles, bricks, most prehistoric	Unglazed, coarse
Earthenware	10–25%	900–1200°C	Coarse: tiles, bricks Fine: wall and floor tile, majolica	Glazed or unglazed, unvitrified
Stoneware	0.5–2.0%	1200–1350°C	Roof tiles, tableware, unglazed artware	Glazed or vitrified
Chinaware	Usually <1%	1100–1200°C	Tableware	White, vitrified
Porcelain	<<1%	1300–1400°C	Fine tableware, fine artware, rings	Hard body when tapped

Source: P. M. Rice, 1987, *Pottery Analysis: A Sourcebook* (Chicago: University of Chicago Press).

where, because of isolation or conservative traditions or both, ancient methods have been preserved.

Pottery is classified into "wares" largely on the basis of firing characteristics and surface treatment (table 13.1). A distinction is made between vitrified ceramics, which have been subjected to high enough firing temperatures to fuse the clay into a glassy matrix, versus unvitrified, porous material, which has been fired at lower temperatures. The former includes stoneware and porcelain; the latter, terra-cotta and earthenware.

In order to make pottery, two important properties of the main constituent, clay, must be known: its plastic properties in working and the physical changes that take place on exposure to firing. Mixtures of clay and water have a range of plastic properties depending on size, composition, and mineralogy. Such a mixture can be pressed into a shape that is retained when the pressure is released. When the shaped mixture is dry, it loses its plasticity and the clay becomes hard and brittle. Upon firing, chemical and mineralogical changes take place over a wide range of temperature. Some clays dry and fire satisfactorily; that is, they shrink uniformly and preserve their shape. Most pure clays, however, dry unevenly and start to crack even before firing. Therefore, pottery is seldom made from pure clay. Other substances that need to be present or added are nonplastic materials called tempers. They facilitate uniform drying and counteract the excessive shrinkage of clay that is normal upon drying and firing. The physical combination of clay (the paste matrix) and filler (the temper) is the ceramic body.

The earliest fired pottery worldwide is terra-cotta. Terra-cotta is fired at relatively low temperatures, usually below 900°C. Terra-cotta vessels, sculptures, and tiles are generally not glazed, although they may be surface treated by roughening, to absorb heat and prevent slipping when wet, or covered with a slip. The slip is a suspension of fine-grained clay and water that may be added to enhance the beauty of the pot by coloring and smoothing, as well as to reduce porosity and retard leaking.

Earthenware bodies are also porous and unvitrified but are fired at higher temperature than terra-cotta, up to about 1100–1200°C. Because of the higher temperatures of firing, material at the surface can fuse and form a natural glaze; earthenware may also be unglazed. When fired at lower temperature, they grade into terra-cotta. The clays used in the manufacture of earthenware are generally coarse, plastic when wet, and, often, red when firing. Earthenwares include a wide variety of products, ranging from "coarse," such as bricks and tiles, to "fine" wall tiles and majolica, which is tin enameled and made from a good-quality white-burning clay.

Stonewares are fired at temperatures that are sufficiently high enough, 1200–1350°C, to achieve partial fusion or vitrification of the clay body. The body itself is medium to coarse grained, opaque, and commonly light brown or gray. Modern stoneware is usually made of sedimentary clays that are transported, rather than residual, deposits, such as ball clays, which are highly plastic and low in iron. Archaeological stonewares may or may not be glazed.

Chinaware and porcelain are generally considered synonymous, chinaware being lower quality than porcelain. Porcelains were first produced in China towards the end of the Han period (206 BC–220 AD). They are fired at 1280–1400°C or even higher, which is necessary to vitrify the highly refractory white clay, kaolin, used in their manufacture. The clay is relatively pure and mixed with quartz and alkali feldspars, which act as a flux, allowing lower firing temperatures than needed to vitrify pure kaolin. At 1150°C, K-feldspar (microcline and orthoclase) melts, and at 1118°C Na-rich plagioclase (albite) melts, giving the porcelain a translucency, hardness, and ring when tapped. Porcelains today are made up of about half kaolin plus roughly equal amounts of feldspar and quartz.

The History of Ceramics

Clay is one of the most abundant raw materials on earth. Many early societies throughout the world recognized the usefulness of clay and the fact that most clay is plastic and can be molded into a desired shape. After drying, the clay retained its shape, so that figurines and vessels could be manufactured. By the late Paleolithic, two other key properties were discovered: fire hardens objects made of clay, and adding various substances improves the properties of clay such as allowing it to preserve its form during firing. The oldest fired clay objects date to 26,000 BP and were found at Dolní Věstonice in Czechoslovakia. They appear 14,000 years before the earliest pottery vessels. Firing temperatures ranging from 500°C to 800°C were attained. The collection includes a bear and a lioness, as well as a well-preserved, 11.2-cm "Venus" figurine. Six other Upper Paleolithic sites have yielded smaller ceramic collections, in the French Pyrenees, in Russia, and Japan, with hand-modeled clay objects that may have also been fire hardened.[1]

A variety of theories have been proposed on how pottery was discovered. Some of the earliest pottery mimics containers made of other materials, such as metal, soapstone, wood, or leather. Logically, the first pots, also containers, would be molded into similar shapes.[2] One interesting theory, proposed by Z. Goffer, is based on the well-known phenomenon of mud cracks that form in soil. Water evaporates more rapidly from the finer-grained surface of loess and clay-rich soils than from the layers underneath. The loss of surface water diminishes the volume of the surface; fissures form to compensate for this loss. The growing fissures plus the rapid near-surface evaporation of water produce irregular, dishlike concave crusts which break away from the moister soil underneath. These dishlike soil crusts may attain an area of about 100 cm², with a thickness ranging from 1 mm to 1 cm. The soil crusts

that form from mud cracks and drying are relatively soft, but firing in small bonfires produces hard, dishlike pieces that can be broken by hand only with some difficulty. The physical properties of pieces produced this way (hardness, porosity, strength) are similar to those
of primitive pottery. Primitive humans may have accidently discovered the properties and
usefulness of molded clay this way. If so, then the idea of molding clay by hand, drying the
form, and then firing it would follow naturally. A distinction could be made between truly
fabricated ancient pots and those made accidentally: the man-made objects should have
been smoothed by hand both inside and outside; the accidental pot would have rough, un-
prepared surfaces.[3]

Clay

Clay is a term applied to a variety of earthy materials of different origins and composition.
The actual definition depends on the field of interest: most archaeologists and potters have
a different idea of what constitutes clay than do geologists and agronomists. To the potter,
clay is any earthy aggregate that (*a*) when pulverized and mixed with water exhibits plasticity, (*b*) becomes rigid on drying, and (*c*) develops hardness and strength when heated to
a high temperature. For the geologist, clay is a fine-grained weathering product of silicate
minerals. The structural breakdown of minerals to form clay is caused by natural geochemical reactions induced by atmospheric and meteorological agents, including water, water vapor, and gases rich in oxygen, CO_2, and SO_x, as well as by various biological agents,
such as lichens and anaerobic bacteria. Depending on the parent rock composition, the end
result of weathering can be a variable mixture of clay, quartz, hydrous ferric and aluminum
oxides, resistant detrital minerals, and micaceous material such as vermiculite. Texturally,
clay is characterized by extremely fine particle size, below $2\mu m$ in diameter, and highly
variable chemical and physical properties.

"Primary" clays were formed in place as residual weathering products and generally
overlie their source rocks. "Secondary" clays are sedimentary clays that formed elsewhere
and were transported by water from their source. In the State of Georgia, one of the world's
principal kaolin sources, the kaolin deposits of Cretaceous age are primary and high grade;
the Tertiary deposits are secondary and have more impurities. The secondary kaolin was
formed by weathering of crystalline rocks on an upland surface to the northwest; the weathered material was transported away from its source, refined to finer grain sizes, picked up
impurities during transport, especially organic matter, and redeposited where it is found today. Other clays such as montmorillonite and bentonite are also found as primary and as
secondary deposits.

Since secondary clays are carried by water, they tend to pick up many impurities which
affect the manner in which the clay may be fashioned, tempered, and fired. Because of abrasion and alteration during transport, the particle sizes tend to be finer than in primary clays.
Their fine grain size and decayed organic matter tend to make them generally more plastic
than primary clays.

Clay Deposits

Most clay deposits are found with an admixture of silt, sand, and other impurities, including organic material, micas, and iron hydroxides. The impurities may constitute over 50%,

so such deposits must be refined before they can be used for pottery making.

The differentiation among clay, silt, and sand is by particle size, but unfortunately, there is no agreement on where the size boundaries should be drawn. Most agronomists, engineers, and geologists accept granulometry scales with clay smaller than either 0.002 or 0.004 (Wentworth Scale) mm in diameter (2–4 μm), silt up to 0.05 mm, sand 0.05–2 mm, and gravel greater than 2 mm. The name assigned to mixtures of the first three depends on the relative amount of each component present (see chapter 3). Soil scientists call clay any mixture with more than about 40% clay; mixtures with more silt and sand are called loam.

The Chemical Composition and Structure of Clay

Clays are composed principally of silica, alumina, and hydroxyl (OH) groups. A theoretical formula given for "pure clay" is $Al_2O_3 \cdot 2SiO_2 \cdot 2H_2O$. Chemical composition alone, however, is not enough to characterize clay. Clays with similar chemical compositions may have completely different physical and technological properties, controlled largely by their crystallographic structure, which is best determined by x-ray diffraction (XRD) analysis.

Clay structures are built up of two basic sheetlike units, building blocks that explain clay crystallography. Each block is composed of ions of silicon combined with oxygen or of ions of aluminum combined with oxygen or hydroxyls (OH):

1. The silica sheet is made up of connected tetrahedral (4-sided) units. Mainly silicon ions with varying amounts of Al^{3+} or Fe^{3+} are located at the centers of the tetrahedra; oxygen or hydroxyl anions form the four corners. The individual tetrahedra are connected with adjacent tetrahedra by sharing three corners with the three basal oxygens. The fourth oxygen is apical, points normal to the sheet, and joins the base of the octahedral sheet (figure 13.1).
2. The alumina sheet consists of connected octahedral (8-sided) units that are larger than the tetrahedral. This allows medium-sized elements, such as Al, Mg, and Fe to occupy the center of the octahedra, with oxygen or hydroxyls forming the six corners.

Elements that can enter into either unit are determined by the element itself, its ionic charge, and its ionic radius. The ionic radius of an element varies according to its distance to surrounding anions, so that in the smaller tetrahedron, ionic radii are much smaller than in the larger octahedron. Oxygen and hydroxyl anions (negative charges) compose the corners of the structures, and the cations (positive charges) are in the center. With tetrahedral coordination (i.e., each cation surrounded by four oxygen anions), silicon has an ionic radius of 0.26 Å and fits more easily into the structure than aluminum, with a radius of 0.39 Å, or ferrous iron, with a radius of 0.63 Å. On the other hand, in the much larger octahedra, where six oxygens make up the corners, both aluminum, now with a radius of 0.54 Å, and ferric iron, with 0.65 Å, are better fits than silicon, which is only 0.40 Å.

The number of oxygens that surround an element in question is called the coordination number (CN); the higher the CN, the larger the space enclosed. In larger spaces, elements with large ionic radii can fit into the crystal lattice. Potassium, a constituent of the clay mineral illite, has a radius of 1.46 Å, with a six-fold coordination. The rarity of potassium in other clay minerals is easily explained by its size: it is much too large to fit into either tetrahedral or octahedral structures. Most of the larger cations such as K actually do not lie within either of the basic sheet structures, but between the sheets, held in place by excess anion

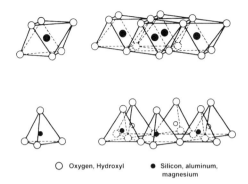

○ Oxygen, Hydroxyl   ● Silicon, aluminum, magnesium

Figure 13.1 Clay structural units (*top*): octohedral sheet, (*bottom*): tetrahedral sheet.

charges. Their presence makes the clay less refractory; that is, it lowers the melting temperature and allows the formation of new mineral phases during firing.

Composition and stability of the clay crystal structure are also strongly dependent on the charges (valences) of the elements involved. In clay minerals, anions are divalent oxygen, O^{2-}, and monovalent hydroxyl, OH^{1-}. The principal cations are Si^{4+}, Al^{3+}, Fe^{3+}, Mg^{2+}, Na^{1+}, and K^{1+}. In structures where all the valences are satisfied (i.e., the total charge of the anions equals the total charge of the cations), there is little room for substitution by exotic elements. Thus, in an ideal kaolinite (table 13.2), the total positive charges of 2Al ($2 \times 3 = 6$) + 2Si ($2 \times 4 = 8$) equal the negative charges of 5O ($5 \times 2 = 10$) + 4OH ($4 \times 1 = 4$). Electrons are shared between the cations and anions and a stable chemical compound is formed. Of all the clay minerals, kaolinite allows the least substitution of other elements and water. If an extra Al^{3+} substitutes anywhere in the lattice for a Si^{4+}, then that tetrahedron will have an excess charge of $+1$. This will allow further substitutions to satisfy the oxygen atoms, each of which needs two electrons to complete its outer shell. Exotic elements in a clay structure will act as fluxes and lower the temperature of firing of a ceramic body. Pure kaolin is highly refractory, with a softening temperature of about 1700°C, quite out of the range of normal ceramic manufacture. That temperature can be brought down if oxides other than Al are present, either in the clay structure or as constituents of the clay-sized or silt-sized fraction.

Virtually any clay can be used for making pottery, depending upon the skill of the potter. The main clay mineral groups include (1) kaolinite, which has a relatively fixed structure (figure 13.2), (2) a group of expandable clays called smectites of overwhelming importance in most soils whose most common member is montmorillonite, and (3) the illite group, a non-expandable mica-like clay.[4] The sum of a layer plus an interlayer is a structural unit in clay. The structural properties of these groups govern many technological choices by potters.

The structural unit for kaolinite is a 1:1 layer, consisting of one tetrahedral sheet with one octahedral sheet. The space between two successive 1:1 layers is an interlayer devoid of any chemical elements if the layers are electrostatically neutral (i.e., with no excess valences). Since this is true in most kaolins, the interlayers remain relatively free of cations or water. The simple rigid structure of kaolin does not allow for much substitution by exotic elements (alkali elements lower melting temperature). In addition, it has the highest

Table 13.2 Idealized General Chemical Formulas for the Three Principal Clay
Mineral Groups

Kaolinite	$Al_2(Si_2O_5)(OH)_4$
Illite	$(K_{1-x},H_2O)(Al,Mg,Fe)_2(Al,Si)_4O_{10}\{(OH)_2H_2O\}$
Smectite (montmorillonite)	$(Na,Ca)_{0.33}(Al,Mg)_2Si_4O_{10} \cdot nH_2O$

Al:Si ratio of the clay minerals (higher Al means higher melting temperature) and is least able to absorb water (lowest plasticity). The use of pure kaolin clays in ceramics is thus restricted to the most advanced ceramic technologies.

The smectite structural unit is a 2:1 layer in which two tetrahedral sheets sandwich an octahedral sheet. The position of the two tetrahedral sheets is inverted so that all apical oxygens point to the octahedral sheet, where they are shared by both sheets. Most 2:1 clay minerals have an excess negative charge produced by unbalanced ionic substitutions, such as Mg^{2+} for Al^{3+} or OH for O. This is neutralized in the interlayer by various cations (mainly K, Na, and Mg), hydrated cations, hydroxyl groups, and water. Abundant substitution for Al and the presence of interlayer cations make smectites less refractory than other clays. They are also expandable because their charge deficiency is in the interior alumina layer. At lower firing temperatures than other clays, about 600°C, major dehydroxylation occurs, which causes irreversible structural changes. Smectites also have higher shrinkage in drying and firing than other clays.

The illite clays have structures similar to micas. In the illites about one-sixth of the Si^{4+} is replaced by Al^{3+}, leading to a charge deficiency that is balanced chiefly by K^{1+} but also by Ca^{2+}, Mg^{2+}, and H^{1+}. The charge deficiency in the illites is primarily in the outer silica layers and therefore closer to the surface than in the smectites. The basic unit is a 2:1 layer, similar to the smectites but nonexpanding, with cations filling in the interlayer. The excess negative charges on the illite interlayers are satisfied by cations, which accounts for a more rigid structure. Because of their fine particle size and low shrinkage, which helps produce a natural luster through minimal mechanical treatment, illite clays are desirable as pottery slips.

Other, less important clay minerals are vermiculite and the chlorite group. Vermiculite, so named because it looks like small worms (Latin *vermiculus*) when heated, is related to the smectites. Vermiculites also have an expanding lattice, but not as great as in the smectites. The chlorite group are light green, mixed-layer minerals common in low-grade metamorphic rocks and sediments. They can form in a variety of ways: as alteration of ferromagnesian silicate minerals or authigenically in sediments. They are commonly mixed with other clay minerals and are thus hard to identify. The three-layered smectite clays, including montmorillonite, are expandable, but the three-layered illites and the two-layered kaolinite are not.

The exact chemical compositions of most clay deposits are quite variable, especially between different beds, which causes their physical properties to vary as well, creating serious problems for the potter. Substitutions of water and different cations in clay structures, as well as indeterminate mixtures of different clay and detrital minerals, can make drying and firing behavior hard to predict. This explains why the choice of clay and the preparation recipe are the most conservative aspects of ceramic production by traditional potters.

Because chemical compositions of clays vary widely in elements present, especially

Figure 13.2 Kaolinite as viewed in the transmission electronmicroscope. (Courtesy Paul Schroeder.)

trace elements at the ppm level, and in water content, a structural determination by XRD analysis is more important than a chemical analysis to identify individual clay minerals. This is especially important for characterization of provenance and technological studies.

Ceramic Properties of Clay

In the making of ceramics, it is crucial that the potter know how the clay body will behave during the different stages of fabrication: aging the clay, forming the matrix, finishing the ceramic, drying, and finally firing. The properties that determine the usefulness and problems for each clay raw material are (1) plasticity in forming, (2) shrinkage by drying, (3) thermal (firing) properties, and (4) esthetic properties, including color.

Plasticity

The addition of water to raw clay will produce a plastic mixture which can readily be shaped without rupture. Clay exhibits plastic properties for primary reasons: the clay particles themselves are platelike, a reflection of their crystalline structure; the particles are small, below 0.002 mm in diameter; and water, called "water of formation," is present between

Table 13.3 Consistency of Clay-Water Mixtures

Consistency	Water %
Dry	5–18
Stiff	10–20
Stiff to plastic	12–25
Plastic	15–30
Oversoft	15–35
Liquid	20–50

Note: The variation is due primarily to the clay species and its crystalline structure.

the clay particles. Water acts as a lubricant and allows the clay particles to slide across each other rather than rupture when a shear force is applied.

The development of plasticity in clay is critical to the manufacture of a ceramic, and depending on the composition of the clay, plasticity will develop over a narrow range of water content. Dry clay is nonplastic; as water is added, the clay becomes plastic, but with too much water, it becomes too fluid to retain its shape when formed (table 13.3). Kaolinite has a nonexpanding two-layer structure and forms relatively large particles, 0.3–0.01 μm in diameter; smectites have an expandable three-layer structure and a finer grain size. Thus, kaolinite has much lower plasticity and undergoes much less shrinkage than smectite upon drying. If a potter has access to clays which are highly plastic, this property can be modified by the addition of nonplastic materials (soil, less plastic clay, crushed rock or pottery, or organic material) as temper.

Many factors control the plasticity of clays, but most important are the shape and size of the clay particles themselves. Clay is very fine grained and crystallographically platelike, with a thickness/diameter ratio of about 1:12. When water is added to clay, it acts as a lubricant, surrounding the clay platelets and allowing them to glide over each other. Surface tension forces acting on the water and the clay bind them weakly together and allow a molded object to retain its shape. When excess water is added, more than the minimum needed to maintain surface tension, the water and the mass will start to flow.

Other important factors which affect plasticity are the nature of adsorbed ions and clay particle size. Large monovalent ions, including $Na+$, $K+$ and ammonia, NH_4^+, diminish plasticity; smaller di- and trivalent ions, including Mg^{2+}, Fe^{2+}, Fe^{3+}, and Ca^{2+}, enhance plasticity. The large monovalent ions are too large to fit into the clay lattice and tend to disrupt the nonliquid water structure, promoting disorder. The smaller, more highly charged ions easily fit into the space available in the lattice and thus do not exercise any disordering effect.

Shrinkage

After the ceramic has been formed, the wet clay must be dried. Water will be lost in two phases. During the first phase, the water of formation will evaporate freely and the ceramic will contract. As the water is removed from around the clay particles, the particles will draw closer together until they eventually touch each other and start to adhere. The ceramic will

Table 13.4 Linear Shrinage and Dry Strength of Clay

Clay mineral	Drying shrinkage (%)	Dry strength (kg/cm2)
Kolinite	3–10	0.5–50
Illite	4–11	15–75
Montmorillonite	12–23	20–60

contract no farther at this stage. The next water loss takes place more slowly and it will come from the pore spaces between the clay particles.

Internal stresses are set up in the ceramic if it loses water too fast at this first stage. The rate of evaporation is critical to prevent the clay from cracking or breaking up. Again, different clays vary greatly in their drying shrinkage (i.e., the reduction in size upon evaporation of water). Clays also vary in their dry strength (i.e., the resistance to breaking when a force is applied after the water of formation has been removed) (table 13.4). This property is important to potters who scrape and smooth vessel walls at this stage, perhaps prior to applying a slip.

The first stage of drying is considered "air drying" and is generally carried out at <150°C. During drying, the mechanically admixed water, which is present as films surrounding clay particles, is lost. The clay mixture shrinks and loses its plasticity. As the clay particles come closer together, they may come into contact with each other. Stresses set up at this stage may later lead to structural damage to the body. During this period of drying, the body will undergo its maximum shrinkage and exhibit its maximum water loss.

Water first evaporates from the surface of the body. Then by capillary action, water rises from the interior to the surface and in turn evaporates. To ease the process of evaporation, nonplastic fillers, called temper, earlier added to the clay paste, help provide pathways for water to escape. Other materials or characteristics may be present or added to help bind the clay together and strengthen it during drying and shrinkage. This includes fine, rather than coarse, particle size clay, the presence of Na^{1+}, and added organic materials such as flour or gums of various trees. All these will help bind the clay particles together and help prevent uneven drying, which may cause cracking or warping.

Thermal (Firing) Properties

Firing the clay body to the hardness of a ceramic body is the next critical stage in manufacture. The ceramic then acquires its hardness, toughness, and durability. Original mineral and clay structures are transformed and chemical compositions change as the body loses its basically sedimentary condition and becomes its metamorphic equivalent. The principal changes that take place are by *(a)* oxidation or reduction; *(b)* loss of volatiles and dehydration, principally loss of hydroxyl ions (OH) from the clay structures; and *(c)* formation of higher-temperature phases and, finally, at very high temperatures of firing, vitrification.

Clay minerals undergo important structural and chemical changes when they are fired at temperatures over 600°C, and particularly at higher temperatures of ~1200°C. These changes are important to convert the clay into a useful ceramic. High temperatures dislodge ions from original positions held in the clay structure into more favorable sites, thus converting the clay into chemically stable and structurally stronger material. Chemical changes

Table 13.5 Thermochemical Changes Induced by the Firing of Clay

Temperature (°C)	Reaction
100–150	Loss of water of formation (plasticity)
450–600	Loss of water of constitution (in clay structure), modification of clay structure
600–950	Clay structure breaks down, new mineral phases formed, incipient vitrification
900–1750	Clay completely converted to new minerals
>1750	Fusion, complete vitrification

that take place during the firing of clay are shown in table 13.5. The exact changes that take place for each stage of firing depend on many variables, including those that are material dependent: (1) the chemical, mineralogical, and structural composition of the clay and the nature of included materials; (2) the wall width of the ceramic body; and (3) the nonplastic temper added by the potter. At 450–500°C, water that is chemically combined in the clay structure is lost. The clay crystalline structure itself will break down above 800°C and be largely converted to new mineral phases. Above 1800°C, a temperature far above the capability of any ancient technology, the ceramic body will fuse (table 13.5). Other changes are dependent on firing conditions. (4) The rate of firing must allow gases, including those formed by the breakdown of organic material and carbonates, to escape. Too fast a rate will trap gases and result in higher porosity. (5) The longer high temperatures are held, the more complete will be mineral reactions forming glass and high-temperature mineral phases, such as spinels and mullite, which lead to a stronger ceramic body. (6) The atmosphere in the kiln can be either reducing or oxidizing. Color is strongly dependent on this: ferrous iron can cause a green color; ferric and unoxidized organic material result in black.

Oxidation and Reduction

If materials such as dung and straw are utilized in the making of the ceramic body, then, when heated to 200–500°C in an oxidizing atmosphere, such organics will oxidize and be given off as CO_2, leaving vacant spaces and a porous body. If, on the other hand, the ceramic body was heated under reducing conditions, in those parts of the kiln where oxygen was not available, then the organic material is only charred and a residue of elemental carbon is left. Under these conditions, a gray or black ware will result. Commonly, firing is timed to allow oxidation to go to completion so that no organic matter is trapped in the fired clay before the stages of dehydration of clay and the formation of new mineral phases and glass begins. However, much archaeological pottery has a gray core, indicating that oxidation did not always go to completion.

As the temperature is raised, and a good draft is obtained in the kiln, the clays themselves will start to oxidize. Clear colors such as red, buff, or yellow suggest an oxidizing atmosphere at firing temperatures of about 700–900°C. If the draft is cut off, preventing outside oxygen from entering, the entire kiln will be filled with reducing gases. The most important oxidation-reduction reaction affecting color is the change from red ferric oxide (hematite, Fe_2O_3) to black ferrous oxide (FeO) and magnetite (Fe_3O_4).

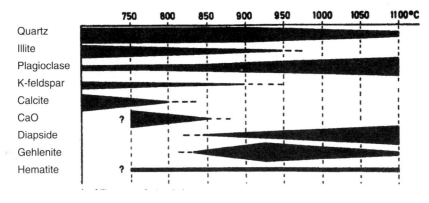

Figure 13.3 Phase changes in pottery.

Loss of Volatiles and Dehydration

When firing temperatures reach about 200–300°C, all mechanically bound or pore water still present after ambient drying is vaporized and driven off. As the kiln temperature rises, hydroxyl ions that are chemically bound within the clay lattice start to be lost. The rate of loss depends on the clay species, with halloysite and kaolinite abruptly losing most hydroxyls at about 400°C (figure 13.3). Other clays, such as smectites, start to lose interlayer water (held between the silicate sheets) at lower temperatures. They then continue to lose water but at a slower rate than kaolinite. It is important to control the rate of these losses because too rapid dehydration may result in cracking and destructive loss of the pot.

In addition to water, as the temperature rises to over 200°C, any carbon or organic matter in the clay will start to oxidize and escape as CO or CO_2. The rate and completeness of oxidation depend on the rate of temperature rise, the length of time the maximum temperature is held, the wall width of the pot, the clay composition, and the nature of the paste. On average, organic carbon is not completely eliminated until temperatures of about 750°C are reached. Other impurities in the clay, including the carbonates calcite and dolomite, the sulfate gypsum, and the iron sulfides pyrite and marcasite, will also start to volatize at about 500–800°C.

Higher-Temperature Phases and Vitrification

The loss of OH^{-1} from the clay molecules causes major structural and chemical changes. Heating kaolinite above about 500°C results in a disordered crystalline form which is still reversible by rehydration back to its ordered kaolinite structure. At 900°C and higher temperatures, all water is lost and the crystal lattice is completely destroyed. If high firing temperatures are maintained for a long enough period, high-temperature aluminum silicates will start to form. Depending on (1) the length of time the temperature is maintained, (2) the temperatures attained, and (3) the other minerals present, such as micas, that may react with the clays, a variety of higher-temperature mineral phases may crystallize. These include, from lower to higher temperature, feldspars, spinels, and above about 1050°C, mullite ($3Al_2O_3 \cdot 2SiO_2$). Mullite is a synthetic aluminosilicate used in the manufacture of high-temperature and high-shock ceramics, such as automobile spark plugs. Identification of such high-temperature minerals can be important clues to a vessel's firing history.

Starting about 600°C, vitrification starts in the more fusible impurities. Feldspars, among the most common impurities and commonly added to the clay paste as a flux by modern potters, will melt: pure Na-feldspars at 1118°C and pure K-feldspars at 1150°C; solid solution mixtures will melt at lower temperatures. Many potters, both ancient and modern, recognize a variety of other materials that can act as fluxes, including illitic clay and iron oxides. When enough vitrification has taken place to give a moderate degree of hardness, the clay is considered baked.

Full firing, or the burning stage for most clay bodies, begins at about 850°C. Two baked vessels struck together will emit a dull sound, whereas when fully burned (for some mixtures fired as low as 850°C, for most about 1000°C), the vessels will emit a ring.

Color

Most relatively pure raw clays are white or light gray. Commercial kaolinite deposits are milk white; in fact, the term comes from the Chinese *kao lin,* meaning "white hill." Colors found in clay deposits are typically due to included iron or manganese minerals or to organic material. Limonite and goethite (hydrous ferric oxides) impart a red color. Magnetite (ferric-ferrous oxide) and manganese oxides give a gray color. Glauconite (a ferrous silicate) green; and organic materials make the clay gray, brown, or black.[5]

Pottery Manufacture

Ceramic studies in archaeology can be broken down into several distinct stages:[6] (1) origin: extraction of the clay from a clay pit; (2) processing: (*a*) preparation of clay paste and shaping, (*b*) drying, and (*c*) slipping, glazing, and decorating; (3) firing: manufacture of a ceramic object; (4) use, reuse, and subsequent breakdown; (5) burial; and (6) analysis.

Origin: The Clay

The paste of ceramic bodies is clay, obtainable in a variety of environments.[7] Potters learn by trial and error which clays yield the most satisfactory ceramics, those that can be formed easily and preserve their shapes under forming, drying, and firing conditions. Commonly, a mixture of clays will be used from different sources and of varying compositions and purity (i.e., percentage of nonclay material such as organics and detrital minerals). The composition of raw clays may differ in chemistry and in mineral content from those of a fired pot due to alteration by refining (accomplished by removing detrital minerals, foreign matter, and coarser fractions), firing, use, and burial. For example, care must be taken in comparing individual sherds petrographically or chemically to clay beds without much additional information, including knowledge of the refining of the original clay, additions of tempers and fluxes, including salts, and mixing of clay sources. Equally serious are the chemical and physical effects of firing, when (1) a selective loss of more volatile elements such as alkalies takes place and (2) oxidation or reduction, especially of iron and manganese occurs. The loss of elements will depend on the positions in which they were held in the clay crystal lattice, the nature of accessory minerals, and the phase changes that occurred under the firing temperatures attained.

Preparation of the Paste and Forming the Ceramic

The addition of a limited amount of water to clay makes it plastic and allows it to be shaped by hand and retain that shape when pressure is removed. The amount of water added varies in each case with the species of clay (table 13.3), as well as with the work habits of the potter. The clay is modeled when wet and plastic. Large pore spaces are filled in by a wet hand. After the pot has dried to some extent, a slip may be applied. This is a thin paste of clay, generally very fine grained, and water. The mixture may or may not also contain pigment. The slip will make the surface smooth and help make the pot impermeable.

Drying

Drying the clay-water mixture is the next step following the forming of the pot. Too rapid drying can set up stresses in the body which may develop into cracks either while in the process of drying or in firing. Water in the system is present either as intergranular (i.e., mechanically combined with the clay matrix) or as chemically combined within the clay molecules. After drying, the plasticity is lost, but it can be restored for reworking by the addition of more water. After the object is sufficiently air-dried, it is baked and becomes hard, permanently losing its plasticity.

Firing: Manufacture of the Ceramic

Much early pottery was fired in open fires, but with the passage of time, more and more firing was done in kilns. The process of firing takes place in distinct phases: setting up the kiln, firing the air-dried clay bodies, cooling down the kiln, and, finally, withdrawing the pottery. Kiln design varies widely through time and place. Most are so-called open-flame kilns, where the heat is used directly. The firebox is in a pit directly below the ware (figure 13.4). The combustion gases generated in the firebox move directly upward into the chamber with the ware by convection.

Temperature control, both in absolute degrees Celsius and in the uniformity of heating throughout the firing chamber, of open-flame kilns is very poor. The temperature throughout the kiln will vary widely, with both vertical and horizontal gradients. Vertical gradients are controlled by the distance to the bottom of the kiln, where it is the hottest, and horizontal gradients develop by the flow of combustion gases toward the vent. Some areas within the kiln will have a reducing atmosphere, especially nearer the bottom, where highly reducing combustion gases enter the firing chamber. Areas nearer the vent will be oxidizing; where a strong draft exists, oxygen is readily available, and the combustion gases themselves become oxidized, for example, CO^{2-} becomes CO_2. Indeed, temperature gradients exist within single vessels and vary according to the size and wall thickness of the vessel.

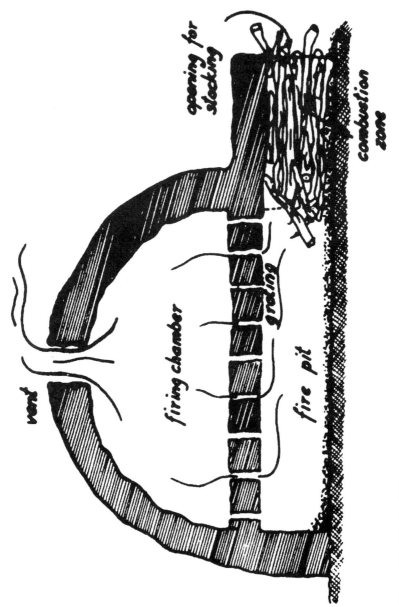

Figure 13.4 Idealized kiln.

Nonplastic Materials

Some physical properties of pure clay, including excess shrinkage during drying and chemical reactions during firing, may have a deleterious effect on the strength of the ceramic body. In order to make a stronger body and counter these effects, nonplastic material (material that does not become plastic when mixed with water) is commonly added or clays are selected that already contain nonplastic material. Depending on the refining processes used after obtaining the clay from its source deposit, some original nonplastic material may still be present, such as silt- and sand-sized quartz and other detrital minerals. In early cultures, this material is more abundant in the clay paste than in later cultures with mass production, which depended more on fast wheel technology requiring refined paste. More important are temper materials added by the potter to improve the workability, shrinkage, and firing behavior of the ceramic. The tempers weaken the body during forming but also reduce the stickiness of clay, making it easier to handle, prevent excess shrinkage and cracking during drying and firing, and can assist in the escape of water during drying. Without the temper, the formed surface will dry too fast and thus shrink more than the interior of the body. This may set up critical stresses that could destroy the ceramic. With the nonplastic fillers, large interparticle spaces are maintained which allow avenues for more rapid water migration and evaporation through the surface. Clay has a low permeability and can otherwise retain too much water.

Most of the fillers used by potters are easily obtained local material. Depending on the site, typical temper includes volcanic ash, quartz sand, crushed limestone or marble, as well as other less chemically reactive rocks, twigs, feathers, straw, dung, sherds, etc. Most temper undergoes thermal changes on firing, especially at elevated temperatures. Carbonates begin to decompose at >750°C, yielding CO_2 and leaving CaO: $CaCO_3 \rightarrow CaO + CO_2\uparrow$. The lime thus produced is unstable and hygroscopic and will, in time, recombine with moisture and form quicklime ($Ca[OH]_2$) in an exothermic reaction (i.e., releasing heat). Quicklime also increases in volume, which sets up possibly destructive stresses in the clay body. Important controls on the exact temperature at which calcite reacts to become lime depend principally on the size of the carbonate particle and the grain size of the calcite crystals. Under the microscope, a carbonate filler will be powdery if fired on average above 750°C (different authors cite 650–900°C) but will preserve its consolidated and crystalline form if fired below that temperature.

Organic materials may be only partially oxidized or may be completely burned away, depending on where they were placed in the kiln. The lower part of the kiln may have a largely reducing atmosphere and preserve organic material, whereas the upper parts or near vents will be oxidizing. Pots fired in the lower part may be gray or black in color due to preserved carbonaceous material, whereas pots fired in the upper part may be reddish due to oxidized (ferric) iron.

To summarize the expected changes in the clay paste during firing:[8]

25–200°C: All mechanically combined adsorbed water is lost from the clay. Much interlayer water will also be lost from smectites.

200–400°C: Organic matter begins to oxidize and migrate to the surface, blackening the clay body before finally being driven off as CO_2.

450–550°C: Chemically bound water is lost from the clay lattice—dehydroxylation. The loss takes place abruptly in kaolin in the structural transformation to metakaolin, with minor shrinkage and a great increase in porosity. In illite and smectite, the loss is more gradual. At 470°C, a red glow can be seen in the dark; at 500°C, all organic matter is oxidized.

550–750°C: Quartz inverts at 573°C from the alpha (low-temperature) form to the beta (high-temperature) form, accompanied by a volume expansion of 2%. Beginning at 600°C, micas lose structural hydroxyls from their lattice.

750–850°C: Most organic water has been successfully burned out of the clay. Kaolin and smectite begin to lose their characteristic crystalline structure.

850–950°C: By 870°C, $CaCO_3$ decomposes to CaO and loses CO_2 gas. Beta quartz inverts to a higher-temperature form, tridymite, but the reaction is very sluggish and will only take place if the firing temperature is held for a relatively long period of time.

950–1100°C: The structure of most kaolin and smectite is irreversibly lost, but some illite may persist. With the breakdown in clays, high-temperature alteration and reaction products can form, accompanied by pronounced shrinkage. Above 1000°C, wollastonite ($CaSiO_3$), calcium ferrisilicates, and, in a reducing atmosphere, hedenbergite ($CaFeSi_2O_6$) with reduced Fe^{2+} iron are typical reaction products. Other high-temperature phases expected to form are spinel ($MgAl_2O_4$) and mullite ($3Al_2O_3 \cdot 2SiO_2$).

>1100°C: Feldspars melt, dissolving silica and initiating a glassy phase of the ceramic. Porosity diminishes rapidly.

Geochemical Analysis for Archaeological Ceramics

Geochemical and petrographic analyses are widely used to characterize archaeological ceramics. Important problems of sourcing raw materials and characterizing manufacturing technology can be resolved by using the complementary techniques of petrographic and geochemical analysis to determine mineral and chemical compositions. The hand lens, binocular microscope, optical (polarizing) microscope, scanning electron microscope (SEM),[9] x-ray diffraction (XRD), and electron microprobe are the most important of these tools. The initial study of a potsherd should begin with a hand lens, followed by thin sections and a petrographic microscope. These studies are done at low magnification (<200×). Textures, including pore spaces, and mineral inclusions or phases can be identified. At this point, unknown fine-grained minerals and phase reactions at mineral boundaries can be labeled for detailed study by electron microprobe or SEM. A preliminary grouping of the potsherd samples can be made and, in conjunction with the archaeologist, decisions made on what other types of analysis, and perhaps field sampling, are needed to obtain further information.

Optical Methods

There are many good arguments for beginning a pottery study with petrographic analysis. Not only is this the cheapest kind of analysis, but information can be obtained that cannot be satisfactorily obtained by any other method. In a study of prehistoric ceramics from Cyprus, S. J. Vaughan used macroscopic observation and optical instrumental techniques and was able to solve many problems associated with the Late Bronze Age pottery called Base Ring Ware. A goal of the study was to establish a fabric typology which could be used by archaeologists to classify sherds in the field. To achieve this, research was designed to characterize the raw material used for the ware, identify individual fabrics, suggest provenance in Cyprus of samples where possible, and establish specific technical attributes which defined the ware in general and which were diagnostic of fabric groups in particular (figure 13.5).[10]

Figure 13.5 Late Bronze Age Canaanite pottery from Palaiokastro-Maa in western Cyprus. Diameter of the photomicrographs is about 2 mm, crossed polars. (Courtesy Sarah J. Vaughn.) *Top:* Fine-grained (0.15–0.25 mm) bimodal inclusions are subrounded-to-rounded argillaceous-calcareous clasts; coarser-grained (0.5–1.0 mm) inclusions are primarily micritic clasts, some sparry or recrystallized calcite, rare detrital chert, feldspar, and amphibole. *Bottom:* Terrigenous clastic sediments, abundant planktonic microfossils (biomicritic), occasionally detrital chert, iron oxides, altered olivine, and clinopyroxine. Possible source is a local Miocene marl.

Macroscopic study was done with a binocular microscope on sherds with fresh fractures to establish a standard list of features with an established range of values for each group. Steps in the analysis were

1. characterization of the raw materials and their preparation, such as the nature and distribution of inclusions and voids;
2. identification of aspects of primary and secondary forming procedures, such as evidence of assembly, scraping, degree of compaction, wall width, and smoothing;
3. identification of finishing techniques, such as wet-hand slurry or slip application.

The information obtained by binocular microscope study enabled a block clustering of archaeologically meaningful groups of samples. Subsamples of each sherd group were then selected for further thin section analysis for mineral and fabric identification. The samples were also analyzed geochemically by ICP (inductively coupled plasma) spectrometry for elemental characterization, by XRD for mineral identification, and by SEM and electron microprobe for micromorphological refined mineral data. A parallel study of Cypriot clays, raw and fired, was carried out to assist in the characterization and provenence assignment of the archaeological samples.

Petrographic analysis showed the use of finer-grained clays that had been subjected to low-temperature metamorphism and contained carbonate impurities. This allowed a correlation to flysch deposits north and east of the Troodos massif. Younger examples of the ware, with a coarser grain size, had been considered a decline in the potter's skill. Petrographic analysis showed merely that a shift in population centers to the south and west opened up other areas with different sources of clay. The new clays formed on steeper slopes than the older and thus had detrital fragments larger than the clays used earlier in the northern valleys. The degree of vitrification seen in thin section was consistent with firing temperatures of 800–900°C (figure 13.6).

In a study of ancient Ubaid-style ware from Ur and central and southern Mesopotomia, ranging in age from about 5500 to 3500 BC, D.C. Kamilli and A. Steinberg used a variety of methods to analyze the material for provenance determination and to decipher the firing history. Their study began with the petrographic microscope and went on to SEM and the electron microprobe. By identifying a variety of refractory minerals in the paste, including quartz, Ca-plagioclase, and augite, they were able to compare the sherds to clay sources to find a provenance. The data suggested that most of the wares were made locally, with good communication between the major pottery-manufacturing sources, suggesting that centralized organizations of pottery production existed in a number of places.

To help determine the temperature of firing, thin sections were studied with the optical microscope, SEM, and the electron microprobe. The SEM revealed samples with unmelted clay particles, fired at relatively low temperatures. At higher temperatures, the clay had become fused and showed glassy breakage. In the paste and painted surface, mullite (an aluminum silicate) and pseudobrookite (an iron-titanium oxide), both high-temperature minerals, had formed. This mineral assemblage suggests firing temperatures above 1000°C.[11]

Chemical Analysis

After petrographic and optical examination, chemical analysis can be carried out. Much chemical information could have been obtained in step 1 using the SEM and electron microprobe—in many cases, enough to answer all important questions of provenance or tech-

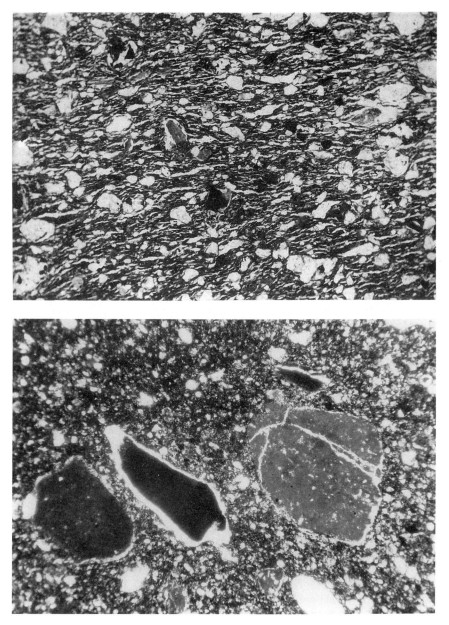

Figure 13.6 Late Bronze Age Base Ring Ware from Cyprus. Diameter of photomicrographs is about 2 mm, plane polarized light. (Courtesy Sarah J. Vaughn.) *Top:* Overfired fabric with elongated voids resulting from shrinkage. Abundant inclusions of relatively well sorted quartz silt, iron oxides, feldspars, and chlorite biotite. *Bottom:* Argillaceous rock fragments (ARF) with shrinkage rims in clay matrix. Polygonal cracking in one fragment diagnostic of ARF and is a distinguishing characteristic of grog (pottery temper). Differential optical densities of clasts due to varying of iron oxides.

nology of firing or both. If more detailed information is needed, especially quantitative chemical analysis of major and trace elements of the paste, temper, slip, and detrital minerals for provenance determination or for technology of firing, then a method of analysis can be decided upon. Many methods of chemical analysis are available and some, such as neutron activation analysis (NAA), are very expensive and may not produce the needed results. This is a critical step in a ceramic study; before one kind of analysis is decided upon, the following should be considered:

1. Some methods can only analyze material within a few millimeters of the surface, such as SEM and the electron microprobe, but give exceptional small-scale detail on chemistry and phase changes.
2. Each method involves a varying degree of destructive or nondestructive sample preparation.
3. If samples are to be compared against an existing database, it is preferable to do the same kind of analysis as that which created the database. The same elements measured by NAA and XRF, for example, will show completely different abundances.
4. Chemical analysis by itself generally does not resolve what parameter is responsible for which element. This is best done by other methods where optical control is maintained, such as SEM and the electron microprobe. However, NAA has been used for a relatively long period, and NAA databases have been established for ceramic provenance studies.[12] ICP spectroscopy is a relatively new technique used in archaeology and has many advantages over other analytical systems. Suites of major elements, as well as trace elements, are routinely determined with a high degree of precision and relatively inexpensively. All principal types of chemical analysis, including their advantages and shortcomings, have been discussed in a previous chapter.

Analysis of Residues

Analysis of organic materials has been carried out to determine reactions that may have taken place between food or perfume and the pot in which it had been stored or cooked. This type of analysis is generally carried out by Fourier transform infrared (FTIR) spectroscopy or by similar methods in specialized labs that can identify organic molecules. A recent publication of the University Museum, Philadelphia, highlights the attempts and successes of many of these studies.[13]

Diagenetic Changes

During burial, elements of the pot will be subject to both infiltration of new material and dissolution of some original constituents by solutions passing through and organisms present in the environment of burial. Unfortunately, in many cases the nature of diagenetic changes is destroyed by archaeological treatment. On most digs, pottery is cleaned with relatively strong acids which dissolve much of the products of diagenesis, including the carbonates and hydrosilicates (zeolites). Since both gypsum and calcite will be leached by this treatment, important information will be lost. Gypsum is a product of cementation during burial; the form of the calcite, either powdery or crystalline, will reveal whether or not temperatures of lime formation were reached and if soil processes caused the formation of secondary calcite.

Diagenetic changes during burial are an especially serious problem for low-temperature ceramics, that is, those fired at about 500–700°C. Such pottery is generally very porous, allowing easy access for outside solutions. Dissolution and neocrystallization will take place: deposition in pore spaces and replacement of primary minerals by secondary carbonates, hydrates, oxides, hydrosilicates (zeolites), and sulfates (gypsum). The initial porosity is due to the manufacturing and firing processes, the chemistry and mineralogy of the starting material, and the loss of outburning substances.

Provenance

Determining the provenance of clay used in the manufacture of a pot is an important problem that is traditionally attacked by geochemical techniques. Many studies have used trace-element analysis to compare the possible sources to a fired ceramic. However, conclusions of such studies must be carefully interpreted since ceramic pastes will not always be similar mineralogically or chemically to their sources, except in the cases of unrefined clay, typical of very ancient potters and very low temperature firing. Most ceramic clay has been refined and mixed, resulting in a great variation in particle size distribution and mineralogical variations. M. J. Blackman found that levigation of the raw clay resulted in major changes in mineralogy and chemistry if silt and finer sizes were removed. If only sand-sized particles were removed, the changes were not as serious, except for trace elements that were concentrated in the accessory minerals.[14]

A study by V. Kilikoglou et al. found that most trace elements showed great variation between raw and fired clay. Whether the elements increased or decreased upon firing and purification depended on many variables, including their (1) physical nature—whether they were held in fine- or coarse-grained material, in organic matter, in the clay, or in detrital minerals; (2) chemistry—whether they were present as oxides, hydroxyls, or carbonates, and their position in the clay lattice, that is, interlayer, adsorbed on the surface, or within a sheet; (3) physical properties—most important, the temperature at which they volatize. The conclusion of their study was that under no circumstances can the chemistry of raw clay be compared to that of a finished pot. To demonstrate the provenance of the clay in the pot, the steps taken by the potter in making the pot must be followed: the clay must be purified and fired under analogous conditions.[15]

The analytical methods then used to compare the synthetic pot to the original should be sensitive to elements present in trace amounts and include optical emission spectroscopy, NAA, XRF, SEM, or the electron microprobe. Then, so that meaningful statistical analysis can be carried out, (1) the number of elements determined should be as large as possible, and (2) a large number of samples should be analyzed.

Determining Firing Temperatures

Many methods have been developed to estimate firing temperatures of pots. Some of the most important include the study of mineral phase reactions in prepared thin sections, texture analysis by SEM, electron spin resonance (ESR) spectra analysis, and Mossbauer spectra analysis.[16]

Mineral Reactions

Data on the temperature of decomposition of existing minerals and the formation of new phases can be obtained depending on what mineral reactions have been observed by thin section microscopy, and followed up by XRD, SEM, or the electron microprobe. Calcite, for example, breaks down at about 700–800°C and potassium feldspar at >900°C. In actual practice, this may be difficult to apply because the thermal stability of minerals is also a function of kiln atmosphere, number of molten phases that form, mineralizers, kinetic retardation of decomposition reactions, and the formation of neominerals by diagenesis. The increase and disappearance of different minerals is, however, a good qualitative indication of the temperatures reached in firing.

Texture Analysis by SEM

Quantitative results are difficult because of inhomogeneous pore distribution and problems in re-creating exactly the same conditions of the original firing. Experimentally, the clay is fired under carefully controlled conditions at different temperatures that span the approximate original temperature of firing. The resulting microtextures from each firing are then compared by the SEM to the sherd. The approximate temperature of firing of the original is considered shown by the texture of the clay paste most like that of the original sherd. In addition to the clay mineral appearance giving evidence for firing at any temperature, vitrification is good evidence that the pot was fired above about 800°C.

Electron Spin Resonance

The ESR signal of the pot is determined. This was presumably set when the pot reached its high temperature of firing and then cooled down. The sherd sample is then irradiated and heated stepwise. At each step of heating, the ESR signal is read again. When the temperature of refiring reaches or is greater than that of the original firing temperature, the ESR spectrum changes.

Another technique is to fire clay taken from the same source as the pot at varying temperatures. When the original temperature of firing is reached, the ESR spectra of the sherd and the synthetic clay will match. This system works best at temperatures below about 600°C.

Mossbauer Spectroscopy

Mossbauer spectra normally yield information on the chemical state and the physical environment of iron and show the changes that take place during firing. Analytical procedures are carried out in a similar way as in ESR analysis. The spectrum of the sherd is first determined. Then clays from the same source as the original are subjected to stepwise heating and their spectra read at each temperature. A match between the spectra of the experimental firings and the original indicates that the original firing temperature was reached. The best temperature range for Mossbauer spectra determination is about 750–1100°C. Clays to be compared should have an iron content of at least 5%.

14

Applications of Stable Isotopes in Archaeological Geology

I sotopic ratios of elements in natural materials on the earth either have been constant in time and space or have varied as a result of radioactive decay or geochemical fractionation. Elements which show variations in isotopic abundances in different samples and the reasons for these variations have helped resolve many geological and archaeological problems. Radioactive decay has provided absolute dating clocks: for archaeology, the most useful systems have been associated with ^{14}C, [40]Ar, and U-disequilibrium series. Variations in isotopic ratios of the stable elements H, C, O, N, S, Sr, and Pb have helped solve problems of provenance, paleoenvironments, and paleodiets.

The rationale for isotopic variations of individual elements will determine the types of applications to archaeological geology. The most important applications are the determinations of artifact signatures, paleodiet, and paleoenvironment.

Isotopic Fractionation of Light Elements

Isotopic fractionation of light elements by physical, chemical, and biological processes is controlled by those thermodynamic properties which are determined by atomic weight and electronic configuration.[1] Thermodynamic properties of molecules that are mass and temperature dependent include energy, which decreases with decreasing temperature, and vibrational frequency, which varies inversely in proportion to the square root of the reduced mass. Easily measurable isotopic separation is generally restricted to the lighter elements, that is, with atomic weights less than 40. Because isotopic fractionation is mass dependent, the separation is greater for elements with the greater mass difference between isotopes. The greatest separation is expected for hydrogen (mass 1) versus deuterium (mass 2); the other light elements commonly have isotopic differences closer to 10%. Thus, the lighter isotopes have higher vibrational energy and their chemical bonds are more easily broken.

The different reactivity of lighter versus heavier isotopes of an element is responsible for their separation during geochemical and biological processes.

Isotopic Variation in Heavier Elements

Thermodynamic behavior has been considered a principal cause for variations, not in isotopic abundances of the heavier elements Sr and Pb, but rather in abundances of their parent radionuclides: Rb for Sr and U and Th for Pb. Recently, however, P. Budd and others suggested that under nonequilibrium conditions, fractionation could theoretically take place among the lead isotopes. In roasting and smelting of lead ores during which evaporation took place, differential loss of lead isotopes might occur. They calculated, for example, that with a 40% lead loss during the refining of a complex Pb-Ag ore, lead isotopic ratios would change from their initial ratios as follows: $^{207}Pb/^{206}Pb$ from 0.8311 to 0.8321, $^{206}Pb/^{204}Pb$ from 18.87 to 18.92, and $^{208}Pb/^{206}Pb$ from 2.058 to 2.063.[2] This theoretical loss of lead isotopes has not yet been demonstrated experimentally.

Strontium Isotopic Variation

Strontium abundance in soils and rocks from a region and isotopic ratios in any object from that region depend purely on local geology.[3] Strontium concentrations in plants and animals are controlled by trophic position, but the isotopic composition is invariant; that is, Sr does not fractionate. Thus, bones and teeth in an individual will have different Sr abundances but identical $^{87}Sr/^{86}Sr$ ratios. Strontium isotopic ratios vary with age and type of rock and are used geologically to detemine the age and source of the rock formation. The isotope ^{87}Sr forms over time by the radioactive decay of ^{87}Rb, with a half-life of 4.7×10^{10} years. Thus, the amount of ^{87}Sr in a rock will depend on (1) the original amount present when the rock formed, (2) the amount of Rb present in the rock, and (3) the age of the original rock.

Lead Isotopic Variation

Lead consists of four stable isotopes: ^{204}Pb, ^{206}Pb, ^{207}Pb, and ^{208}Pb. The isotope ^{204}Pb is not the product of any known radioactive decay series; ^{206}Pb results from the decay of ^{238}U; ^{207}Pb from ^{235}U; and ^{208}Pb from ^{232}Th. Variations found in nature in lead isotopic ratios can be explained in the following ways:[4]

1. Each of the parents has a different half-life, resulting in different rates of productivity of the daughter lead isotopes. This abundance factor is thus controlled by the age of the original source material and the half-lives of the parent radionuclides: $^{238}U = 4.5 \times 10^9$ yr, $^{235}U = 0.7 \times 10^9$ yr, and $^{232}Th = 14.0 \times 10^9$ yr.
2. The relative abundance of the parent radionuclides differs in the original source material. Today, ^{238}U is 99.27% of all uranium, ^{235}U is 0.72%, and Th/U ratios in igneous rocks vary from 2.8 in high Ca-granite to 5.7 in low Ca-granite, and in sediments from 0.77 (in carbonates) to 5.4 (in deep-sea clay).
3. As discussed in chapter 5, intermediate daughter products formed during the decay se-

ries from the original parent to the stable daughter lead may be lost. Daughter products may be separated from their parents through geological processes such as weathering, adsorption on clay minerals, fractionation, and crystallization of magma, etc. During the processes leading to the formation of lead ore minerals, generally galena, lead is separated from its parent U and Th nuclides, which ends the processes responsible for variations in isotopic ratios. The ratios are then frozen in and form distinctive isotopic fingerprints. Pb isotopic ratios have been determined in ores for many geological sites and mining districts and found to vary systematically from place to place. This information has been used to explain the genesis of many occurrences as well as to supply isotopic fingerprints for many ores in North America and the eastern Mediterranean.[5] Lead isotopic signatures have been used to determine provenance for a variety of archaeological objects, including metallic lead, leaded bronze, and lead glass from the Early Bronze Age through classical times.[6]

Methodology

Measurements of stable isotopic ratios are carried out with a mass spectrometer, an instrument that measures proportions in very small samples of different isotopic masses of several elements. In the newer, state-of-the-art machines, less than 5 mg of sample are needed for an analysis.

A mass spectrometer consists of three essential parts: (1) a source of a positively charged, monoenergetic beam of ions; (2) a magnetic analyzer; and (3) an ion collector (figure 14.1). The mass spectrometer is evacuated to a high vacuum, of the order of 10^{-6}–10^{-9} mm Hg. Gaseous samples, such as CO_2, or solid samples containing lead can be analyzed depending on the design of the machine. For analysis of a gas, the sample is allowed to leak into the source through a small orifice. The molecules are ionized by bombardment with electrons. The positively charged ions are then accelerated by an adjustable voltage and collimated into a beam by slit plates. For analysis of solid samples, a salt of the element is deposited on a filament in the source composed of Ta, Re, or W, heated electrically sufficient to ionize the element to be analyzed, and collimated into a beam, as with the gaseous analysis.

The beam enters a magnetic field whose field lines are perpendicular to the direction of travel of the ions. The magnetic field deflects the ions into circular paths whose radii are proportional to the masses of the isotopes. The heavier ions are deflected less than the lighter ones. The separated ion beams continue through the analyzer tube to the collector.

The mass analysis is obtained by varying either the magnetic field or the accelerating voltage in such a way that the separated ion beams are focused into the collector in succession. The resultant signal, traced out on a strip recorder, shows a series of peaks and valleys that form the mass spectrum of the element. Each peak represents a discrete isotope whose abundance is proportional to the relative height of the peak in the spectrum.

The data are reduced by comparison to an accepted international standard. For example, the precise measurement of the ratios of oxygen and carbon isotopes, $^{18}O/^{16}O$ and $^{13}C/^{12}C$, in marble is carried out after suitable chemical treatment has separated these elements in the form of CO_2 from the calcium carbonate. The results are expressed as a deviation from a conventional standard, the Pee Dee belemnite, a carbonate fossil from South Carolina, and form the isotopic signature. This deviation, called δ, is expressed as $\delta^{13}C$ or $\delta^{18}O$, mea-

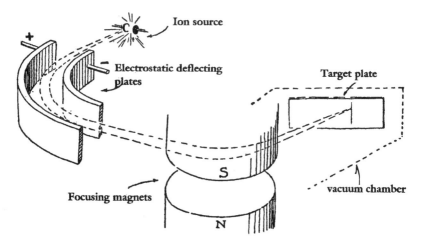

Figure 14.1 Schematic diagram of a 60° sector mass spectrometer showing arrangement of ion source, electromagnet, and collector (modified from Pauling, RH9).

sured in parts per thousand (or per mil, ‰), and calculated as follows: $\delta(‰) = 1000(R$ sample/R standard $- 1$), where $R = {}^{13}C/{}^{12}C$ or ${}^{18}O/{}^{16}O$. Thus, if marble has a $\delta^{18}O = +10‰$, the isotopic ratio of the oxygen is 10 parts per thousand enriched in the heavy isotope ${}^{18}O$ compared to the standard.   The isotopic variability data usually are expressed as a scatter plot of $\delta^{18}O$ and $\delta^{13}C$ values. The precision of $\delta^{18}O$ values is of the order of $\pm0.2‰$ or better, including instrumental, analytical, and sampling errors.[7] For the heavier isotopes, such as ${}^{87}Sr/{}^{86}Sr$ and ${}^{206}Pb/{}^{204}Pb$, results are expressed directly as ratios and generally not normalized to a conventional standard.

Artifact Signatures

The earliest attempt to use isotopes as signatures of provenance was by Robert Brill and his colleagues starting in 1965 on glass artifacts.[8] In their studies, they found evidence for local sources (lead in Egyptian glass vessels was locally mined) and for early trade in lead (some yellow and red pigments decorating glass panels in Greece were from outside sources).

H. Craig and V. Craig first suggested using isotopic patterns plotted on a $\delta^{18}O$–$\delta^{13}C$ diagram to identify quarry sources of classical Greek and Roman artifacts. They tested 10 archaeological marbles and found that 5 from Athens, Delphi, and Naxos could be assigned a provenance, but that 5 others, from Delphi, Epidaurus, and Caesarea, Israel, could not. The small amounts needed for analysis could readily be acquired without causing harm to museum specimens.[9] An extensive isotopic database for the principal classical quarries is now available and many marble objects can now be related to their sources.[10] Isotopic analysis can also routinely be used to associate broken fragments from inscriptions and statues, allowing only those fragments with similar isotopic compositions to be associated.[11] The marble portrait of Antonia Minor in the Fogg Museum of Harvard University, for ex-

Figure 14.2 Marble bust of Antonia Minor: height 58 cm, acquisition no. 1972.306. (Photograph courtesy Fogg Art Museum, Harvard University.)

ample, was found to consist of five fragments from different marble statues, three fragments made of Greek Parian marble (the head, right shoulder and breast, and upper left shoulder) and two of Roman Carrara (the ponytail and lower left shoulder) (figure 14.2).

For any signature to be viable, its values must be uniform throughout an artifact, should be uniform over a quarry, and, hopefully, show only small variations within the limits of a mining district. Isotopes that show such a restricted variation have been used successfully as artifact signatures to determine source and to help in assembly of broken pieces. These include O and C for marble and Pb for metal, glass, and glazes.[12] Databases of Mediterranean and Near Eastern sources are available for these isotopes, and a database of North American sources is available for lead. Sr may also prove to be helpful for carbonates and obsidian, but the databases are still incomplete.

Oxygen and Carbon

Variations in $\delta^{18}O$ and $\delta^{13}C$ have been used with great success in studies of carbonates, including marble and shells, and with lesser success on malachite in patinas of ancient bronzes.[13] Controls of the isotopic composition of oxygen and carbon in the carbonate of marble are principally through temperature, chemical composition, and isotopic ratios of water. The processes involved are as follows:

1. mode of origin, as either a chemical precipitate or a "hash" of organic shell fragments or a mixture of both, and composition of all subsequent diagenetic cements;
2. isotopic composition of water associated with the carbonate minerals during their formation, diagenesis, and later history;
3. temperature of metamorphism which converted the limestone into marble; and
4. later weathering history.[14]

In this way, marble from a given region that was formed at a particular time with its own geological history may develop unique isotopic characteristics.

In the Carrara district of Italy variations in $\delta^{13}C$ and $\delta^{18}O$ within statuary blocks were less than 0.5‰; they were less than 2‰ within an outcrop in most quarries. The two principal marble districts, the classical Roman Carrara and the Renaissance Serraveza, could also be distinguished from each other, but individual quarries within each district could not be told apart.[15]

Weathering may change the original isotopic composition of a marble sample. If fragments of an artifact have different weathering histories—for example, one piece buried in soil, another in a well, and a third in a wall—then each could exchange oxygen with waters that are of different isotopic compositions and also quite different from those present at the time of formation. Since many artifacts can only be sampled near or on their surfaces, changes due to weathering must be assessed (table 14.1). Weathering commonly causes little change in $\delta^{13}C$ but decreases $\delta^{18}O$ if exchanged with meteoric waters. In one study, about 50 years of weathering had caused a decrease of about 0.6‰.

Biodeterioration is an important source of fractionation, especially of carbon. Lichens, bacteria, and algae can penetrate the surface of marble to a depth of greater than 40 mm.[16] The microflora colonies will then, over time, change the normal isotopic composition of the fresh marble following the fractionation of each specific species. Deposits of calcium oxalate will form, replacing calcite as a result of this microfloral activity. $\delta^{13}C$ will decrease; $\delta^{18}O$ will be less affected.

Table 14.1 δ^{13}C and δ^{18}O Analysis of Fresh and Weathered Marble

Sample no.[a]	Condition	δ^{13}C	δ^{18}O
1	Fresh	+1.17	−7.24
	Weathered	+1.12	−7.93
2	Fresh	+1.37	−6.93
	Weathered	+1.20	−7.45
3	Fresh	+2.57	−7.83
	Weathered	+2.64	−8.16

Source: N. Herz, 1987, Carbon and oxygen isotopic ratios: A data base for classical Greek and Roman marble, Archaeometry 29:35–43. M. Coleman and S. Walker, 1979, Stable isotope identification of Greek and Turkish marbles, Archaeometry 21:107–12.

Note: In mils, relative to Pee Dee belemnite.

[a] Samples 1 and 2 are from the Tate Quarry, Georgia; 3 is from a sarcophagus in the British Museum (Coleman and Walker, 1979).

A stable isotopic database of multiple marble samples from the major ancient quarries of Greece, Turkey, Italy, and Tunisia is used to determine the source of artifacts.[17] The provenance of samples is determined by comparison to the database either by visual inspection of δ^{13}C versus δ^{18}O plots (figure 14.3) or by statistical analysis, such as discriminant analysis.

Knowing the difference in isotopic composition of possible source sites has also been used to determine the provenance of shells. N. J. Shackleton and C. Renfrew studied Neolithic to Bronze Age sites in the Balkans where *spondylus* shells had been used for ornaments. Possible sources could have been either the Black Sea or the Aegean, which have similar temperatures but differ by ±5.3‰ in δ^{18}O. Isotopic measurements showed the source for all the shells in ornaments was the Aegean, even for sites that were close to the Black Sea.[18]

N. R. Shaffer and K. B. Tankersley, in eastern North America, and D. N. Stiles, R. L. Hay, and J. R. O'Neil, in Olduvai Gorge, Tanzania, used δ^{18}O to determine the source of chert artifacts.[19] In Tanzania, the sources were paleolakes with varying salinities. Salinity is another important control of the oxygen isotopic ratio, with δ^{18}O enrichment associated with higher salinities.[20]

Strontium and Sulfur

Strontium isotopes have been used to distinguish the provenance of Mediterranean marble (table 14.2), obsidian, and, together with S isotopic ratios, gypsum.[21] The strontium contained in carbonate shells of marine organisms and in directly precipitated carbonate should reflect the composition of seawater at the time of deposition. Since the approximate ^{87}Sr/^{86}Sr variation in seawater at different geologic times (figure 14.4) is known,[22] and Sr isotopes do not fractionate during metamorphism, the measurement of this ratio in marble could theoretically fix time constraints for the formation of the original carbonate. If there is a negligible contribution from river discharge, the principal control of Sr isotopic composition of the marble is seawater; seawater Sr isotopic composition in turn is controlled by seawater-oceanic basalt exchange.

Since Sr is not fractionated by magmatic processes, Sr isotopic ratios in obsidian are

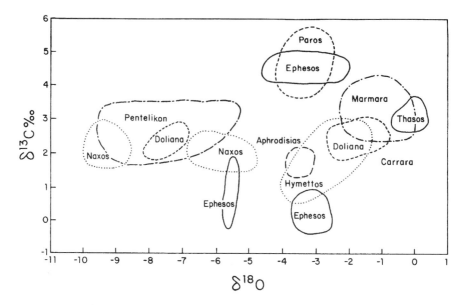

Figure 14.3 δ^{13}C vs. δ^{18}O for classical marble quarries (Herz, 1985).

controlled by magmatic compositions. Lower crust and upper mantle sources generally have a lower ratio than those in the upper crust. Contamination by crustal materials can raise an initially lower isotopic ratio derived from the mantle. In a study of 14 known Mediterranean obsidian sources, N. H. Gale found poor discrimination using Sr isotopic ratios alone. However, when the abundance of Rb was introduced as a second variable, the discrimination was excellent. The ratios of ^{87}Sr/^{86}Sr varied from a low of 0.70336 ± 0.00006, with Rb 219 ppm, at Pantelleria, to a high of 0.72027 ± 0.00009 with Rb 395 ppm at Antiparos in the Cyclades.[23]

Gypsum was widely used in the interiors of buildings in Minoan Crete and Thera and in Mycenean Greece from the beginning of the Bronze Age. By comparing Sr and S isotopes, Gale et al. found that all the gypsum came from Cretan sources.[24] This long-distance trade in gypsum from Crete northward paralleled the trade in lead and silver from Laurion southward to Crete.

Lead

There are numerous sources of lead in the eastern Mediterranean, commonly associated with gold and silver mines. Many of these mines were noted by Herodotus (490–425 BC) and Strabo (63 BC–AD 21), among others.[25] Some of the best known are the rich lead-silver deposits of Laurion, 40 km southeast of Athens. When the mines of Laurion went into a decline, after the fifth century BC, the power of Athens was also diminished (see chapter 12). Mining in the eastern Mediterranean region must have started in the Early Bronze Age; some of the oldest lead artifacts, from Siphnos, date to about 3000 BC.[26] Analysis of Bronze Age artifacts from the Aegean region has revealed a very simple picture; the overwhelm-

Table 14.2 Means of Sr, O, and C Isotopic Analysis: Classical Marble Quarries

Quarry (no. of samples)	$^{87}Sr/^{86}Sr$		$\delta^{18}O$		$\delta^{13}C$	
	Mean	s.d.[a]	Mean	s.d.	Mean	s.d.
Paros ($n = 4$)	0.70755	15	−3.78	0.41	5.09	0.22
Naxos ($n = 3$)	0.70832	0	−8.75	0.66	1.93	0.14
Thasos ($n = 3$)	0.70779	12	−0.78	1.63	3.34	0.21
Marmara ($n = 4$)	0.70796	14	−1.67	0.75	2.78	0.43
Penteli ($n = 4$)	0.70846	24	−6.84	0.75	2.65	0.21
Hymmetus ($n = 4$)	0.70727	11	−2.71	0.35	1.98	0.58
Doliana, group I quarries ($n = 3$)	0.70783	9	−1.50	0.69	2.51	0.27
Carrara, archaeological quarries ($n = 6$)	0.70789	8	−2.02	0.98	2.16	0.31

Source: S. C. Carrier, N. Herz, and B. Turi, 1996, Isotopic strontium: its potential for improving the determination of the provenance of ancient marble, International Symposium on Archaeometry, program and abstracts (Urbana: University of Illinois) p. 21.
Note: C and O in mils, relative to Pee Dee belemnite.
[a]Times 10^{-5}.

ing sources of lead for the Cretan and Theran cultures were mines at Laurion and Siphnos (figure 14.5).[27]

Lead isotopes have been used for provenance studies of North American Indian cultures where galena itself was used as a raw material. Metallurgical techniques for obtaining lead from ores were apparently unknown to the Indians. R. M. Farquhar and I. R. Fletcher used lead isotopic signatures to study trading patterns. Galenas found at some aboriginal sites

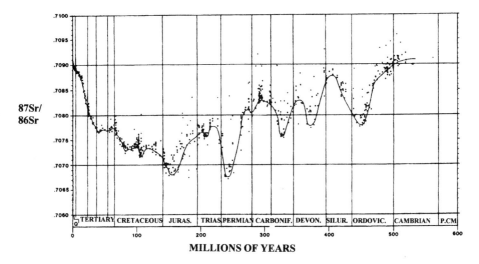

Figure 14.4 Sr isotopic variation with time (modified from Burke et al., 1982).

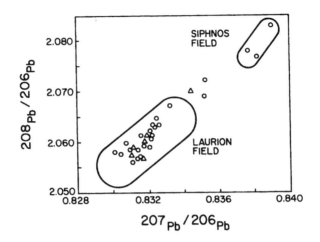

Figure 14.5 Lead artifacts from Thera (*circles*) and Crete (*triangles*) plotted on a $^{208}Pb/^{206}Pb$ vs. $^{207}Pb/^{206}Pb$ diagram. The Laurion and Siphnos fields are outlined (modified from Gale and Stos-Gale, 1981).

were primarily obtained from only two veins within a few hundred kilometers of the sites. During the Late Archaic–Early Woodland periods (3500–2500 BP), trading took place over much greater distances. Thus, shells from the Gulf of Mexico were apparently brought to the Great Lakes area, where they may have been traded for native copper. Galenas analyzed at five sites in the Great Lakes area apparently came from veins near Rossie, New York; Ottawa; and the mineral district of southern Wisconsin–northwestern Illinois (figure 14.6).[28]

Paleodiets

Information on diets of prehistoric people can be obtained indirectly from isotopic analysis of carbon and nitrogen in collagen of bones and dentine of teeth, both human and animal, and directly from food, either refuse or foodstuffs burned onto pots. Since bone recovered from archaeological sites has been subjected to a variety of diagenetic effects, consideration must be given in the interpretation of an analysis to the possible addition of contaminants such as humic acids and the loss of original portions due to the action of fungi or bacteria.[29] The differential loss of certain amino acids from collagen is another potentially serious problem since each amino acid fractionates carbon and nitrogen differently. The loss of glycine, for example, which contributes 30% of the total nitrogen to protein, would increase the $\delta^{15}N$ of the residual and lead to a mistaken interpretation of diet.[30] Information on residence in a site or immigration into the site can be obtained from strontium isotopic information, which gives information on the regional geology.[31] Because

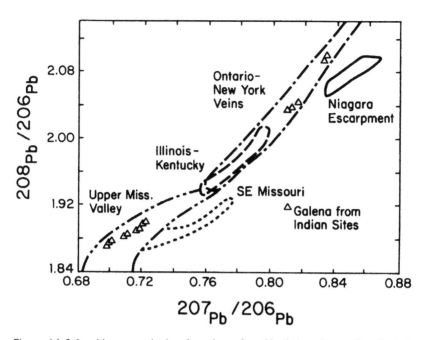

Figure 14.6 Lead isotope ratio data for galenas from North American archaeological sites compared to data from mineral deposits (modified from Farquhar and Fletcher, 1984).

plants and animals have distinctive isotopic signatures, isotopic composition of humans can therefore be correlated with diets.

The first attempt to use isotopic information for figuring paleodiets used $^{13}C/^{12}C$.[32] The C isotopic composition of corn is distinct from that of other crop plants, such as beans or squash. The difference in $\delta^{13}C$ is due to the operation of a different photosynthetic pathway to fix CO_2 (C_4 photosynthesis) in corn relative to that operational in most other higher terrestrial plants (C_3 photosynthesis). Of much less importance for paleodiet studies is the third pathway, taken by the CAM (crassulacean acid metabolism) plants, typical of cacti and other succulent desert plants. The introduction of corn into the diet of prehistoric North American Indians caused a change in the stable C isotopic ratios of the protein collagen preserved in fossil bones. Now nitrogen, oxygen, and strontium are also used,[33] although carbon remains the most important for determining paleodiet.

Analysis of the collagen portion of bone has yielded isotopic evidence on the diets of ancient humans because the $^{13}C/^{12}C$ and $^{15}N/^{14}N$ ratios are passed along, with modification, through the food chain from the base (plants) to the consumers (animals and humans). Although with time some amino acids of collagen break down, making it debatable if the original isotopic ratios of ancient bone have been preserved, M. Fizet and others have shown that animal and human bones from the 40,000–45,000 BP Neanderthal site of Marillac did preserve collagen. They found no significant differences in isotopic ratios in bone collagen when they compared the modern species to its ancestor. Bone also showed that the Neanderthals ate meat.[34] Thus, the isotopic ratios in bone are affected by important items of the diet (e.g., meat vs. grains, cultivated vs. wild plants, and even corn vs. wheat).

Carbon and Nitrogen

The photosynthetic cycle strongly fractionates carbon isotopes when plants metabolize atmospheric carbon dioxide.[35] Plants convert CO_2 into more complex molecules by three pathways, called C_3, C_4, and CAM.[36] The effects of these different metabolic pathways are seen in the stable isotopic ratios found in plant tissue (table 14.3). For example, C_3 plants, which include most North American and European species, first convert CO_2 to a 3-carbon molecule called 3-phosphoglycerate. The 4-carbon dicarboxylic C_4 plants, such as maize and sugarcane, convert CO_2 more efficiently into plant tissue and need less time and water. They end up less depleted in $\delta^{13}C$, with average values of $-12.5‰$, compared to the $-26.5‰$ of C_3 plants, but with a near overlap in variability. The CAM pathway, with average $\delta^{13}C = -17.7‰$, is of little importance for archaeology (figure 14.7).

Stable nitrogen isotopic ratios $^{15}N/^{14}N$ are expressed as $\delta^{15}N$ values relative to the air standard (atmospheric N_2, $\delta^{15}N = 0‰$). Nitrogen isotopic ratios differ in terrestrial plants,

Table 14.3 Typical Biosphere $\delta^{13}C$ Values

Air (CO_2)	
Normal atmosphere	-7
Urban atmosphere, Los Angeles (diluted by fossil fuel consumption)	-8
Amazon rain forest at ground level (diluted by rotting leaf litter)	-15.5
C3 Plants (leaves)	
Beta vulgaris (sugar beet)	-30.1
Raphanus sp. (radish)	-28.8
Pisum sativum (pea)	-26.1
Triticum aestivum (bread wheat)	-23.7
C4 Plants (leaves)	
Atriplex vesicara (goosefoot herbs)	-15.1
Zea mays (maize)	-14.0
Saccharum sp. (sugarcane)	-13.9
Thallasia testudinum (water grass)	-9.3
Herbivores (bone collagen)	
Browsers	
Grey duiker	-23.8
Tropical grysbuck	-22.6
Kudu	-21.1
Giraffe	-20.8
Mixed Feeders[a]	
Sable antelope	-13.6
Mountain zebra	-11.8
Hippopotamus	-9.1
Tsesabe	-6.8

Source: N. J. van der Merwe, 1982, Carbon isotopes, photosynthesis, and archaeology. *American Scientist* 70:596–606.
Note: In mils, relative to Pee Dee blemenite.
[a]Group arranged from truly mixed feeders at top to pure grazers at bottom.

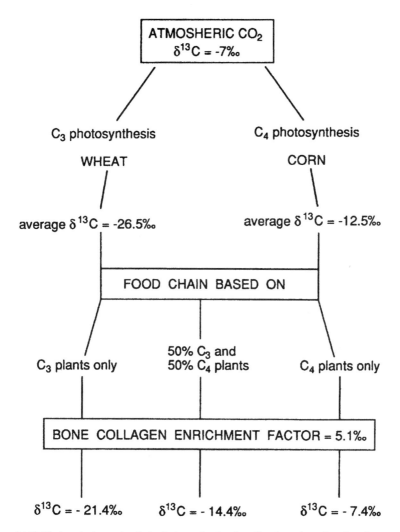

Figure 14.7 Carbon isotopes in diet: photosynthetic plant fractionation of carbon isotopes and its effects on ^{13}C in human bone collagen (modified from van der Merwe, 1982). Average values are shown; assumption is that the consumers are vegetarians.

which is in turn reflected in the diet and tissues of animal consumers.[37] Two groups of plants have distinctive nitrogen isotopic ratios: legumes and nonlegumes. The legumes use atmospheric N_2 and nitrate and ammonium ions in the soil as nitrogen sources, and nonlegumes use only soil nitrogen, since they cannot fix N_2. The difference in $^{15}N/^{14}N$ ratios is not due to fractionation by different photosynthesis pathways but merely reflects different isotopic compositions of the sources: soil nitrate and ammonia have $\delta^{15}N$ values more positive than that of the atmosphere. Thus, peas, peanuts, beans, and other legumes have $\delta^{15}N$ values lower than those of nonleguminous plants. Prehistoric legumes probably had values typical of modern legumes, that is, about $+1‰$, because the isotopic composition of the atmosphere has not changed over the past few million years. Prehistoric nonlegumes, on the other hand, with $\delta^{15}N$ values of about $+9‰$, are more positive than modern because of the widespread use of chemical fertilizers today that are synthesized from atmospheric nitrogen and so average around $0‰$.[38] Protein in animal tissue is enriched in $\delta^{15}N$ relative to the $\delta^{15}N$ in their diet by $\pm3‰$.

The difference between $\delta^{13}C$ of atmospheric CO_2 and maize is $-7.0‰$ (Table 14.3). The mean $\delta^{13}C$ value of C_3 plants is $-26.5‰$; of C_4 plants, $-12.5‰$ (table 14.3).[39] J. C. Vogel studied the way large animals, in this case ungulates, fractionate carbon isotopes. In the case of the kudu, which feeds almost exclusively on C_3 plants with an average $\delta^{13}C$ of $-26.5‰$, he found fractionation factors of $+5.3‰$ for bone collagen, $+3‰$ for flesh, and $-3‰$ for fat. He also noted that there was little difference in isotopic composition of bone collagen in carnivores and the animals consumed.[40]

M. J. DeNiro and S. Epstein traced isotopic ratios up the food chain and found that experimentally controlled diets affected both the carbon and the nitrogen isotopic content of bone in laboratory animals.[41] In order to assess the applications to human bones, it was necessary to know how human metabolism affects isotopic ratios. The initial carbon isotopic ratios of plants are fractionated differently by each body organ. Human bone consists of collagen (a soft protein tissue) and apatite. Soft tissues and bone collagen are apparently enriched about $5‰$ relative to the food intake.[42] The carbonate radical of bone hydroxyapatite is enriched about $13‰$.[43] In adults, collagen has a slow turnover rate and its carbon probably accumulated from food eaten over several decades.[44] Collagen is also inert and does not exchange carbon with air or other organic materials in archaeological deposits.

An early application of stable isotopes to the study of human skeletons was on American Indian bones in New York State. The point in question was the time that maize agriculture had been introduced to the region. The $\delta^{13}C$ values of the Archaic and Woodland (2500–100 BC) remains clustered around $-21‰$, about the same as prehistoric western European values. The sites dated from AD 1000 to 1500 showed a slow change over time from $-16.5‰$ to $-13.5‰$, which was correlated with an increase in maize consumption. Conclusions from this study were not only that the arrival of maize in a C_3 environment could be dated but also that the amount of maize consumed in human diets could be measured (figure 14.8). Over a 200-year period, from AD 1000 to 1200, the proportion of maize in the diet increased from zero to more than half the carbon in the bones. The findings suggested that farming very rapidly overtook hunting and gathering as a basis for food production, displacing a tradition that had lasted for thousands of years.[45]

The biogeochemical group of the Geophysical Laboratory has suggested that greater problems exist than were realized in reconstructing paleodiets based only on a shift in $\delta^{13}C$ bone collagen. Different amino acids in collagen have variations in $\delta^{13}C$ values ranging up to $10‰$ within a single bone. Variations in the amount and isotopic composition of amino acids can be produced by starvation, which produces changes due to autodigestion, and by degradation in soil. The relative amounts of amino acids, as well as the $\delta^{13}C$ values, are changed during differential processing by bacteria.[46]

In the Tehuacan Valley of Mexico, carbon isotope analysis of skeletal material demonstrated the increasing domestication of maize and other crops from ca. 6000 BC to AD 1000.[47] Maize became paramount in the diet by ca. 4000 BC and continued to be consumed at roughly the same level throughout the rest of the period studied. By analysis of nitrogen

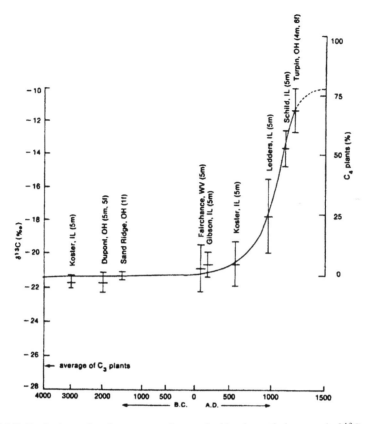

Figure 14.8 Beginnings of maize consumption marked by dramatic increase in $\delta^{13}C$ values of the Archaic and Woodland skeletal remains (modified from van der Merwe and Vogel, 1977).

isotopes in skeletal material, DeNiro and Epstein showed that the consumption of beans actually decreased slowly between 6000 BC and AD 1000.[48]

B. S. Chisholm, E. E. Nelson, and H. P. Schwarcz compared carbon isotope values for indigenous populations in Canada and their food chain to discern the effects of a marine versus a terrestrial protein diet. On the Pacific coast, the flesh of marine animals differs from that of terrestrial animals by 7.9‰, reflecting the 7‰ difference between oceanic and atmospheric carbon. Marine phytoplankton, the base of the marine food chain, fractionate carbon to the same extent as terrestrial C_3 plants—that is, about 19‰ relative to their carbon source. Thus, the 7‰ difference between air and ocean carbon is maintained in the plants. A direct relationship was found in comparing $\delta^{13}C$ values in collagen of bones of the indigenous populations to the mammals, fish, and littoral species they consumed. The study concluded that carbon isotopic measurements could determine the relative proportions of marine- and terrestrial-based protein in the aboriginal diets.[49]

Using both carbon and nitrogen isotopes, P. L. Walker and M. J. DeNiro were able to figure the contributions of marine and terrestrial resources in prehistoric diets. They studied burial sites in southern California and found the mean $\delta^{13}C$ and $\delta^{15}N$ values in collagen changed progressively from the Channel Islands ($\delta^{13}C = -14.0, \delta^{15}N = +16.3$) to the mainland coast ($\delta^{13}C = -14.5, \delta^{15}N = +14.9$) to the interior ($\delta^{13}C = -17.2, \delta^{15}N = +10.9$). This parallels the changes in ratios of the two isotopes in marine compared to terrestrial organisms and shows a declining marine contribution to the diet in inland localities.[50]

In many instances, organic residues scraped from potsherds yield highly accurate dietary information.[51] Larger postmortem isotopic alteration can occur in uncarbonized plants than in burned plant remains.[52] Analysis showed that relative to modern plants, uncarbonized plant remains may undergo large shifts in $\delta^{15}N$ and, in some cases, in $\delta^{13}C$ as well. One of the highest values ever reported for a terrestrial specimen, $\delta^{15}N = +46‰$, was found in uncarbonized maize. On the other hand, carbonized plants retain their original isotopic ratios. Apparently, burning seals the original isotopic ratios for both carbon and nitrogen, preserving the distinctions between the three groups in prehistoric plants. Based on studies of the carbonized food remains in the highlands of Peru, nonleguminous C_3 plants predominated from 200 BC to AD 1000 and from AD 1200 to 1470. In contrast, during the intermediate period (AD 1000–1200), a trend toward C_4 plants appeared, suggesting that nonleguminous plants and maize both were cooked. Combined analysis of both carbon and nitrogen isotopic ratios clearly has many advantages over the analysis of one element alone. Agricultural communities have lower $\delta^{15}N$ values than foodstuffs of marine origin and this is independent of the photosynthetic mode of the crops as reflected in the $\delta^{13}C$. The present state of knowledge on relations between bone collagen isotopic values and diet is shown in figure 14.9.[53] For example, the area within the triangle shows a diet of maize, gastropods, bivalves, and terrestrial herbivores that ate C_3 plants for tropical island dwellers.

Combined radiocarbon and AMS dating and stable isotope analysis of human bone collagen from Mesolithic and Neolithic skeletons from Portugal showed a marked diet change just prior to 7000 BP, at the Mesolithic-Neolithic transition.[54] The transition from the Mesolithic to Neolithic was characterized by a massive change in societies, from hunting and gathering to settled agriculture and husbandry. This change should be reflected in the diet, with the Mesolithic diet more varied and consisting of a food mix of marine and terrestrial origins. With the development of a settled society, the Neolithic diet, based largely on agriculture, should be more dependent on terrestrial food sources.

Comparing age of sample to nitrogen and carbon isotopic ratios in bone collagen revealed highly variable ratios for the Mesolithic, suggesting a varied diet, but by about 7000 BP and for the entire Neolithic, a uniform diet based on terrestrial sources was the norm. Correlating $\delta^{15}N$ to $\delta^{13}C$ showed a linear relationship. This implies that the marine component of the Mesolithic diet was isotopically homogeneous and consisted of either a single species or a well-defined mix of several species. The proposed end points of the trend line are

	$\delta^{13}C$	$\delta^{15}N$
Low end	−20.0%	8.5%
High end	−5.0%	13.0%

The low end is approximately the composition expected for the collagen of a human consuming a mixture of two possible foods: flesh of herbivores feeding on C_3 plants and C_3

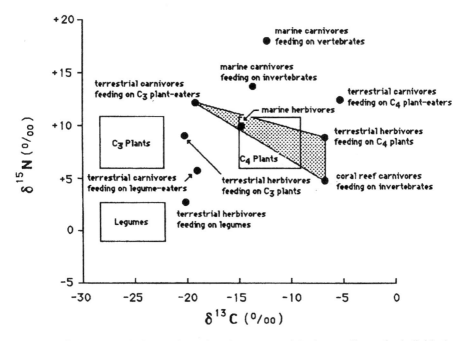

Figure 14.9 Carbon and nitrogen isotopic values expected for bone collagen for individuals feeding on a single type of food. The triangle outlines theoretical tropical island dwellers (see text for explanation). Also shown are the fields for C_3 and C_4 plants and legumes (modified from DeNiro, 1987; DeNiro and Hastorf, 1985). Diagram is highly schematic; points shown should be areas.

plants themselves.[55] This value represents a basically terrestrial diet. The upper value represents consumers of a wholly marine diet.[56]

Van der Merwe and others have shown how dietary information obtained from isotopic analysis can be used to help police the illegal ivory trade. The $\delta^{13}C$ and $\delta^{15}N$, as well as $^{87}Sr/^{86}Sr$ isotopic ratios, analyzed in ivory tusks of elephants can pinpoint their diet, and from that information, their environment (i.e., rain forest, savannah, desert) can be determined. Tusks seized as contraband had distinctive isotopic signatures. These signatures were similar to those of elephant populations living in different parts of Africa.[57]

Ivory and bone samples were obtained from more than 100 adult elephants from over 20 game refuges in 10 African countries. Carbon and nitrogen extracted from bone and from ivory dentine showed no significant differences. Some of the results of analysis are shown in table 14.4. The results turned out as predictable from the different regions of Africa. The $\delta^{13}C$ is controlled by a diet of browsing (preferred by elephants) or by grazing of grasses (if they are pushed). The first is a C_3 pathway to convert CO_2 to sugars; tropical grasses use the more efficient C_4 pathway. C_3 discriminates against ^{13}C, so trees and shrubs have lower ratios than C_4. Thus the C ratios will give an indication of the amount of grass versus leaves eaten by elephants, showing the environment where they lived.

Table 14.4 Isotopic Analysis of Collagen from Elephant Bone and Dentine

Country	Region	$\delta^{13}C$	$\delta^{18}N$	$^{87}Sr/^{86}Sr$
Liberia	Sapo	−26.80	7.77	0.71191
Sierra Leone	Gola Forest	−24.00	8.80	0.73314
Burkina Faso		−21.60	5.20	0.71493
Kenya	Shimba Hills	−21.70	9.04	0.71238
	Tsavo	−16.10	12.58	0.71000
Malawi	Kasunga	−19.62	4.03	0.72683
Zambia	Luangwa Valley	−18.16	6.78	0.72155
Botswana	(various)	−19.22	8.66	0.72126
	Tuli Block	−17.20	10.80	0.70649
Namibia	Etosha	−18.02	10.55	0.71646
South Africa	Kruger Park, east	−18.40	8.98	0.71171
	Kruger Park, west	−20.94	6.77	0.72406

Source: N. J. van der Merwe et al., 1990, Source-area determination of elephant ivory by isotopic methods, *Nature* 346:744–46.

1. In woodland savannahs, elephants eat primarily C_3 plants and have collagen $\delta^{13}C$ values around −21.5‰ (Kruger Park W, Shimba Hills)
2. In dense forests, isotopically light CO_2 is recycled under the forest canopy.[58] Thus, the normal atmospheric $\delta^{13}C$ value of about −7‰ will become more negative in the forest; the denser the forest, the more the atmosphere will be recycled, resulting in more negative values. Elephant $\delta^{13}C$ values more negative than −22‰ identify those from the rain forest environment (Sapo, Gola).
3. In unforested regions, the elephants will graze on grasses which follow the C_4 pathway and have higher $\delta^{13}C$ values than the plants of the rain forest. Bone collagen values more positive than −21.5‰ indicate a significant amount of grass in the elephants' diet (eastern Tsavo).
4. In arid regions, the browse diet may include an unusually high amount of CAM plants with mean $\delta^{13}C$ values of −17.7‰ rather than either C_3 (−26.5‰) or C_4 (−12.5‰) plants (northwestern Namibia, eastern Tsavo).

In addition to carbon isotopic ratios, $^{15}N/^{14}N$ correlates closely with rainfall and type of vegetation, giving another fix. Low $\delta^{15}N$ values indicate abundant rainfall and C_3 vegetation (bush/tree), which fixes its nitrogen from the air. High $\delta^{15}N$ values, on the other hand, indicate arid regions, with lower rainfall, and C_4 plants, which can fix nitrogen from the soil as well as from the atmosphere. This is seen in values from the desert regions of Damaraland and Kaokoveld in northwestern Namibia.

Strontium Isotopic Ratios

When the nitrogen and carbon values overlapped in the elephant ivory, $^{87}Sr/^{86}Sr$ was used to give information on the nature of the geological substrate. The Sr isotopic values depend on the age of the parent rock and the ratio of Sr/Rb in the rock. Young volcanic soils in the Rift Valley developed on basaltic rocks have $^{87}Sr/^{86}Sr$ similar to that of seawater, 0.70906 (Liwonde, Tuli Block). Soils developed on old granitic rocks, on the other hand, have values greater than 0.715 (Kruger Park, South Africa). Using all three isotopic ratios (table

14.4), 27 different areas of elephant habitation could be distinguished.

Strontium isotopic ratios in human bone and tooth enamel have also been used to investigate patterns of migration in the late prehistoric period (14th century) of east-central Arizona. Grasshopper Pueblo, a 500-room masonry pueblo, is underlain by Upper Paleozoic limestone which has relatively high Sr concentrations (325 ppm) and low $^{87}Sr/^{86}Sr$ ratios (0.70893). Comparative data were gathered from Walnut Creek, a site less than 20 km west of Grasshopper on a terrain of Precambrian intrusive rocks, some with relatively high Rb and exceptionally high $^{87}Sr/^{86}Sr$ ratios (0.71523 in soil). These great variations allowed for testing of migration models for Walnut Creek, such as proposed for a Hohokum migration.[59]

Measurable differences were found between the bones and the tooth enamel of the same individuals and among communities in the area. The Sr isotopic ratios in the bones at each community remained relatively constant, reflecting diet. For some, however, the teeth enamel ratios were different from bone. Since the teeth ratios are set early and bone will change according to the diet, a clear separation could be made between those who were locals (i.e., with teeth matching bone) and those who had migrated. In this latter case, teeth represented the original site of habitation and the bone the new migratory site. A typical migrant to Grasshopper Pueblo showed enamel $^{87}Sr/^{86}Sr$ = 0.71192 and bone = 0.71029, compared to a local with enamel = 0.71066 and bone = 0.71006. This isotopic evidence supported the contention made on the basis of ceramic variation and building episodes that the site was settled to a great extent by immigrants.

Paleoenvironments

Valuable information on past climates can be obtained from isotopic ratios, especially those isotopes in water, $^{18}O/^{16}O$ and D/H. These ratios in water will change over time, fractionated by evaporation conditions, which, in turn, are controlled by climate. The lighter isotopes, with a higher vapor pressure, will evaporate at a faster rate than the heavier. In other words, the process of evaporation discriminates against the heavier isotopes so that the vapor will always be "lighter" than the water source. The water will have a higher $\delta^{18}O$ and D/H than the vapor, and the higher the temperature during evaporation, the greater the differences. During condensation, the same effect will occur, with the condensate becoming "heavier" than the vapor. Thus, through time, the isotopic pool of water present in the sea, rainwater, and the atmosphere will change. This can be clearly measured in ice cores from continental glaciers of Greenland and Antarctica and in deep-sea drilling cores (figure 14.10).[60]

Oxygen

Oxygen isotopic fractionation is thus strongly temperature dependent. The isotopic ratios in the carbonate fraction of a shell can act as a paleothermometer. Shells that were harvested by inhabitants of ancient sites from nearby bodies of water can give valuable information on paleoclimatic conditions of the time.[61] Some possible sources of error in these determinations are the following:

1. Chemical and mineral composition of the shell material: organic protein, aragonite,

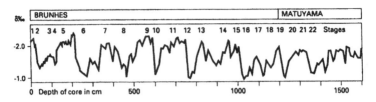

Figure 14.10 Oxygen isotopic trace from deep-sea cores (modified from Bell & Walker, 1992, fig. 2.27).

and dolomite may be present, each with different ^{18}O enrichment factors than calcite;

2. Isotopic composition of water at the time: should be known, as well as the effects of evaporation and precipitation, especially of small parental aquifers, and the presence or absence of global glaciation;
3. Postdepositional isotopic exchange: diagenetic and weathering processes involving re-crystallization can result in a change in the oxygen isotopic composition of the original carbonate.

Oxygen isotopic measurements have yielded information on the times of occupation of temporary sites. Measurements of incremental growth in mollusk shells have clearly shown seasonal influences due to temperature control of oxygen fractionation. N. J. Shackleton found differences from prehistoric midden sites in southern Africa of 1.4‰ in shell increments, suggesting a difference of about 7°C from winter to summer. The last growth in the shells took place in the winter, clear evidence that they were consumed at that time and that the sites were occupied only during the winter months. Thus, oxygen isotopic variation in entire mollusk shells can yield information on the yearly average and individual shell layers the seasonal variations in temperature. Seasonality problems are resolved by analyzing the growth increments in the shell.[62] The method and results are shown in figure 14.11, where variations in $\delta^{18}O$ between winter and summer temperatures in molluscan shells from New Guinea are clearly demonstrated.[63]

Long-term climatic variations have been worked out by oxygen isotopic studies of shells. Shells formed during glacial episodes yield higher ^{18}O values than interglacial ones. Details of sea-level change, as well as temperature, have been worked out in the South Pacific. At times of maximum glaciation, sea level was much lower and coral reefs were exposed. In addition, tectonic movements in the region, an active plate tectonic regime, have also uplifted coral reefs above sea level. Combining knowledge of sea level changes and oxygen isotope data allows a quantitative evaluation of ice volume and temperature. From figure 14.12, the Last Interglacial Phase includes a slow warming followed by a rapid cooling during the sea-level rise at VIIB, which has been attributed to an Antarctic surge. The cooling event, centered at the onset of the Ice Age at 120,000 BP, has unique characteristics, and its manifestation in the Tropics waned while the northern glaciation was initiated. Warm interstadials generated by brief retreats of incipient ice caps at 107,000 and 85,000 BP suggest that the subsequent chilling of the climate was confined to the high latitudes. In contrast, later oscillations in the expanded ice caps affected global climate, because they are correlated to the cold interstadials at 60,000, 45,000–40,000, and 28,000 BP. The relationship between higher sea level at times of chilling and ice cap formation is well shown

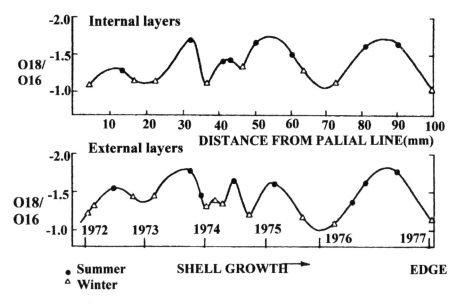

Figure 14.11 $\delta^{18}O$ isotope seasonal variation in mollusk shells of a modern giant clam *Tridacna gigas* from northeastern New Guinea (modified from Aharon, 1982).

by oxygen isotopes, radiometric dating, and relative sea-level changes. This type of information has been used to help work out the climatic history of the Holocene, yielding evidence that temperatures were 2°C higher than the present during the climatic optimum in Italy and 5°C above the present in Australia.[64]

Carbon

Holocene rock varnish formed in different environments shows the influence of plants on the $\delta^{13}C$ composition.[65] Plants with different photosynthetic pathways have distinct $\delta^{13}C$ signatures which apparently control the isotopic composition of the varnish. Isotopic determinations in late Pleistocene desert landforms revealed $\delta^{13}C$ ranging from -12 to $-19‰$, whereas samples from more humid environments averaged around $-23‰$. This was interpreted as reflecting the abundance of C_4 and CAM plants in warm, arid environments versus C_3 plants in cool, moist ones.

Deuterium

Natural water is a mixture of molecules of oxygen and hydrogen with different isotopic compositions. $H_2{}^{16}O$, $H_2{}^{18}O$, and $HD^{16}O$ are the most abundant and are the ones whose abundance is generally measured.[66] Their distribution in a water sample is a function of the geological history and environmental conditions undergone by the water. An important control is geography, including latitude, longitude, and elevation. Thus the D/H ratios as well as $^{18}O/^{16}O$ can give information on present and past climates. Isotopic ratios of D/H in wood

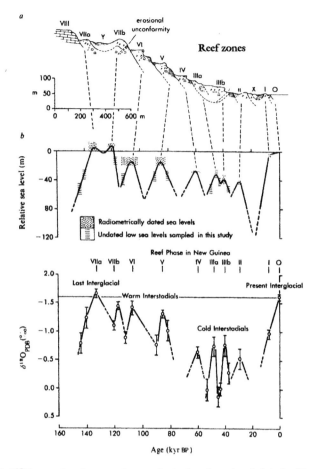

Figure 14.12 $\delta^{18}O$ record and supporting geological and sea-level data for New Guinea reefs at Sialum for the past 150,000 years (modified from Aharon, 1983).

reflect the climate in which the tree grew; D represents the water used during growth.[67] Because of fractionation effects, δD values in nitrated cellulose of trees average −20‰ below that of associated meteoric waters. C. J. Yapp and S. Epstein[68] examined samples of wood from a number of trees from North America and concluded that meteoric waters that precipitated over ice-free regions from 22,000 to 14,000 BP were enriched in D by an average of 19‰ compared to present-day precipitation in the same regions (figure 14.13).

Conclusions from the study were that the enrichment was caused by lower ocean temperatures and smaller temperature gradients between the ocean and the precipitation sites than exist at present. Winters were warmer and summers cooler during the Glacial age.

A problem in using D/H ratios in speleothems to determine paleoenvironment was demonstrated in a study in Iowa, West Virginia, Kentucky, and Missouri.[69] D/H ratios were measured in speleothems deposited over the past 250,000 years. At each site, glacial age waters were depleted in D relative to those of the Interglacial age. The average shift in

Figure 14.13 δD values in cellulose of modern plants. Inferred water δD values are underlined. (modified from Yapp and Epstein, 1977; in Bradley fig. 10.33).

Interglacial/Glacial precipitation in D over ice-free areas of east-central North America was about −12‰. To explain this apparent discrepancy, climatic controls over D/H ratios in meteoric precipitation were considered:

1. With the decrease in temperature, the D of modern precipitation at coastal marine and continental interior sites increases from 1.2 to 5.6‰.
2. During the transition from Glacial to Interglacial conditions, the D composition of the oceans decreased 4–13‰.
3. During Glacial times, Arctic air circulation patterns were displaced to the south, resulting in a southerly shift in iso-D contours.

4. Summer temperatures over interior North America were lower during Glacial times than at present. Estimates of the decrease range from 4–5°C to 11–14°C for 18,000 BP. During that time, average global sea-surface temperature decreased by 2°C. Thus, the summer temperature gradient between ocean source and continental precipitation site would have increased by 3–12°C.

Factors 1 and 4 would tend to decrease D in summer precipitation by 4–63‰; factor 2 would increase it by 4–13‰. Within the range of uncertainty, Glacial age continental precipitation might have differed from Interglacial waters by +7 to −63‰. The average Glacial/Interglacial difference in D of −12‰ found in this study is well within the limits imposed by these models.

Discrepancies in data between measurements of speleothems and ancient trees may also arise. This is due to the fact that trees use shallow groundwater, representing precipitation that fell during the growing season, whereas cave deposits use a deeper groundwater that may be hundreds or thousands of years old. In addition, in some areas, glacial meltwaters may have contributed to the deep groundwater.

Applications of stable isotopic geochemistry to solve geoarchaeological problems have been growing exponentially during the past 20 years. As different applications are suggested, the growth in research and publications in that field becomes explosive. No attempt has been made here to completely review the entire geological literature. Rather, we have introduced some of the important applications of stable isotopic ratios as artifact signatures or for determining paleodiet, paleoclimate, and paleoenvironment.

Notes

Foreword

1. C. Renfrew and J. S. Peacey, 1968, *Aegean Marble: A Petrological Study,* Annual British School in Athens 63:45–66.
2. B. Ashmole, 1970, *Aegean Marble: Science and Common Sense,* Annual British School in Athens 65:1–2.
3. J. Spier, 1990, Blinded with science: the abuse of science in the detection of false antiquities. *The Burlington Magazine* 132:623–31.
4. N. Herz and C. J. Vitaliano, 1983, Archaeological geology in the eastern Mediterranean, *Geology* 11:49–52.
5. M. R. Waters, 1992, *Principles of Geoarchaeology: A North American Perspective* (Tucson: University of Arizona Press).
6. B. M. Fagan, 1994, *Archaeology: A Brief Introduction* (New York: Harper Collins).

Chapter 1

1. G. Torraca, 1982, The scientist's role in historic preservation with particular reference to stone conservation, in *Conservation of Historic Stone Buildings and Monuments* (Washington: National Academy Press), pp. 13–21.
2. C. Renfrew, 1976, Archaeology and the earth sciences, in *Geoarchaeology: Earth Science and the Past,* ed. D. A. Davidson and M. L. Shackley (London: Duckworth), pp. 1–5.
3. E. R. Caley, 1967, The early history of chemistry in the service of archaeology, *Journal of Chemical Education* 44:120–23.
4. G. Rapp, Jr., and J. A. Gifford, 1982, Archaeological geology, *American Scientist* 70:45–53.
5. G. R. Lepsius, 1890, *Griechische Marmorstudien* (Berlin: Akademie der Wissenschaften).
6. H. S. Washington, 1898, The identification of the marbles used in Greek sculpture, *American Journal of Archaeology* 2:1–18.

7. M. Moltesen, N. Herz, and J. Moon, 1992, The Lepsius marbles, in *Ancient Stones: Quarrying, Trade, and Provenance,* ed. M. Waelkens, N. Herz, and L. Moens, Acta Archaeologica Lovaniensia Monographiae 4, Leuven: Leuven University Press, pp. 277–81.

8. M. R. Waters, 1994, *Principles of Geoarchaeology* (Tucson: University of Arizona Press).

9. N. Panin, S. Panin, N. Herz, and J. E. Noakes, 1983, Radiocarbon dating of Danube delta deposits, *Quaternary Research* 19:249–55.

10. J. C. Kraft, I. Kayan, and S. E. Aschenbrenner, 1985, Geologic studies of coastal change applied to archaeological settings, in *Archaeological Geology,* ed. G. Rapp, Jr., and J. A. Gifford (New Haven: Yale University Press), pp. 57–84.

11. Z. Goffer, 1980, *Archaeological Chemistry* (London: Wiley-Interscience), p. 356.

12. D. Kamilli and A. Steinberg, 1985, New approaches to mineral analysis of ancient ceramics, in *Archaeological Geology,* ed. G. Rapp, Jr., and J. A. Gifford (New Haven: Yale University Press), pp. 313–30.

13. H. Craig and V. Craig, 1972, Greek marbles: Determination of provenance by isotopic analysis, *Science* 176:401–3.

14. N. J. van der Merwe, 1982, Carbon isotopes, photosynthesis, and archaeology, *American Scientist* 70:596–606.

15. M. J. DeNiro and S. Epstein, 1978, Influence of diet on the distribution of carbon isotopes in animals, *Geochimica et Cosmochimica Acta* 42:495–506.

Chapter 2

1. M. R. Waters, 1992, *Principles of Geoarchaeology: A North American Perspective* (Tucson: University of Arizona Press); J. K. Stein, and W. R. Farrand, eds., 1985, *Archaeological Sediments in Context: Peopling the Americas,* Vol. 4 (Orono: Center for the Study of Early Man, Institute of Quaternary Studies, University of Maine).

2. D. A. Davidson, 1985, Geomorphology and archaeology, in *Archaeological Geology,* ed. G. Rapp, Jr., and J. A. Gifford (New Haven: Yale University Press), pp. 25–55.

3. K. W. Butzer and C. L. Hansen, 1968, *Desert and River in Nubia* (Madison: University of Wisconsin Press).

4. J. F. Hoffecker, 1988, Applied geomorphology and archaeological survey strategy for sites of Pleistocene age: An example from central Alaska, *Journal of Archaeological Science* 15:683–713.

5. Davidson, Geomorphology and archaeology.

6. David J. Hally, 1995, personal communication.

7. M. Blackburn and M. Fortin, 1994, Geomorphology of Tell 'Atij, northern Syria, *Geoarchaeology* 9:57–74.

8. W. H. Hoyt, J. C. Kraft, and M. J. Chrzastowski, 1990, Prospecting for submerged archaeological sites on the continental shelf: Southern Mid-Atlantic Bight of North America, in *Archaeological Geology of North America,* ed. N. P. Lasca and J. Donahue, Geological Society of America, Centennial Special Volume 4 (Boulder), pp. 147–60.

9. Ibid.

10. A. Holmes, 1978, *Principles of Physical Geology,* 3rd ed. (New York: John Wiley).

11. M. J. Stright, 1990, Archaeological sites on the North American continental shelf, in *Archaeological Geology of North America,* ed. N. P. Lasca and J. Donahue, Geological Society of America, Centennial Special Volume 4 (Boulder), pp. 439–65; L. R. Johnson, 1990, *Paleoshorelines* (Boca Raton: CRC Press).

12. C. B. DePratter and J. D. Howard, 1981, Evidence for a sea level low stand between 4500

and 2400 bp on the southeast coast of the United States, *Journal of Sedimentary Petrology* 51:1287–95.

13. Stright, Archaeological sites.

14. J. A. Clark, W. E. Farrell, and W. R. Peltier, 1978, Global changes in postglacial sea level: A numerical calculation, *Quaternary Research* 9 (3):265–87; T. R. Nardin, R. H. Osborne, D. J. Bottjer, and R. C. Scheidmann, Jr., 1981, Sea-level curves for Santa Cruz shelf, California continental borderland, *Science* 213 (4505):331–33.

15. L. A. Carbone, 1984, Early Holocene environments and paleoecological contexts on the central and southern California coast: Hunters and gatherers of early Holocene coastal California, in *Perspectives in California Archaeology*, ed. J. M. Erlandson and R. H. Colten (Los Angeles: Institute of Archaeology, UCLA), pp. 11–17.

16. J. G. Hutton, A. C. Hine, M. W. Evans, E. B. Osking, and D. F. Belknap, 1984, Influence of a karstified limestone surface on an open-marine, marsh-dominated coastline: West central Florida, in *Sinkholes: Their Geology, Engineering and Environmental Impact,* ed. Barry Beck (Boston: A. A. Belkema), pp. 35–42.

17. J. F. Donoghue, 1993, Late Wisconsin and Holocene depositional history, northeastern Gulf of Mexico, *Marine Geology* 112 (1):185–206.

18. Davidson, Geomorphology and archaeology.

19. J. Donahue and J. M. Adovasio, 1990, Evolution of sandstone rockshelters in eastern North America: A geoarchaeological perspective, in *Archaeological Geology of North America,* ed. N. P. Lasca and J. Donahue, Geological Society of America, Centennial Special Volume 4 (Boulder), pp. 231–252.

20. C. V. Constantinidis, J. Christdoulias, and A. I. Sofianos, 1988, Weathering processes leading to rockfalls at Delphi archaeological site, in *The Engineering Geology of Ancient Works, Monuments, and Historical Sites,* ed. P. G. Marinos and G. C. Koukis (Rotterdam: A. A. Balkema), vol. 1, pp. 201–6; J. A. Gusmão Filho, J. M. Justino da Silva, and J. F. Thomé Jucá, 1988, The movements in the hills of Olinda, Brazil, in *The Engineering Geology of Ancient Works, Monuments, and Historical Sites,* ed. P. G. Marinos and G. C. Koukis (Rotterdam: A. A. Balkema), vol. 1, pp. 191–200.

Chapter 3

1. I. S. Allison and D. F. Palmer, 1980, *Geology,* 7th ed. (New York: McGraw-Hill), p. 112.

2. B. W. Avery, 1990, *Soils of the British Isles* (Wellingford: C.A.B. International), pp. 37–80.

3. C. F. Marbut, 1935, Soils of the United States, in *Atlas of American Agriculture,* Part 3, (Washington, D.C.: U.S. Department of Agriculture).

4. J. K. Stein and W. R. Farrand, eds., 1985, Context and geoarchaeology: An introduction, in *Archaeological Sediments in Context: Peopling of Americas* (Orono: Center for the Study of Early Man, Institute for Quaternary Studies, University of Maine), vol. 1, p. 1.

5. E. P. Odum, 1954, *Principles of Ecology* (Philadelphia: W. B. Saunders Co.), p. 260.

6. J. K. Stein, 1985, Interpreting sediments in cultural settings, *supra,* p. 6; W. C. Krumbein and L. L. Gloss, 1963, *Stratigraphy and Sedimentation* (San Francisco: Freeman); H. Blatt, G. Middleton, and R. Murray, 1972, *Origin of Sedimentary Rocks* (Englewood Cliffs, N. J.: Prentice Hall); W. H. Twenhofel, 1950, *Principles of Sedimentation* (New York: McGraw-Hill).

7. Allison and Palmer, *Geology,* p. 114.

8. Ibid., pp. 187–88.

9. B. Arnold and C. Monney, 1978, Les amas de galets du village littoral d'Auvernier-Nord

(Bronze Final; lac de Neuchâtel): Études, geologique et archéologique, *Bulletin de la Société Neuchâteloise des Sciences Naturelle,* pp. 153–66.

10. J.-P. Portmann, 1954, Petrographie des moraines du glacir wurmier du Rhône dans la region des lacs subjurassions (Suisse), *Bulletin de la Societe de Neuchâtel Geographie* 51(5):13–55.

11. P. Hadorn, 1995, *Saint-Blaise/Bains des Dames, 1, Palynolgie d'un site néolithique et histoire de la végétation des derniers 16000 ans,* Archéologie Neuchâteloise 18 (Neuchâtel: Musée Cantonal d'Archéologie), pp. 58–62, 115.

12. J.-C. Gall, 1983, *Ancient Sedimentary Environments and the Habitats of Living Organisms: Introduction to Palaeoecology* (Berlin: Springer-Verlag), p. 51.

13. R. L. Folk, 1954, The distribution between grain-size and mineral composition in sedimentary rock nomenclature, *Journal of Geology* 62(4):344–59.

14. M. E. Tauggi and R. R. Andrier, 1971, Micromorphological recognition of paleosolic features in sediments and sedimentary rocks, in *Paleopedology: Origin, Nature, and Dating of Paleosols,* ed. Dan H. Yaalon (Jerusalem: International Society of Soil Science and Israel Universities Press), p. 163.

15. T. Gé, M.-A. Courty, W. Matthews, and J. Wattez, 1993, Sedimentary formation processes of occupation surfaces, in *Formation Processes in Archaeological Context,* ed. P. Goldberg, P. T. Nash, and M. D. Petraglia, Monographs in World Archaeology, no. 17 (Madison: Prehistory Press), pp. 149–63.

16. M. B. Schiffer, 1976, *Behavioral Archaeology* (New York: Academic Press), pp. 27–39; M. B. Schiffer, 1987, *Formation Processes of the Archaeological Record* (Albuquerque: University of New Mexico Press).

17. F. Bordes, 1975, Sur la notion de sol d'habitatien préhistorie paléolithique, *Bulletin de la Société Préhistorique Francaise* 72:139–44; F. Bordes, 1972, *A Tale of Two Caves* (New York: Harper and Row); F. David, M. Julein, and C. Karlin, 1973, Approche d'un niveau archéologique en sédiment homogene, in *L'homme, hier et aujourd'hui,* ed. Cujas (Paris: C.N.R.S.), pp. 65–72; H. L. Movius, Jr., 1977, *Excavation of Abri Pataud, Les Eyzies,* Stratigraphy. American School of Prehistoric Research, bulletin 31 (Harvard: Peabody Museum).

18. Gé et al., Sedimentary formation processes, p. 150.

19. Schiffer, *Behavioral Archaeology.*

20. Farrand, Rockshelters and cave sediments, pp. 21–39.

21. A. Sjöberg, 1976, Phosphate analysis of anthropic soils, *Journal of Field Archaeology* 3:447–54.

22. T. Webb III and R. A. Bryson, 1972, Late and post-glacial change in the northern midwest USA: Quantitative estimates derived from fossil pollen spectra by multivariate statistical analysis, *Quaternary Research* 2:20–215. See also V. M. Bryant, Jr., and R. G. Holloway, 1983, The role of palynology in archaeology, in *Advances in Archaeological Method and Theory* (New York: Academic), vol. 6, p. 208.

23. K. Faegri and J. Iversen, 1975, *Textbook of Pollen Analysis,* 3rd ed. (New York: Hafner Publishing), p. 256.

24. V. M. Bryant, Jr., 1989, Pollen: Nature's fingerprints of plants, in *1990 Yearbook of Science and the Future* (Chicago: Encyclopedia Britannica, Inc.), pp. 96–97.

25. Ibid., p. 104.

26. P. V. Glob, 1969, *The Bog People: Iron Age Man Preserved* (Ithaca: Cornell University Press).

27. J. Iversen, 1956, Forest clearance in the Stone Age, *Scientific American* 194(3):36–41; W. G. Solheim II, 1972, An earlier agricultural revolution, *Scientific American* 226:34–41; J. R. Harlan and D. Zohany, 1966, Distribution of world wheats and barley, *Science* 153:1074–80;

A. Leroi-Gourham, 1969, Pollen grains of Gramineae and Cerealia from Shendiar and Zawi Chemi, in *The Domestication and Exploitation of Plants and Animals,* ed. P. Ucko and G. Dimbleby (Chicago: Aldine), pp. 143–48; M. Van Campo and R. Cogne, 1960, Palynologie et géomorphologie dans le sud tunésian, *Pollen et Spores* 2:275–84; E. M. Van Zinderen Bakker, 1962, A pollen diagram from Equatorial Africa (Cherangani, Kenya), *Nature* 194:201–3.

28. V. M. Bryant, Jr., 1978, Palynology: A useful method for determining paleoenvironments, *Texas Journal of Science* 30:25–42.

29. R. H. Tshudy, 1969, Relationship of palynomorphs to sedimentation, in *Aspects of Palynology,* ed. R. H. Tshudy and R. A. Scott (New York: John Wiley and Sons), pp. 79–96.

30. A. J. Havinga, 1964, Investigations into the differential corrosion susceptibility of pollen and spores, *Pollen et Spores* 6:621–35.

31. Havinga, Differential corrosion susceptibility.

32. J. Overpeck, 1992, Mapping eastern North American vegetation changes of the past 18 ka: No analogs and the future, *Geology* 20:1071–74; C. R. Janssen, 1970, Problems in the recognition of plant communities in pollen diagrams, *Vegetatio* 20:187–95; H. J. Birks, T. Webb III, and A. A. Beitz, 1975, Numerical analysis of pollen samples from central Canada: A comparison of methods, *Review of Paleobotany and Palynology* 20:133–69; P. E. O'Sullivan and D. Riley, 1974, Multivariate numerical analysis of surface pollen spectra from a native Scots pine forest, *Pollen et Spores* 16:239–64; A. M. Soloman et al., 1980, Testing a simulation model for reconstruction of prehistoric forest-stand dynamics, *Quaternary Research* 14:275–93.

33. Bryant and Holloway, The role of palynology in archaeology, pp. 204–5.

34. F. A. Barkley, 1934, The statistical theory for pollen analysis, *Ecology* 15:283–89.

35. Bryant and Holloway, The role of palynology in archaeology.

36. H. Richard, 1993, Palynological micro-analysis in Neolithic lake dwellings, *Journal of Archaeological Science* 20:241–62.

37. C. Kunth, 1826, Examen botanique, in *Catalogue raison et historique de antiquités découvertes en Égypte* (Paris: J. Passalacqua).

38. O. Heer, 1865, Die Pflanzender Pfahlbauten, *Neujahr-blatt für Naturforschung Gesellschaft* (Zurich) 68:1–54.

39. J. M. Renfrew, 1973, *Paleoethnobotany* (New York: Columbia University Press).

40. V. H. Jones, 1957, Botany, in *The Identification of Non-artifactual Archaeological Materials,* ed. W. Taylor, National Academy of Sciences—National Research Council Publication, no. 565 (Washington, D.C.: National Academy of Sciences), pp. 35–38.

41. V. M. Bryant, Jr., and D. G. Williams, 1975, The coprolites of man, *Scientific American* 232:100–105.

42. G. F. Fry, 1977, *Analysis of Prehistoric Coprolites from Utah,* University of Utah Anthropological Papers, no. 97 (Salt Lake City: University of Utah Press); A. K. G. Jones, 1986, Parasitological examination of Lindow Man, in *Lindow Man: The Body in the Bog,* ed. I. Stead, J. Bourke, and D. Brothwell (London: British Museum Publications), pp. 136–39; K. J. Reinhard, 1990, Archaeoparasitology in North America, *American Journal of Physical Anthropology* 82:142–63; K. J. Reinhard and V. M. Bryant, Jr., 1992, Coprolite analysis: A biological perspective on archaeology, in *Archaeological Method and Theory,* vol. 4., ed. Michael B. Schiffer (Tucson: University of Arizona Press), p. 250.

43. Reinhard and Bryant, Coprolite analysis, p. 263.

44. D. R. Piperno and D. M. Pearsall, 1993, Current research in phytoliths: Applications in archaeology and paleoecology, in *MASCA Research Papers in Science and Archaeology,* ed. D. M. Pearsall and D. R. Piperno (Philadelphia: University Museum of Archaeology and Anthropology), vol. 10, p. 9.

45. H. C. Schellenberg, 1908, Wheat and barley from the North Kurgan, Anau, in *Explorations*

in Turkestan, Expedition of 1904: Prehistoric Civilizations of Anau; Origins, Growth and Influences of Environment, Vol. 2, ed. R. Pumpelly (Washington, D.C.: Carnegie Institution), pp. 471–73.

46. F. Netolitzky, 1900, Mikroskopische Untersuchung Ganzlich verkohlter vorgeschicht-licher Nahrungsmitted aus Tirol, *Zeitschrift für Untersuchung der Nahrungs-und Genussmittel* 3:401–7; F. Netolitzky, 1914, Die Hirse aus antiken Funden, *Sitzbuch der Kaiserliche Akademie für Wissenschaft der Mathematisch-Naturwissen-Schaften* 123:725–59.

47. V. M. Bryant, Jr., 1993, Phytolith research: A look toward the future, in *MASCA Research Papers in Science and Archaeology,* ed. D. M. Pearsall and D. R. Piperno (Philadelphia: University Museum of Archaeology and Anthropology), vol. 10, pp. 175–76.

48. H. Helbaek, 1959, Domestication of food plants in the Old World, *Science* 130:365–73; H. H. Helbaek, 1961, Studying the diet of ancient man, *Archaeology,* 14:95–101; H. Helbaek, 1969, Paleo-ethnobotany, in *Science in Archaeology,* ed. D. Brothwell, E. Higgs, and G. Clarke (New York: Basic Books), pp. 177–185; N. Watanabe, 1955, Ash in archaeological sites, *Rengo-Taikai Kiji: Proceedings of the Joint Meeting of the Anthropological Society of Nippon and Japan Society of Ethnology* (in Japanese) 9:169–71; N. Watanabe, 1968, Spodographic evidence of rice from prehistoric Japan, *Journal of the Faculty of Science of the University of Tokyo,* Section V, 3:217–35.

49. D. R. Piperno, 1984, A comparison and differentiation of phytoliths from maize and wild grasses: Use of morphological criteria, *American Antiquity* 49(2):361–83; D. R. Piperno, 1988, *Phytolith Analysis: An Archaeological and Geological Perspective* (New York: Academic); D. M. Pearsall, 1978, Phytolith analysis of archaeological soils: Evidence for maize cultivation in Formative Ecuador, *Science* 199:177–78; D. M. Pearsall, 1989, *Paleoethnobotany: A Handbook of Procedures* (New York: Academic); I. Rovner, 1971, Potential of opal phytoliths for use in paleoecological reconstruction, *Quaternary Research* 1:343–59; I. Rovner, 1983, Plant opal phytolith analysis: Major advances in archaeobotanical research, in *Advances in Archaeological Method and Theory,* ed. Michael B. Schiffer (New York: Academic), vol. 6, pp. 225–66.

50. S. C. Mulholland, 1993, A test of phytolith analysis at Big Hidatsa, North Dakota, in *MASCA Research Papers in Science and Archaeology,* ed. D. M. Pearsall and D. R. Piperno (Philadelphia: University Museum of Archaeology and Anthropology), vol. 10, p. 138; D. A. Brown, 1984, Prospects and limits of a phytolith key for grasses in the central United States, *Journal of Archaeological Science* 11:345–68; A. M. Rosen, 1993, Phytolith evidence for early cereal exploitation in the Levant, in *MASCA Research Papers in Science and Archaeology,* ed. D. M. Pearsall and D. R. Piperno (Philadelphia: University Museum of Archaeology and Anthropology), vol. 10, p. 163; J. C. Russ, and I. Rovner, 1989, Stereological identification of opal phytolith populations from wild and cultivated *Zea mays, American Antiquity,* 54(3): 784–792.

51. S. Bozarth, 1993, Biosilicate assemblages of boreal forests and aspen parklands, in *MASCA Research Papers in Science and Archaeology,* ed. D. M. Pearsall and D. R. Piperno (Philadelphia: University Museum of Archaeology and Anthropology), vol. 10, p. 96.

52. E. G. Garrison, 1991, Archaeogeophysical studies at Cortaillod and St. Blaise, Lac de Neuchâtel, Switzerland, *Geoarchaeology: An International Journal* 6(2):111–32.

53. W. Ludi, 1935, Das grosse Moos in Westschweizerischen Seelande und die Geschichte seiner Entstenung, *Veröffentlichungen des Geobotanischen Institutes Rübel in Zürich* 11; R. Müller, 1973, in Les nivaux des lacs du Jura: Contribution aux researche archéologiques de la 2e correction des eaux du Jura, ed. A. Schwab and R. Müller (Fribourg: Editions Universitaires), pp. 154–76; R. Kasser, 1975, Yverdon: Histoire d'un sol et d'un site avec la cité qu'ils ont fait naître, *Eburodunum* 1; J. L. Brochier, 1986, La séquence sédimentaire lacustre, in

Cortaillod-Est, un village du Bronze final, 4, Nature et environment, Archéologie neuchâteloise 4 (Saint-Blaise: Éditions du Ruau), p. 31; B. Moulin, 1991, *Hauterive-Champréveyres, 3, La dynamique sédimentaire et lacustre durant le Tardiglaciaire et le Postglaciaire,* Archéologie neuchâteloise 9 (Saint-Blaise: Éditions du Ruau), p. 100; Garrison, Archaeogeophysical studies.

54. Brochier, Séquence sédimentaire lacustre, pp. 11–12.

55. B. Arnold, 1990, Cortaillod-Est et les villages du lac de Neuchâtel au Bronze final, in *Structure de l'habitat et proto-urbanisme, 6,* Archéologie neuchâteloise 6 (Saint-Blaise: Éditions du Ruau), pp. 20–23.

56. Brochier, Séquence sédimentaire lacustre, pp. 11–35.

57. Garrison, Archaeogeophysical studies.

58. Brochier, Séquence sédimentaire lacustre, p. 31.

59. It is important to note at this point that at the time of this survey, 1987, subbottom surveys in extremely shallow water were not easily accomplished. The system used to obtain the seismic records at Cortaillod was a highly modified Edo Western pulse-echo subbottom profiler system (4–7 kHz) with an early digitally controlled chart recorder (EPC Laboratories model 614). The large, heavy, cone-shaped transducer greatly reduced acoustical "ringing" in the outgoing pulse, thus providing a highly efficient receiver of the relatively weak returning pulses from the shallow sediments. The use of this system and a carefully modulated, low-power transmittal pulse and postreturn digital expansion of the first few microseconds of the signal was successful in the very shallow water depths. Today, only a few years later, digital processing technology in systems such as multiple-frequency CHIRP and swept frequency sonars has made such shallow-water surveys routine.

60. Moulin, Hauterive-Champréveyres, 3.

61. J. Affolter, M.-I. Cattin, D. Leesch, P. Morel, N. Plumettaz, N. Thew, and G. Wendling, 1994, Monruz-une nouvelle station magdalénienne au bord du lac de Neuchâtel, *Archéologie Suisse* 17(3):94–104; V. Markgraf, 1974, Paleoclimatic evidence derived from timberline fluctuations, in *Les méthods quantitative d'étude des variations du climat an cours du Pléistocène* (Gif-sur-Yvette, 5–9 juin 1973), Colloques Internationaux du CNRS 219 (Paris: C.N.R.S.), pp. 67–77; H. Kerschner, 1980, Outlines of climate during the Egesen advance (Younger Dryas, 11,000–10,000 BP) in the central Alps of the Western Tyrol, Austria, *Zeitschrift für Gletscherkunde und Glazialgeologie* 16(2):229–40.

62. U. Eiches, U. Siegenthalen, and S. Wagmüller, 1981, Pollen and oxygen isotope analyses on Late and Post Glacial sediments of the Tourbière de Chirens (Dauphiné, France), *Quaternary Research* 15:160–70.

63. K. J. Hsü, F. Giovanoli, and K. R. Kelts, 1984, Introduction: The Zübo 80 Project, *Quaternary Geology of Lake Zürich: An Interdisciplinary Investigation of Deep-Lake Drilling,* ed. K. J. Hsü and K. R. Kelts (Stuttgart: E. Schweizerbartsche Verlagbuchandlung), pp. 1–2.

64. K. J. Hsü, 1984, Quaternary geology of the Lake Zürich region, in *Quaternary Geology of Lake Zürich,* p. 203.

65. C. Sidler, 1984, Palynological investigations of Zübo sediments, in *Quaternary Geology of Lake Zürich,* pp. 112–13.

66. M. Magny and P. Olive, 1981, Origine climatique des variations du riveau du lac Léman au cours de l'Holocène: La crise de 1700 à 700 BC, *Archives Suisses d'Anthropologie Générale* 45(2):159–69.

67. Hadorn, *Saint-Blaise/Bains des Dames, 1.*

68. Avery, *Soils of the British Isles,* p. 385.

69. Avery, *Soils of the British Isles,* p. 387.

70. P. Rowsome, 1987, Roman street life, *Archaeology Today* 8(9):24.

71. J. Schofield, 1983, Piecing it all together, *Popular Archaeology* 5(4):9.

72. E. A. Bettis, 1992, Soil morphological properties and weathering zone characteristics as age indicators in Holocene alluvium in the upper Midwest, in *Soils in Archaeology: Landscape Evolution and Human Occupation,* ed. V. T. Holiday (Washington, D.C.: Smithsonian Institution Press), pp. 119–44.

73. S. Limbrey, 1979, *Soil Science and Archaeology* (London: Academic Press).

74. R. B. Cate, Jr., 1964, New data on the chemistry of submerged soils, *Economic Geology* 59:161–62.

Chapter 4

1. D. H. Thomas, 1991, *Archaeology: Down to Earth* (Fort Worth: Harcourt, Brace, Jovanovich), pp. 102–4.

2. A. Bartsiokas and A. P. Middleton, 1992, Characterization and dating of recent and fossil bone by x-ray diffraction, *Journal of Archaeological Science* 19:63–72.

3. P. Hille, K. Mais, G. Rabeder, N. Vavra, and E. Wild, 1981, Über Aminosäuren- und Stickstoff/Fluor-Datierung fossiler Knochen aus österreichischen, *Höhle* 32:74–91.

4. Bartsiokas and Middleton, Characterization and dating of recent and fossil bone by x-ray diffraction.

5. M. R. Schurr, 1989, Fluoride dating of prehistoric bones by ion selective electrode, *Journal of Archaeological Science* 16:265–70.

6. K. P. Oakley, 1969, Analytical methods of dating bones, in *Science in Archaeology,* ed. D. Brothwell and E. Higgs (London: Thames and Hudson), pp. 35–45; R. Millar, 1972, *The Piltdown Men* (London: Gollanz).

7. K. L. Feder, 1990, Dawson's Dawn Man: The hoax at Piltdown, in *Frauds, Myths, and Mysteries* (Mountain View, Calif.: Mayfield Publishing), pp. 26–35.

8. P. E. Hare and P. H. Abelson, 1968, Racemization of amino acids in fossil shells, *Carnegie Institution of Washington Yearbook* 65:362–664.

9. J. Shreeve, 1992, The dating game, *Discover* 82:76–83.

10. Hare and Abelson, Racemization of amino acids in fossil shells.

11. J. L. Bada, B. P. Luyendyk, and J. B. Maynard, 1970, Marine sediments: Dating by the racemization of amino acids, *Science* 170:730–32.

12. T. D. Dillehay and D. J. Meltzer, eds., 1991, *First Americans: Search and Research* (Boca Raton: CRC Press); V. M. Bryant, Jr., 1992, In search of the first Americans, in *1993 Yearbook of Science and the Future* (Chicago: Encyclopaedia Britannica), pp. 8–27; J. F. Hoffecker, W. R. Powers, W. Rodgers, and T. Goebel, 1993, The colonization of Beringia and peopling of the New World, *Science* 259:46–53.

13. J. L. Bada, K. A. Kuenholden, and E. Peterson, 1973, Racemization of amino acids in bones, *Nature* 245:308–10; J. L. Bada and P. M. Helfman, 1975, Amino acid racemization of fossil bones, *World Archaeology* 7:160–73.

14. J. L. Bada, 1972, The dating of fossil bones using the racemization of isoleucine, *Earth and Planetary Science Letters* 15:223–31.

15. Bada and Helfman, Amino acid racemization of fossil bones.

16. J. L. Bada, R. Gillespie, J. A. J. Gowlett, and R. E. M. Hedges, 1984, Accelerator mass spectrometry radiocarbon ages of amino acid extracts from California paleoindian skeletons, *Nature* 312:442–44.

17. Shreeve, The dating game.

18. B. A. Blackwell, H. P. Schwarcz, and A. Debénath, 1983, Absolute dating of hominids and paleolithic from the cave of La Chaise-de-Vouthon (Cherente), France, *Journal of Archaeological Science* 10:493–513; B. A. Blackwell and H. P. Schwarcz, 1993, Archaeo-

chronology and scale, in *Effects of Scale on Archaeological and Geoscientific Perspectives,* ed. J. K. Stein and A. R. Linse, Geological Society of America Special Paper 283 (Boulder), p. 56.

19. Shreeve, The dating game, pp. 76–83.

20. C. A. Prior, P. J. Ennis, E. A. Noltmann, P. E. Hare, and R. E. Taylor, 1986, Variations in D/L aspartic acid ratios in bones of similar age and temperature history, *Proceedings of the 24th International Archaeometry Symposium,* ed. J. S. Olin and M. J. Blackman (Washington: Smithsonian Institution).

21. G. H. Miller, 1993, Chronology of hominid occupation at Bar Tarfawi and Bir Sahara East, based on the epimerization of isoleucine in ostrich eggshells, in *Egypt during the Last Interglacial,* ed. F. Wendorf, R. Schild, A. E. Close, and associates (New York: Plenum Press), pp. 218–23.

22. Hare and Abelson, Racemization of amino acids in fossil shells.

23. D. Q. Bowen, G. A. Sykes, A. Reeves, G. H. Miller, G. T. Andrews, J. S. Brew, and P. E. Hare, 1985, Amino acid geochronology of raised beaches in southwest Britain, *Quaternary Science Reviews* 4:279–318.

24. E. N. Powell, J. A. King, and S. Boyles, 1991, Dating time-since-death of oyster shells by the rate of decomposition of the organic matrix, *Archaeometry* 33:51–68; E. N. Powell, A. Logan, J. Stanton, R. J., D. J. Davies, and P. E. Hare, 1989, Estimating time-since-death from the free amino acid content of the mollusc shell: A measure of time averaging in modern death assemblages? Description of the technique, *Palaios* 4:16–31.

25. Powell, King, and Boyles, Dating time-since-death of oyster shells.

26. I. Friedman and R. L. Smith, 1960, A new dating method using obsidian, part 1: The development of the method, *American Antiquity* 25:476–93.

27. I. Friedman, R. L. Smith, and D. Clark, 1963, Obsidian dating, in *Science in Archaeology,*

28. J. J. Mazer, C. M. Stevenson, W. L. Ebert, and J. K. Bates, 1991, The experimental hydration of obsidian as a function of relative humidity and temperature, *American Antiquity* 56:504–13.

29. I. Friedman, F. Tembour, F. L. Smith, and G. I. Smith, 1994, Is obsidian hydration dating affected by relative humidity? *Quaternary Research* 41:185–90.

30. I. Friedman and F. Trembour, 1983, Obsidian hydration dating update, *American Antiquity* 48:544–47.

31. Friedman, Smith, and Clark, Obsidian dating.

32. I. Friedman and W. Long, 1970, Hydration rate of obsidian, *Science* 191:347–52.

33. J. W. Michels, 1986, Obsidian hydration dating, *Endeavor* 10:97–100.

34. R. Lee, D. Leich, T. Tombrello, J. Ericson, and I. Friedman, 1974, Obsidian hydration profile measurements using a nuclear reaction technique, *Nature* 250:44–47.

35. Norris, using hydrogen profiling, where a ^{15}N ion beam measured the hydration layer of obsidian, reports a resolution to this level. Norris, S. 1995, New Methods in Hydration Dating: Results from Yautepec, Morelos, Mexico. Paper presented at the 60th Annual Meeting of the Society of American Archaeology. Minneapolis.

36. J. W. Michels, I. S. T. Tsong, and G. I. Smith, 1983, Experimentally derived hydration rates in obsidian dating, *Archaeometry* 25:107–17; C. M. Stevenson, J. Carpenter, and B. E. Scheetz, 1989, Obsidian dating: Recent advances in the experimental determination and application of hydration rates, *Archaeometry* 31:193–206; F. Trembour and I. Friedman, 1984, The present status of obsidian hydration dating, *Quaternary Dating Methods,* ed. W. C. Mahaney (Amsterdam: Elsevier), pp. 141–51; Michels, Obsidian hydration dating.

37. J. E. Francis, L. L. Loendorf, and R. I. Dorn, 1993, AMS radiocarbon and cation-ratio dating of rock art in the Bighorn Basin of Wyoming and Montana, *American Antiquity* 58(4):711–37.

38. R. I. Dorn, 1991, Rock varnish, *American Scientist* 79:542–53; P. R. Bierman, A. R. Gillespie, and S. Kuehnr, 1991, Precision of rock-varnish chemical analyses and cation-ratio ages, *Geology* 19:135–38; P. R. Bierman and A. R. Gillespie, 1991, Accuracy of rock-varnish chemical analyses: Implications for cation-ratio dating, *Geology* 19:196–99; C. D. Harrington and J. W. Whitney, 1987, Scanning electron microscope method for rock-varnish dating, *Geology* 15:967–70. R. I. Dorn, A. J. T. Jull, D. J. Donahue, T. W. Linick, and L. J. Toolin, 1989, Accelerator mass spectrometry radiocarbon dating of rock varnish, *Bulletin of the Geological Society of America* 101:1363–72.

39. K. G. Krauss and T. L. Ku, 1980, Desert varnish: Potential for age dating via uranium-series isotopes, *Journal of Geology* 88:95–100.

40. R. I. Dorn, 1983, Cation-ratio dating: A new rock varnish age-determination technique, *Quaternary Research* 20:49–73.

41. R. I. Dorn, 1989, Cation-ratio dating of rock varnish: A geographical assessment, *Progress in Physical Geography* 13(4):559

42. Francis, Loendorf, and Dorn, AMS radiocarbon and cation-ratio dating of rock art.

43. Ibid.

44. S. L. Reneau and C. D. Harrington, 1988, Age determinations for rock varnish formation within petroglyphs: Cation-ratio dating of 24 motifs from the Olary Region, South Australia, *Rock Art Research* 5(2):141–42.

45. P. Bierman and A. Gillespie, 1991, Range fires: A significant factor in exposure age determination and geomorphic surface evolution, *Geology* 19:641–44.

46. M. P. Lanteigne, 1991, Cation-ratio dating of rock-engravings: A critical appraisal, *Antiquity* 65:292–95.

47. Ibid.

48. Francis, Loendorf, and Dorn, AMS radiocarbon and cation-ratio dating of rock art.

49. K. G. Harry, 1995, Cation-ratio dating of varnished artifacts, *American Antiquity* 60(1):118–30.

50. Dorn, Cation-ratio dating: A new rock varnish age-determination technique; R. I. Dorn and D. S. Whitley, 1983, Cation-ratio dating of petroglyphs from the western Great Basin, *Nature* 302:816–18.

51. R. I. Dorn and D. H. Krinsley, 1992, Cation-leaching sites in rock varnish, *Geology* 20(11):1051.

Chapter 5

1. R. Siegel, 1993, Exploring mesoscopia: The bold new world of nanostructures, *Physics Today* 46(1):64–68.

2. A. O. Nier, 1936, A mass spectrographic study of the isotopes of argon, potassium, rubidium, zinc, and cadmium, *Physics Review.* 50:1041.

3. C. Patterson, 1956, Age of meteorites and the earth, *Geochimica et Cosmochemica Acta* 10:230–37.

4. M. Ivanovich and R. S. Harmon, eds., 1982, *Uranium Series Disequilibrium: Applications to Environmental Problems* (Oxford: Clarendon Press), pp. 104–44.

5. H. P. Schwarcz, 1986, Geochronology and isotopic geochemistry of speleothems, in *Handbook of Isotope Geochemistry: The Terrestrial Environment,* ed. J. C. Fontes and P. Fritz (Amsterdam: Elsevier), pp. 271–303; M. Gascoyne, 1992, Paleoclimate determination from cave calcite deposits, *Quaternary Science Reviews* 11(6): 609–32; J. A. Dorale, L. A. Gonzalez, M. K. Reagan, D. A. Pickett, M. T. Murell, and R. G. Baker, 1992, A high resolution record of

Holocene climate change in speleothem calcite from Cold Water Cave, northeast Iowa, *Science* 258:1626–30.

6. N. J. Shackleton and N. D. Opdyke, 1973, Oxygen isotope and paleomagnetic stratigraphy of equatorial Pacific Core 28–238: Oxygen isotope temperatures and ice volumes on a 10^5 year and 10^6 year scale, *Quaternary Research* 3:39–55.

7. H. P. Schwarcz and B. Blackwell, 1985, Uranium-series disequilibrium dating, in *Dating Methods of Pleistocene Deposits and Their Problems,* ed. N. W. Rutler (Toronto: Geological Association of Canada), Geoscience Canada Reprint Series 2, pp. 9–18.

8. M. J. Aitken, 1990, *Science-Based Dating in Archaeology* (London: Longman), p. 130.

9. T. Chen and S. Yuan, 1988, Uranium series dating of bones and teeth from Chinese palaeolithic sites, *Archaeometry* 30:59–76; A. M. Rae, M. Ivanovich, and H. P. Schwarcz, 1987, Absolute dating by uranium series disequilibrium of bones from the cave of La Chaise-de-Vouthon (Charente), *Earth Surface Processes and Land Forms* 12:543–50.

10. C. T. Williams and P. J. Potts, 1988, Element distribution maps in fossil bones, *Archaeometry* 30:237–47.

11. Aitken, *Science-Based Dating in Archaeology,* p. 131.

12. P. Lo Bello, G. Feraud, C. M. Hall, D. York, P. Lavina, and M. Bernet, 1987, $^{40}Ar/^{39}Ar$ step-heating and laser fusion dating of a Quaternary pumice from Neschers, Massif Central, France: The defeat of xenocrystic contamination, *Chemical Geology* (Isotope Geoscience) 66:61–77.

13. R. D. Dallmeyer, 1979, $^{40}Ar/^{39}Ar$ dating: Principles, techniques, and applications in orogenic terranes, *Lectures in Isotope Geology,* ed. E. Jäger and J. C. Hunziker (Berlin: Springer-Verlag), pp. 77–104.

14. $J = \dfrac{\exp(\lambda_{total}\, t_{monitor}) - 1}{(^{40}Ar/^{39}Ar)_{monitor}}$ where $t_{monitor}$ is a sample of known age.

15. R. C. Walter, 1994, Age of Lucy and the first family: Single-crystal $^{40}Ar/^{39}Ar$ dating of the Denen Dora and lower Kada Hadar members of the Hadar Formation, Ethiopia, *Geology* 22:6–10; C. C. Swisher, G. H. Curtis, T. Jacob, A. G. Getty, and A. Suprijo Widiasmoro, 1994, Age of the earliest known hominids in Java, Indonesia, *Science* 263:1118–21.

16. D. C. Johanson, T. D. White, and Y. Coppens, 1978, A new species of genus *Australopithecus* (Primates: Hominidae), *The Pliocene of Eastern Africa: Kirtlandia* 28:1–14.

17. M. H. Day, 1987, Lucy jilted? *Nature* 300:574.

18. R. C. Walter and J. L. Aronson, 1982, Revisions of K/Ar ages for the Hadar hominid site, Ethiopia, *Nature* 296:122–27.

19. I. McDougall, 1985, K-Ar and $^{40}Ar/^{39}Ar$ dating of the hominid-bearing Pliocene-Pleistocene sequence at Koobi Fora, Lake Turkhana, northern Kenya, *Geological Society of America Bulletin* 96:159–75.

20. M. Leakey and J. M. Harris, eds., 1987, *Laetoli: A Pliocene Site in Northern Tanzania* (Oxford: Clarendon Press).

Chapter 6

1. C. W. Naeser and N. D. Naeser, 1988, Fission-track dating of Quaternary events, in *Special Paper 227, Geological Society of America* (Boulder), pp. 1–11.

2. R. L. Fleisher, 1979, Where do nuclear tracks lead? *American Scientist* 67:194–203; C. W. Naeser, N. D. Briggs, J. D. Obradovich, and G. A. Izett, 1981, Dating of tephra, geochronology of Quaternary tephra deposits, in *Tephra Studies,* ed. S. Self and R. S. J. Sparks (Dordrecht: Reidel Publishing), pp. 13–47.

3. W. H. Huang and R. M. Walker, 1967, Fossil alpha-recoil tracks: A new method of age determination, *Science* 155:1103–6.

4. R. L. Fleischer, P. B. Price, and R. M. Walker, 1964, Techniques for geological dating of minerals by chemical etching of fission fragment tracks, *Geochimica et Cosmochimica Acta* 28:1705–14; R. L. Fleischer, P. B. Price, and R. M. Walker, 1964, Glass dating by fission fragment tracks, *Journal of Geophysical Research* 69:331–39; R. L. Fleischer, P. B. Price, and R. M. Walker, 1964, Fission track ages of zircons, *Journal of Geophysical Research* 69:4885–88; R. L. Fleischer, P. B. Price, and R. M. Walker, 1975, *Nuclear Tracks in Solids* (Berkeley: University of California Press).

5. N. Bohr, 1940, Velocity-range relation for fission fragments, *Physics Review,* 59:270–75.

6. Ibid.

7. Naeser and Naeser, Fission-track dating of Quaternary events, p. 2.

8. H. B. Lück, 1974, A plastic track detector with high sensitivity, *Nuclear Instruments Methods* 114:139–40.

9. E. G. Garrison, O. H. Zinke, and C. R. McGimsey III, 1975, Alpha-recoil tracks in archaeological ceramic dating, *Archaeometry* 20:39–46; D. Wolfman and T. M. Rolniak, 1979, Alpha-recoil track dating: Problems and prospects (Proceedings of the 18th International Symposium on Archaeometry and Archaeological Prospection, Bonn, 1978), *Archeo-physika* 10:512–21.

10. W. H. Huang and R. M. Walker, Fossil alpha-recoil tracks.

11. Wolfman and Rolniak, Alpha-recoil track dating.

12. Wolfman and Rolniak, Alpha-recoil track dating, p. 512.

13. M. S. Tite, 1981, *Methods of Physical Examinations in Archaeology* (London: Seminar Press).

14. C. Lalon and G. Valladas, 1984, Thermoluminescence dating, in *Nuclear Methods of Dating,* ed. Etienne Roth and Bernard Poty (Dordrecht: Kluwer Academic Publishers), pp. 243–45.

15. M. J. Aitken, M. S. Tite, and J. Reid, 1964, Thermoluminescence dating of ancient ceramics, *Nature* 202:1032; M. J. Aitken, D. W. Zimmerman, and S. J. Fleming, 1968, Thermoluminescence dating of ancient pottery, *Nature* 219:442; D. W. Zimmerman, 1971, Thermoluminescence using fine grains from pottery, *Archaeometry* 13:29–52; S. J. Fleming, 1970, Thermoluminescent dating: Refinement of the quartz inclusion method, *Archaeometry* 12:53–55.

16. Lalon and Valladas, Thermoluminescence dating, pp. 254–62; M. J. Aitken, 1985, *Thermoluminescence Dating* (London: Academic).

17. M. J. Aitken, 1990, Luminescence dating, *Science-based Dating in Archaeology* (London: Longman), pp. 262–66; Lalon and Valladas, Thermoluminescence dating, pp. 260–70; F. Albarède, A. Michard, and M. Cumey, 1989, The uranium-thorium-lead dating, in *Nuclear Methods of Dating,* ed. Etienne Roth and Bernard Poty (Dordrecht: Kluwer Academic Publishers), p. 121; Geyh and Schleicher, *Absolute Age Determination,* pp. 262–64.

18. M. J. Aitken, 1974, *Physics and Archaeology,* 2nd ed. (Oxford: Clarendon Press).

19. Zimmerman, Thermoluminescence using fine grains from pottery; S. R. Sutton and D. W. Zimmerman, 1976, Thermoluminescent dating using zircon grains from archaeological ceramics, *Archaeometry* 18:125–34.

20. S. J. Fleming, 1966, Study of thermoluminescence of crystalline extracts from pottery, *Archaeometry* 9:170–73.

21. Zimmerman, 1976, *supra;* Geyh and Schleicher, *Absolute Age Determination,* p. 265.

22. R. H. Templer, 1986, Auto-regenerative TL dating of zircon inclusions, *Radiation Protection Dosimetry* 17:235–39; B. W. Smith, 1988, Zircon from sediments: A combined OSL and TL auto-regenerative dating technique, *Quaternary Science Reviews* 7:401–6.

23. G. Guérin, 1983, La thermoluminescence des plagiocases, méthode de datation du volcanisme, Applications du domaine volcanique français: Chaîne des Puys, Mont-Dore et Cezaillier, Bas Vivarais (Thèse Dr. Sc. Université Paris VI).

24. H. Y. Göksu, J. H. Fremlin, H. T. Irwin, and R. Fryxell, 1974, Age determination of burnt flint by a thermoluminescence method, *Science* 183:651–54; A. G. Wintle and M. J. Aitken, 1977, Thermoluminescence dating of burnt flint: Application to a Lower Palaeolithic site, Terra Amata, *Archaeometry* 19:111; H. Valladas, 1978, Thermoluminescence dating of burnt stones from prehistoric sites, *PACT* 2:180.

25. A. G. Wintle, 1987, Thermoluminescence dating of loess, *Catena Supplement* 9:103–15.

26. V. A. Zubakov, ed., 1973, *Chronology of Pleistocene and Climatic Stratigraphy* (in Russian) (Leningrad: Geographical Society of the USSR, Pleistocene Commission), p. 286.

27. R. P. Feynman, P. B. Leighton, and M. Sands, 1964, *The Feynman Lectures on Physics* (Palo Alto: Addison-Wesley).

28. J. E. Wertz and J. R. Bolton, 1972, *Electron Spin Resonance, Elementary Theory, and Practical Applications* (New York: McGraw-Hill), pp. 16–17.

29. C. P. Poole, Jr., 1967, *Electron Spin Resonance* (New York: Interscience—John Wiley and Sons), p. 825.

30. Wertz and Bolton, *Electron Spin Resonance,* p. 154.

31. M. Ikeya, 1988, Dating and radiation dosimetry with electron spin resonance, *Magnetic Resonance Review* 13:19–134.

32. Ibid.

33. E. J. Zeller, P. W. Levy, and P. L. Mattern, 1967, Geological dating by electron spin resonance: Proceedings of the Symposium on Radioactive Dating and Low-Level Counting. *International Atomic Energy Agency* (Vienna), pp. 531–40; Garrison et al., ESR dating of ancient flints; E. G. Garrison, 1989, Characterization of an ESR geochronological dating center in flints, *Journal of Physics and Chemistry of Minerals* 16:767–73; D. R. Griffiths, N. J. Seeley, H. Chandra, and M. C. R. Symons, 1983, ESR dating of heated chert, *PACT* 9:399–409.

34. M. Stapelbroek, D. L. Griscom, E. J. Friebele, and G. H. Sigel, Jr., 1979, Oxygen-associated trapped holes in high-purity fused silicas, *Journal of Non-crystalline Solids* 32:313; D. L. Griscom and E. J. Friebele, 1981, Fundamental defect centers in glass: ^{29}Si hyperfine structure of the non-bridging oxygen hole center and peroxy radical in α-SiO_2, *Physics Review B,* 8:4896–98.

35. K. S. V. Nambi, 1979, On ESR dating of minerals, *Japan Journal of Applied Physics* 18:2319–20; G. V. Robins, N. J. Seeley, D. A. C. McNeil, and M. R. C. Symons, 1978, Identification of ancient heat treatment in flint artefacts by ESR spectroscopy, *Nature* 276:703–4.

36. Geyh and Schleicher, *Absolute Age Determination,* p. 280.

37. G. J. Hennig and R. Grün, 1983, ESR dating in Quaternary geology, *Quaternary Science Reviews* 2:157–238.

38. R. Grün and H. P. Schwarcz, 1987, Some remarks on ESR dating of bones, *Ancient TL* 5(2):1–9.

39. Geyh and Schleicher, *Absolute Age Determination,* p. 281.

40. G. Poupeau and A. M. Rossi, Electron spin resonance dating, in *Nuclear Methods of Dating,* ed. E. Roth and B. Pody (Paris: CEA), p. 287.

41. M. G. Seitz and R. E. Taylor, 1974, Uranium variations in a dated fossil in bone series from Olduvai Gorge, Tanzania, *Archaeometry* 16:129–35.

42. M. Ziaei, H. P. Schwarcz, C. M. Hall, and R. Grün, 1990, Radiometric dating of the Mousterian site at Quneitra, in *Quneitra: A Mousterian Site on the Golan Heights,* ed. N. Goren-Inbar, QEDEM 31 (Jerusalem: Hebrew University of Jerusalem), pp. 232–35.

43. R. Grün, H. P. Schwarcz, and S. Zymela, 1987, Electron spin resonance dating of tooth enamel, *Canadian Journal of Earth Science* 24:1022–37.

44. Aitken, Luminescence dating, pp. 193–94.

45. Poupeau and Rossi, Electron spin resonance dating, p. 283.

46. Y. Yokoyama, J. P. Quagebear, R. Bibron, C. Leger, N. Chappaz, C. Michelot, G.-J. Shen, and H.-V. Nguyen, 1983, ESR dating of stalagmites of the Caune de l'Arago, the Grotte du Lazaret, the Grotte du Vallonnet, and the Abri pie Lombard: A comparison with the U-Th method, *PACT* 9:381–89.

47. M. Ikeya, 1982, Petralona cave dating controversy, *Nature* 299:281; G. J. Hennig and F. Hours, 1982, Dates pour les passages entre l'Acheuleen et le Paléolithique Moyen à El Kown (Syrie), *Palaeorient* 8:81–83; R. Grün, H. P. Schwarcz, D. C. Ford, and B. Hentzch, 1988, ESR dating of spring deposited travertines, *Quaternary Science Reviews* 7:429–32; H. P. Schwarcz, R. Grün, A. G. Latham, D. Mania, and K. Brunnacker, 1988, New evidence for the age of the Bilzingsleben archaeological site, *Archaeometry* 30:5–18.

48. G. Fauve, 1986, Principles of Isotope Geology, 2nd ed. (New York: John Wiley).

49. W. F. Libby, 1946, Atmospheric helium-three and radiocarbon from cosmic radiation, *Physics Review* 69:671–72. C. L. Bennett, 1978, Radiocarbon dating with accelerators, *American Scientist* 66:450.

50. Activity Calculation for Radiocarbon

AVOGADRO'S NUMBER: 6.023×10^{23} atoms/mole

Atoms / Mole = Avogadro's #/12 grams/mole

1950 Radiocarbon Activity = 13.56 DPM/Gram Carbon

Instead of the difficult-to-determine N, N_0 values, we use activity, I, I_0, and assume a reasonable correlation between the two measures, N and I, such that we can calculate reliable ages.

51. H. E. Suess, 1970, Bristle-cone-pine calibration in time, 5200 BC to present, in *Radiocarbon Dating*, ed. R. E. Berger and H. E. Suess (Berkeley: University of California Press), pp. 777–84.

52. R. E. Lingenfelter, 1963, Production of carbon-14 by cosmic-ray neutrons, *Journal of Geophysical Research* 78:5902–3.

53. C. Renfrew, 1973, *Before Civilization: The Radiocarbon Revolution and Prehistoric Europe* (London: Jonathan Cape).

54. H. E. Suess, 1986, Secular variations of cosmogenic ^{14}C on earth—their discovery and interpretation, *Radiocarbon* 28:259–65.

55. Bennett, Radiocarbon dating with accelerators.

56. Ibid., p. 450.

57. P. E. Damon, D. J. Donahue, B. H. Gore, A. L. Hathaway, A. J. T. Jull, T. W. Linick, P. J. Sercel, L. J. Toolin, C. R. Bronck, E. T. Hall, R. E. M. Hedges, R. Housley, I. A. Law, C. Perry, G. Bonani, S. Trumbone, W. Woefli, J. C. Ambers, S. G. E. Bowman, M. N. Leese, and M. S. Tite, 1989, Radiocarbon dating the Shroud of Turin, *Nature* 337:611–15.

58. D. Roberts, 1993, The Iceman, *National Geographic* 183(6):36–67.

59. J. E. Francis, L. L. Loendorf, and R. L. Dorn, 1993, AMS radiocarbon and cation-ratio dating of rock art in the Bighorn Basin of Wyoming and Montana, *American Antiquity* 58(4):711–37.

60. J. Clottes, 1993, Paint analysis from several Magdalenian caves in the Ariege region of France, *Journal of Archaeological Science* 20:223–35.

61. T. H. Loy, J. Rhys, D. E. Nelson, B. Meehan, J. Vogel, J. Southon, and R. Cosgrove, 1990, Accelerator radiocarbon dating of human blood proteins in pigments from late Pleistocene art sites in Australia, *Antiquity* 64:110–16.

62. H. Valladas, H. Cachier, P. Maurice, F. Bernaldo de Quiros, J. Clottes, V. C. Valdes, P. Uzquiano, and M. Arnold, 1992, Direct radiocarbon dates for prehistoric paintings at the Altamira, El Castillo, and Niaux Caves, *Nature* 357:68–70.

63. H. Valladas, et al. Direct radiocarbon dates for prehistoric paintings at the Altamira, El Castillo, and Niaux caves; J. Clottes, 1993, Paint analyses from several Magdalenian caves in the Ariège region of France, *Journal of Archaeological Science* 20:223–35.

64. J. Clottes, 1995, Rhinos and lions and bears. *Natural History* 104(5):30–36.

65. S. P. M. Harrington, 1990, Dating rock art, *Antiquity,* May/June, p. 24.

66. Valladas, et al., Direct radiocarbon dates.

67. M. L. Kunz and R. E. Reanier, 1994, Paleoindians in Beringia: Evidence from Arctic Alaska, *Science* 263:660–62. V. M. Bryant, Jr., 1992, In search of the first Americans, in *1993 Yearbook of Science and the Future* (Chicago: Encyclopaedia Britannica), pp. 8–27.

68. D. W. Clark, 1984, Northern fluted points: Paleo-Eskimo, Paleo-Arctic or Paleo-Indian?, *Canadian Journal of Anthropology* 4:65–81.

69. Kunz and Reanier, Paleoindians in Beringia, p. 660.

70. H. L. Alexander, 1987, *Putu: A Fluted Point Site in Alaska,* Department of Archaeology, Simon Fraser University, Publication no. 17 (Burnaby, B.C.).

71. J. F. Hoffecker, W. R. Powers, and T. Gabel, 1993, The colonization of Beringia and peopling of the New World, *Science* 259:46–53.

72. D. Lal and B. Peters, 1967, Cosmic-ray produced radioactivity on the earth, *Handbook of Physics* 46:551–612.

73. T. E. Cerling and H. Craig, 1994, Geomorphology and *in situ* cosmogenic isotopes, *Annual Review of Earth and Planetary Science* 22:273–317. E. W. Cliver, 1994, Solar activity and geomagnetic storms: The corpuscular hypothesis, *EOS* 75(52):609–12.

74. F. M. Phillips, M. G. Zreda, S. S. Smith, D. Elmore, P. W. Kubik, and P. Sharma, 1990, Cosmogenic chlorine-36 chronology for glacial deposits at Bloody Canyon, eastern Sierra Nevada, *Science* 248:1529–32; M. G. Zreda, F. M. Phillips, D. Elmore, P. W. Kubik, P. Sharma, and R. I. Dorn, 1991, Cosmogenic chlorine-36 production rates in terrestrial rocks, *Earth and Planetary Science Letters* 105:94–109; B. Liu, F. M. Phillips, D. Elmore, and P. Sharma, 1994, Depth dependence of soil carbonate accumulation based on cosmogenic ^{36}Cl dating, *Geology* 22:1071–74; M. G. Zreda, F. M. Phillips, P. W. Kubik, P. Sharma, and D. Elmore, 1993, Cosmogenic ^{36}Cl dating of a young basaltic eruption complex, Lathrop Wells, Nevada, *Geology* 21:57–60.

Chapter 7

1. D. Currey, 1965, An ancient bristlecone pine stand in eastern Nevada, *Ecology* 46:564–66.

2. H. Michael and E. Ralph, 1971, *Dating Techniques for the Archaeologist* (Cambridge: MIT Press), pp. 49–56.

3. Ibid.

4. M. G. L. Baillie, 1992, Dendrochronology and past environmental changes, in *New Developments in Archaeological Science,* ed. A. M. Pollard (Oxford: Oxford University Press), pp. 5–24.

5. M. G. L. Baillie, 1982, *Tree-Ring Dating and Archaeology* (London: Croom Helm), pp. 31–45.

6. Ibid.

7. J. S. Dean, 1986, Dendrochronology, in *Dating and Age Determination of Biological Materials,* ed. M. R. Zimmerman and J. L. Angel (London: Croom Helm), pp. 126–65.

8. M. J. Aitken, 1990, *Science-Based Dating in Archaeology* (London: Longman), pp. 42–43.

9. F. H. Schweingraber, O. U. Bräker, L. G. Drew, and E. Schär, The x-ray technique as applied to dendroclimatology, *Tree-Ring Bulletin* 38:68–91.

10. Dean, Dendrochronology, p. 143.

11. D. H. Thomas, 1991, *Archaeology: Down to Earth* (Fort Worth: Harcourt Brace Jovanovich), pp. 60–62.

12. Dean, Dendrochronology, p. 127.

13. Ibid., p. 128; Thomas, *Archaeology,* p. 62.

14. Baillie, *Tree-Ring Dating and Archaeology,* p. 39.

15. J. R. Pilcher, M. G. L. Baillie, B. Schmidt, and B. Becher, 1984, A 7,272 year tree-ring chronology for western Europe, *Nature* 312:150–52.

16. H. Egger, 1983, Die absolute Datierung der Saöne-Rhöne Kultur und der Bronzezeit der Westschweiz, *Dendrochronologia* 1:37–44.

17. H. E. Weakley, 1949, Dendrochronology in Nebraska. In *Proceedings of the 5th Plains Conference for Archaeology, Notebook No. 1* (Laboratory of Anthropology, University of Nebraska), pp. 111–14; W. F. Weakly, 1971, Tree-ring dating and archaeology in South Dakota, *Plains Anthropologist* 16(2), *Memoir 8.*

18. D. W. Stable, 1979, Tree-ring dating of historic buildings in Arkansas, *Tree-Ring Bulletin* 39:1–28.

19. J. I. Giddings, 1941, Dendrochronology in northern Alaska, *Laboratory of Tree-Ring Research Bulletin* (Tucson), no. 1; J. I. Giddings, 1948, Chronology of the Kotzebue sites, *Tree-Ring Bulletin* 14(4):20–32; W. Oswalt, 1952, The archaeology of the Hooper Bay Village, Alaska, *Anthropological Papers of the University of Alaska* 1(3):47–91; J. W. Van Stone, 1953, Notes on Kotzebue dating, *Tree-Ring Bulletin* 20(1):6–8.

20. Dean, Dendrochronology, p. 155.

21. N. W. Thompson, 1967, *Novgorod the Great: Excavations at the Medieval City Directed by A. V. Artsikhovsky and B. A. Kolchin* (New York: Praeger).

22. B. Bannister, 1970, Dendrochronology in the Near East: Current research and future potentialities, *Proceedings of the Seventh International Congress of Anthropological and Ethnological Sciences* 5:336–40.

23. H. L. DeVries, 1958, Variation in concentration of radiocarbon with time and location on earth, *Kongress Nederland Akademie Wetter Proceedings,* ser. B, 61:94–102; M. Stuiver, and H. E. Suess, 1966, On the relationship between radiocarbon dates and true age samples, *Radiocarbon* 8:534–40.

24. H. N. Michael, and E. K. Ralph, 1970, Correction factors applied to Egyptian radiocarbon dates from the era before Christ, in *Nobel Symposium 12: Radiocarbon Variations and Absolute Chronology,* ed. I. U. Olsson (Stockholm: Almquist and Wilsell), pp. 109–19.

25. H. E. Suess, 1970, Bristlecone-pine calibration of the radiocarbon time-scale 5200 BC to the present, in *Nobel Symposium 12: Radiocarbon Variations and Absolute Chronology,* ed. I. U. Olsson (Stockholm: Almquist and Wilsell), pp. 308–9.

26. G. Folgerhaiter, 1899, Sur les variations séculaires de l'inclinaison magnétique dans l'antiquité, *Archives des Sciences Physiques et Naturelles* 8:5–116.

27. R. S. Sternberg, and R. H. McGuire, 1990, Techniques for constructing secular variation curves and for interpreting archaeomagnetic dates, in *Archaeomagnetic Dating,* ed. J. L. Eighmy and R. S. Sternberg (Tucson: University of Arizona Press), p. 112.

28. Ibid.

29. R. S. Sternberg, and R. H. McGuire, 1981, Archaeomagnetic secular variation in the American Southwest, *EOS: Transactions of the American Geophysical Union* 62:852.

30. J. L. Eighmy, and D. R. Mitchell, 1994, Archaeomagnetic dating at Pueblo Grande, *Journal of Archaeological Science* 21:445–53.

31. D. Wolfman, 1990, Retrospect and prospect, in *Archaeomagnetic Dating,* ed. J. L. Eighmy and R. S. Sternberg (Tucson: University of Arizona Press), pp. 313–64.

32. Ibid., p. 336.

33. N. Watanabe, 1959, The direction of remanent magnetization of baked earth and its application to chronology for anthropology and archaeology in Japan, *Journal of the Faculty of Sciences, University of Tokyo* 2:1–188.

34. M. Creer, P. Tucholka, and C. E. Barton, eds., 1983, *Geomagnetism of Baked Clays and Recent Sediments* (Amsterdam: Elsevier); A. Cox, ed., 1973, *Plate Tectonics and Geomagnetic Reversals* (San Francisco: W. H. Freeman).

35. Wolfman, Retrospect and prospect, p. 322.

36. D. E. Champion, 1980, Holocene geomagnetic secular variation in the western United States: Implications for the global geomagnetic field, U.S. Geological Survey Open-File Report 80–824 (Washington, D.C); R. S. Sternberg, 1990, The geophysical basis for archaeomagnetic dating, in *Archaeomagnetic Dating,* ed. J. L. Eighmy and R. S. Sternberg (Tucson: University of Arizona Press), p. 18.

37. S. Thorarinsson, 1944, Tefroknonologiska studies p island, *Geografiska Annaler* 26:1–217; S. Thorarinsson, 1974, The terms "tephra" and "tephrachronology," in *World Bibliography and Index of Quaternary Tephrochronology,* ed. J. A. Westgate and C. M. Gold (Edmonton: University of Alberta), pp. xvii—xviii.

38. V. Stern-McIntyre, 1985, *Archaeological Geology,* ed. G. Rapp, Jr., and J. A. Gifford (New Haven: Yale University Press), pp. 266–68.

39. J. G. Nixon, 1985, The volcanic eruption of Thera and its effect on the Mycenean and Minoan civilizations. *Journal of Archaeological Science* 12:9–24.

40. R. C. Walter, 1994, Age of Lucy and the first family, *Geology* 22:6–10.

41. S. D. Stihler, D. B. Stone, and J. E. Beget, 1992, "Varve" counting vs. tephrachronology and ^{137}Cs and ^{210}Pb dating: A comparative test at Skilack Lake, Alaska, *Geology* 20:1019–22.

42. J. P. Grattan and D. D. Gilbertson, 1994, Acid-loading from Icelandic tephra falling on acidified ecosystems a key to understanding archaeological and environmental stress in northern and western Britain, *Journal of Archaeological Science* 21(6):851–60.

43. C. W. Naeser, Briggs, J. D. Obradovich, and G. A. Izett, 1981, Geochronology of tephra deposits, in *Tephra Studies,* ed. S. Self and R. S. J. Sparks (Dordrecht: D. Reidel), pp. 13–47.

44. G. W. Naeser and N. D. Naeser, 1988, Fission-track dating of Quaternary events, *Geological Society of America Special Paper* 227; G. A. Gunnlaugsson, G. M. Gudbergson, S. Thorarinsson, S. Raffinson, and T. Einarsson, 1984, *Skóftareldar 1783–1784* (Icelandic Norse with English summaries) (Reikjavik: Mal Og Memming); E. L. Jackson, 1982, The Laki eruption of 1783: Impacts on population and settlement of Iceland, *Geography* 67(1):42–50; A. E. J. Oglivie, 1986, The climate of Iceland, 1701–1784, *Jökull* 36:57–73.

45. Grattan and Gilbertson, Acid-loading from Icelandic tephra; V. A. Hall, J. R. Pilcher, and F. G. McCormac, 1994, Icelandic volcanic ash and the mid-Holocene Scots pine (*Pinus sylvestris*) decline in the north of Ireland, no correlation, *Holocene* 4(1):79–85.

46. H. Sigurdsson, H. Jóhannesson, and G. Larsen, 1993, *Iceland Geology: A Field Guide* (Boulder: Geological Society of America), p. 22.

47. A. J. Dugmore, and A. J. Newton, 1992, Thin tephra layers in peat revealed by x-radiography, *Journal of Archaeological Science* 19:163–70.

48. Naeser et al., Geochronology of tephra deposits, p. 35.

49. Westgate and Naeser, Tephro-chronology and fission-track dating, p. 34.

50. Steen-McIntyre, *Archaeological Geology,* p. 290.

51. Naeser et al., Geochronology of tephra deposits, p. 37; Steen-McIntyre, *Archaeological Geology,* p. 291.

52. Naeser and Naeser, Fission-track dating of Quaternary events, p. 5.

53. Westgate and Naeser, Tephro-chronology and fission-track dating, p. 34.

54. P. Sheets, 1995, personal communication; Steen-McIntyre, *Archaeological Geology,* p. 272.

55. W. J. E. Hart, 1983, Classic to Post-Classic tephra layers exposed in archaeological sites: Eastern Zapotitán Valley, in *Archaeology and Volcanism in Central America: The Zapotitán Valley of El Salvador,* ed. Payson Sheets (Austin: University of Texas Press), pp. 44–51; W. J. E. Hart and V. Steen-McIntyre, 1983, Tierra blanca joven tephra from the A.D. 260 eruption of Ilopango Caldera, in *Archaeology and Volcanism in Central America: The Zapotitán Valley of El Salvador,* ed. Payson Sheets (Austin: University of Texas Press), pp. 14–34.

56. G. Bosinski, 1992, *Eiszeitjäger im Neuwieder Becken,* Archäologie an Mittelrhein und Mosel 1 (Koblenz: Archäologische Denkmalpflege Amt), p. 12.

57. Ibid., p. 86.

58. Ibid., pp. 89, 111.

59. P. V. D. Bogaard and H.-U. Schminke, 1985, Laacher See tephra—A widespread isochronous Late Quaternary ash layer in central and northern Europe, *Geological Society of America Bulletin* 96:1554–71.

Chapter 8

1. K. Butzer, 1980, Context in archaeology: An alternative approach, *Journal of Field Archaeology* 7:417–22.

2. G. Rapp, Jr., and J. A. Gifford, 1982, Archaeological geology, *American Scientist* 70:45–71.

3. B. M. Fagan, 1994, *Archaeology: A Brief Introduction,* 5th ed. (New York: Harper Collins), pp. 70–71.

4. C. Renfrew, 1973, *Before Civilization* (New York: Alfred Knopf).

5. M. B. Dobrin, 1960, *Introduction to Geophysical Prospecting* (New York: McGraw-Hill).

6. J. W. Weymouth, 1986, Geophysical methods of archaeological site surveying, in *Advances in Archaeological Method and Theory,* ed. M. B. Schiffer (New York: Academic), vol. 9, pp. 315–17; M. J. Stright, 1986, Evaluation of archaeological site potential on the Gulf of Mexico continental shelf using high resolution seismic data, *Geophysics* 51(3):605–22.

7. Weymouth, Geophysical methods of archaeological site surveying, pp. 315–17.

8. W. M. Telford, L. P. Geldart, R. E. Sheriff, and D. A. Keys, 1976, *Applied Geophysics* (New York: Cambridge University Press).

9. P. Hoekstra, L. Raye, C. R. Bates, and D. Phillips, 1992, Case histories of shallow time domain electromagnetics in environmental site assessment, *Ground Water Monitoring Review,* 12(4):110–17.

10. P. Sheets et al., 1985, Geophysical exploration for ancient files at Ceren, El Salvador, *National Geographic Society Research Reports.*

11. D. Chapellier, 1974, Une mèthode seismique au service de l archéologie, *Prospezioni Archaeologiche* 9:93–95.

12. A. Adlung, 1987, *Archäologie und Geophysike, Möglichkeiten und Perspektiven ihier Anwendung in der DDR,* Kleine Schriften des Landersmuseums für Vorgeschichte, vol. 6. (Dresden).

13. W. Neubauer, 1990, *Geophysikalische Prospektion in der Archäologie Institut für Ur- und Frühgeschichte* (Vienna), pp. 54–57.

14. A. J. Witten, T. E. Levy, J. Ursic, and P. White, 1995, Geophysical diffraction tomography: New views on the Shiqmim prehistoric subterranean village site (Israel), *Geoarchaeology* 10:185–206.

15. E. G. Garrison, 1991, Archaeogeophysical studies of paleoshorelines in the Gulf of Mexico, in *Paleoshorelines: An Investigation of Method,* ed. Lucy Johnson (Boca Raton: CRC Press), pp. 103–16; M. J. Stright, 1986, Human occupation of the continental shelf during the late Pleistocene/early Holocene: Methods for site location, *Geoarchaeology* 1(4):347–64.

16. M. J. Stright, 1990, Archaeological sites on the North American continental shelf, in *Archaeological Geology of North America,* ed. N. P. Lasca and J. Donahue, Geological Society of America, Centennial Special Volume 4 (Boulder), pp. 439–65.

17. S. M. Gagliano, 1970, Archaeological and geological studies at Avery Island, 1968–1970 (Baton Rouge: Coastal Studies Institute, Louisiana State University).

18. S. M. Gagliano, C. E. Pearson, R. A. Weinstein, D. E. Wiseman, and C. M. McClendon, 1982, *Sedimentological Studies of Prehistoric Archaeological Sites* (Baton Rouge: Coastal Environments, Inc.).

19. P. M. Masters, and N. C. Flemming, eds., 1983, *Quaternary Coastlines and Marine Archaeology: Towards a Prehistory of Land Bridges and Continental Shelves* (New York: Academic Press).

20. L. W. Patterson, 1981, OCS Prehistoric site discovery difficulties, *Journal of Field Archaeology* 8(2):231–32.

21. J. R. Curray, 1960, Sediments and history of Holocene transgression, continental shelf, northwest Gulf of Mexico, in *Recent sediments, Northwest Gulf of Mexico,* ed. F. P. Shapard, E. B. Phlezar, and T. H. van Andel (Tulsa: American Association of Petroleum Geologists); H. F. Nelson and E. E. Bray, 1970, Stratigraphy and history of the Holocene sediments in the Sabine-High Island area, Gulf of Mexico, in *Deltaic Sedimentation, Modern and Ancient,* ed. J. P. Morgan (Tulsa: Society of Economic Paleontologists and Mineralogists), pp. 48–72.

22. R. L. Edwards, J. W. Beck, D. J. Burr, D. J. Donahue, J. M. A. Chappell, A. L. Bloom, E. R. M. Drufel, and F. W. Taylor, 1993, A large drop in atmospheric $^{14}C/^{12}C$ and reduced melting in the Younger Dryas, documented with ^{230}Th ages of corals, *Science* 260:962–68.

23. I. M. Whillans and R. B. Alley, 1991, Changes in the West Antarctic ice sheet, *Science* 254:959–63.

24. Stright, Human occupation of the continental shelf.

25. CHIRP Sonar, *Sea Technology* 31(9):35–43.

26. P. G. Hallof, 1980, Electrical IP and resistivity, in *Practical Geophysics for the Exploration Geologist,* comp. R. van Blaricom (Spokane: Northwest Mining Association), pp. 40–42.

27. M. J. Aitken, 1974, *Physics and Archaeology,* 2d ed. (Oxford: Clarendon Press); A. Clark, 1990, *Seeing beneath the Soil: Prospecting Methods in Archaeology* (London: Batsford).

28. T. Herbich and K. Misiewicz, 1994, Multilevel resistivity prospecting of architectural remains: The Schwarzach case study (paper presented at the 29th International Symposium on Archaeometry, 9–14 May, Ankara, Turkey); T. Herbich, 1993, The variations of shaft fills as the basis of the estimation of flint mine extent: A Wierzbica case study, *Archaeologia Polonia* 31:71–82.

29. I. Scollar, A. Tabbagh, A. Hesse, and I. Herzog, 1990, *Prospecting, Image Processing, and Remote Sensing* (New York: Cambridge University Press); I. Klein and J. J. Lajoie, 1980, Electromagnetics, in *Practical Geophysics for the Exploration Geologist,* comp. Richard van Blaricom (Spokane: Northwest Mining Association), p. 242.

30. M. S. Tite and C. Mullins, 1969, Electromagnetic prospecting: A preliminary investigation, *Prospezioni Archaeologiche* 4:95–102.

31. M. B. Dobrin and C. H. Sauit, 1988, *Introduction to Geophysical Prospecting,* 4th ed. (New York: McGraw-Hill), p. 168.

32. Klein and Lajoie, Electromagnetics, p. 243.

33. Ibid.; U. Leute, 1987, *Archaeometry* (Weinheim: VCH), p. 21.

34. Aitken, *Physics and Archaeology,* pp. 196–98.

35. D. D. Scott, R. A. Fox, Jr., M. A. Connor, and D. Harmon, 1987, *Archaeological Perspectives on the Battle of the Little Bighorn* (Norman: University of Oklahoma Press); D. D. Scott, 1987, *Archaeologial Insights into the Custer Battle* (Norman: University of Oklahoma Press); C. L. Bond, 1981, Archaeological Investigations at Palo Alto Battlefield (College Station: Cultural Resources Laboratory, Texas A&M University).

36. Klein and Lajoie, Electromagnetics, pp. 270–72.

37. M. S. Tite, 1972, *Methods of Physical Examinations in Archaeology* (New York: Academic Press), pp. 17–18.

38. Leute, *Archaeometry,* p. 21.

39. S. Breiner, 1973, *Applications Manual for Portable Magnetometers* (Sunnyvale: Geometrics), pp. 17–54; M. J. Alden, 1988, Locating archaeological features in magnetic data by cross correlation, *Archaeometry* 30(1):145–54; R. E. Linington, 1966, A first use of linear filtering techniques on archaeological prospecting results, *Prospezioni. Archaeologische* 5:43–54; I. Scollar, B. Weidner, and K. Segeth, Display of archaeological magnetic data, *Geophysics* 51(3):623–33; L. J. Peters, 1949, The direct approach to magnetic interpretation and its practical application, *Geophysics* 14:290–319.

40. P. Hänninen and S. Autio, 1992, Fourth International Conference on Ground-Penetrating Radar, June 8–13, 1992 (Rovaniemi, Finland), *Geological Survey of Finland,* Special Paper 16 (Espoo: Geologian tutkimuskaskus); J. W. Weymouth, 1986, Geophysical methods of archaeological site surveying, in *Advances in Archaeological Method and Theory,* ed. Michael B. Schiffer (New York: Academic Press), vol. 9, pp. 311–91; T. Inai, T. Sakayama, and T. Kanemori, 1992, Use of ground probing radar and resistivity surveys for archaeological investigations, *Geophysics* 52(2):137–50.

41. D. Goodman, 1994, Ground-penetrating radar simulation in engineering and archaeology, *Geophysics* 59(2):224–32; D. Goodman and Y. Nishimura, 1992, 2-D synthetic radargrams for archaeological investigations, in *Proceedings of the 4th International Conference on Ground Penetration Radar, Special Paper 16* (Espoo: Geologian tutimuskaskus) (Finland Geological Survey), pp. 339–43; G. Turner, 1992, Propagation deconvolution, in *Proceedings of the 4th International Conference on Ground Penetration Radar,* pp. 85–93; E. Fisher, G. A. McMehan, and A. P. Annan, 1992, Acquisition and processing of wide-aperture ground-penetrating radar data, *Geophysics* 57:497–504; S. Duke, 1990, Calibration of Ground-Penetrating Radar and Calculation of Attenuation and Dielectric Permittivity versus Depth (M.S. thesis, Colorado School of Mines).

42. M. Powers, and G. Olhoeft, 1995, Modeling dispersive GPR data (paper presented at University of Colorado Workshop on GPR in Archaeology, Boulder).

43. Goodman, Ground-penetrating radar simulation in engineering and archaeology.

44. D. Goodman, and Y. Nishimura, 1993, A ground-radar view of Japanese burial mounds, *Antiquity* 67:349–54; D. Goodman, Y. Nishimura, T. Uno, and T. Yamamoto, 1993, A ground radar survey of medieval kiln sites in Suzu City, western Japan, *Archaeometry* 36(2):317–26.

45. *Supra.* Goodman and Nishimura, A ground-radar view of Japanese burial mounds.

46. G. D. Jones, 1987, The ethnohistory of the Guale Coast through 1684, in The Anthropology of St. Catherines Island: I. Natural and Cultural History, by D. H. Thomas et al., *Anthropological Papers of the American Museum of Natural History* 55(2):178–210.

47. D. H. Thomas, 1988, *St. Catherines: An Island in Time* (Atlanta: Georgia Humanities Council), p. 18.

48. Ibid.; D. H. Thomas, 1987, The archaeology of Mission Santa Catalina de Guale: Search and Discovery, *Anthropological Papers of the American Museum of Natural History* 63(2) p. 108.

49. Thomas, *St. Catherines.*

50. Ibid.; E. G. Garrison, J. G. Baker, and D. H. Thomas, 1985, Magnetic prospection and discovery of Mission Santa Catalina de Guale, Georgia, *Journal of Field Archaeology* 12(3):299–313.

51. D. J. Guion, 1980, Gravity, in *Practical Geophysics for the Exploration Geologist,* comp. Richard van Blaricom (Spokane: Northwest Mining Association), pp. 153–63.

52. Y. Lemoine and J.-P. Baron, 1991, La microgravimetrie, *Dossiers d'Archéologie* 162:78.

53. P. Deletie and Y. Lemoine, 1991, Le capteur thermographique, *Dossiers d'Archéologie* 162:78.

54. SQUIDS, *Scientific American* 271(2):46.

55. E. Le Borgne, 1960, Influence du fer sur les propriétés magnétiques du sol et sur velles du schiste et du granite, *Annales de Geophysique* 16(2):159–95; E. Le Borgne, 1965, Les propriétés magnétiques du sol: Application à la prospection des sites archéologiques, *Archaeophysika* 1:1–20.

56. I. G. Hedley, J. J. Wagner, and D. Ramseyer, 1994, Magnetic susceptibility and human occupation (paper presented at the 29th International Symposium on Archaeometry, 9–14 May, Ankara, Turkey); B. B. Ellwood, D. E. Peter, W. Balsam, and J. Schieber, 1995, Magnetic and geomagnetic variations of paleoclimate and archaeological site evolution: Examples from 41TR68, Fort Worth, Texas, *Journal of Archaeological Science* 22(3):409–16.

57. Aitken, *Physics and Archaeology;* Le Borgne, Influence du fer; Tite, *Methods of Physical Examinations in Archaeology;* Hedley et al., Magnetic susceptibility and human occupation.

58. P. Spoerry, 1992, *Geoprospection of the Archaeological Landscape* (London: Oxbow Books).

Chapter 9

1. P. Bethell, and I. Máté, 1989, The use of soil phosphate analysis in archaeology: A critique, in *Scientific Analysis in Archaeology,* ed. J. Henderson, Oxford University Communications for Archaeology Monograph, 19:1–29.

2. O. Arrhenius, 1931, Die bodenanalyse im dienst der Archäologie, *Zeitschrift für Pflanzenernährung, Düngang und Bodenkunde,* B10:427–39.

3. S. F. Cook and R. F. Heizer, 1965, *Studies on the Chemical Analysis of Archaeological Sites,* University of California Publications in Archaeology 2 (Berkeley).

4. S. Weiner, G. Goldberg, and O. Bar-Yosef, 1993, Bone preservation in Kebara Cave, Israel, using on-site Fourier transform infrared spectrometry, *Journal of Archaeological Science* 20:613–27.

5. Bethell and Máté, Soil phosphate analysis in archaeology.

6. A. H. Johnson and D. Nicol, 1949, Analysis of soil samples from old ground surface under cairn Cairnholy, I, in The excavation of three Neolithic tombs in Galloway, ed. S. Piggott and T. G. E. Powell, *Proceedings of the Society of Antiquaries, Scotland,* 83:161.

7. A. H. Johnson, 1956, Examination of soil from Corrimony chambered cairn, *Proceedings of the Society of Antiquaries, Scotland,* 88:200–207.

8. D. A. Davidson, 1973, Particle size and phosphate analysis: Evidence for the evolution of a tell, *Archaeometry* 15:143–52.

9. W. C. Cavanagh, S. Hirst, and C. D. Litton, 1988, Soil phosphate, site boundaries, and change point analysis, *Journal of Field Archaeology* 15:67–83.

10. P. Craddock, D. Gurney, F. Pryor, and M. Hughes, 1985, The application of phosphate analysis to the location and interpretation of archaeological sites, *Archaeological Journal* 142:361–76.

11. R. D. Lippi, 1988, Paleotopography and P analysis of a buried jungle site in Ecuador, *Journal of Field Archaeology* 15:85–97.

12. R. C. Eidt, 1977, Detection and examination of anthrosols by phosphate analysis, *Science* 197:1327–33; R. C. Eidt, 1984, *Advances in Abandoned Settlement Analysis: Application to Prehistoric Anthrosols in Columbia, South America* (Milwaukee: Center for Latin America, University of Wisconsin—Milwaukee).

13. K. T. Lillios, 1993, Phosphate fractionation of soils at Agroal, Portugal, *American Antiquity* 57(3):495–506.

14. Eidt, *Advances in Abandoned Settlement Analysis.*

Chapter 10

1. R. V. Dietrich and B. J. Skinner, 1979, *Rocks and Rock Minerals* (New York: John Wiley and Sons); D. R. C. Kempe and A. P. Harvey, eds., 1983, *The Petrology of Archaeological Artefacts* (Oxford: Clarendon Press).

2. G. A. Wagner and G. Weisgerber, eds., 1988, *Antike Edel-und Buntmetallgewinnung auf Thasos* (Bochum: Bergbau Museums).

3. T. Zoltai, and J. H. Stout, 1984, *Mineralogy: Concepts and Principles* (Minneapolis: Burgess Publishing Co.).

4. M. Waelkens, N. Herz, and L. Moens, eds., 1992, *Ancient Stones: Quarrying, Trade, and Provenance* (Leuven: Leuven University Press), Acta Archaeologica Lovaniensia Monographiae 4.

5. D. P. S. Peacock, 1992, Rome in the desert: A symbol of power (Inaugural Lecture, University of Southampton).

6. Z. Hawass and M. Lehmer, 1994, Remnant of a lost civilization? *Archaeology* 47(5):44–47; K. Lal Gauri, J. J. Sinai, and J. K. Bandyopadhyay, 1995, Geologic weathering and its implications for the age of the Sphinx, *Geoarchaeology* 10:119–33.

7. Gauri et al., *supra.*

8. J. A. Harrell, 1992, Ancient Egyptian limestone quarries: A petrological survey, *Archaeometry* 34:195–211.

9. R. J. Atkinson, 1979, *Stonehenge* (Harmondsworth: Penguin Books).

10. O. Williams-Thorpe and R. S. Thorpe, 1992, Geochemistry sources and transport of the Stonehenge bluestones, in *New Developments in Archaeological Science,* ed. A. M. Polland (Oxford: Oxford University Press), pp. 133–61.

11. Atkinson, *Stonehenge.*

12. Williams-Thorpe and Thorpe, Stonehenge bluestones, p. 139.

13. H. Howard, 1982, A petrological study of the rock specimens from excavation at Stonehenge, 1979–80, in On the road to Stonehenge: Report on investigations beside the A344 in 1968, 1979, and 1980, by M. W. Pitts, *Proceedings of the Prehistoric Society* 48:104–26.

14. A. Burl, 1979, *The Stone Circles of the British Isles* (New Haven: Yale University Press); D. P. Dymond, 1966, Ritual monuments at Rudston, E. Yorkshire, England, *Proceedings of the Prehistoric Society* 32:86–95; A. Burl, 1985, Geoffrey of Monmouth and the Stonehenge bluestones, *Wiltshire Archaeological and Natural History Magazine* 79:178–83.

15. M. J. O'Kelly, 1982, *Newgrange* (New York: Thames and Hudson), p. 118.

16. R. Bollin and M. Maggetti, 1994, Petrographic and isotopic arguments in provenance

studies of "tessarae" from Swiss Gallo-Roman mosaic, *Abstracts of the 29th International Symposium on Archaeometry, Arkara, Turkey, 9–14 May* (Ankara: Middle Eastern Technical University).

Chapter 11

1. B. Gratuze, A. Goivagnoli, J. N. Barrandon, P. Telouk, and J. L. Imbert, 1993, Aport de la méthode ICP-MS couplée à l'ablation laser pour la caractérisation des archéomatériaux, *Revue d'archaeometrie* 17:89–104.

2. M. J. Aitken, 1974, *Physics in Archaeology,* 2nd ed. (Cambridge: Clarendon), p. 154.

3. R. M. Rowlett, 1988, Titelberg, a Celtic Hillfort in Luxembourg, *Expedition* 30(2):31.

4. Ibid., p. 34.

5. R. F. Tylecotte, 1962, The methods of use of early Iron Age moulds, *Numismatic Chronicle,* pp. 108–9; J. Tournaire, D. Buchsenschutz, J. Henderson, and J. Collis, 1982, Iron Age moulds from France, *Proceedings of the Prehistoric Society* 48:417–35; D. Allen, 1971, The early coins of the Treveri, *Germania* 49:91–110; J. Metzger, and R. Weiller, 1977, Beiträge zur Archäologie und Numismatik des Titelbarges, Schrötlingsformen und Schrötlinge, *Publications Section Historique, Luxembourg* 91:137–41; L. R. Laing, 1969, *Coins and Archaeology* (London: Weidenfeld and Nicolson); L. Reding, 1972, *Les monnaies gauloises du Titelberg,* Publications Nationales du Ministère des Arts et des Sciences (Luxembourg); A. Walker, 1976, Worn and corroded coins: Their importance for the archaeologist, *Journal of Field Archaeology* 3:329–35.

6. Debord, 1989, L'atelier monétaire gauloise de Villeneuve-St. Germain (Aisne) et sa production, *Revue Numismatique,* 6th ser., 31:7–24.

7. Rowlett, Titelberg, p. 36.

8. P. Chevallier, F. Legrand, K. Gruel, I. Porissaud, and A. Tanats-Saugnac, 1993, Étude par rayonnement synchrotron de moules á aveoles de La Tène finale trouves à Villeneuve-St. Germain et au Mont Beuvray, *Revue d'Archéometrie* 17:75–88.

9. E. V. Sayre and R. W. Dodson, 1957, Neutron activation study of Mediterranean potsherds, *American Journal of Archaeology* 61:35–41; I. Perlman and F. Asaro, 1971, Pottery analysis by neutron activation, *Archaeometry* 13:21–52; G. Harbottle, 1982, Provenance studies using neutron activation analysis: The role of standardization, in *Archaeological Ceramics,* ed. J. S. Olin and A. O. Franklin (Washington, D.C.: Smithsonian Press), pp. 67–77.

10. D. E. Moore and R. C. Reynolds, 1989, *X-Ray Diffraction and the Identification and Analysis of Clay Minerals* (New York: Oxford University Press), pp. 291–307.

11. A. Minzoni-Déroche, 1981, X-ray diffraction analysis and petrography as useful methods for ceramic typology, *Journal of Field Archaeology* 8:511–13.

12. C. Yang, N. P.-O. Homman, K. G. Malmquist, L. Johannson, N. M. Haldin, and V. Barbin, in press, Ionoluminescence: A new tool for the nuclear microprobe in geology, *Scanning Microscopy.*

13. A. A. Katsanos, 1986, Applications of ion beam analysis in archaeology and the arts, *Nuclear Instruments and Methods in Physics Research,* B14, 86: 82–85.

14. B. Welz, 1985, *Atomic Absorption Spectroscopy,* 2nd ed., Trans. Christopher Skegg (Weinheim: VCH), p. 1.

15. Ibid., pp. 12–14.

16. Ibid., p. 254.

17. Ibid., p. 27.

18. F. A. Hart and S. J. Adams, 1983, The chemical analysis of Romano-British pottery from the Alice-Holt Forest, Hampshire, by means of inductively coupled plasma emission spec-

troscopy, *Archaeometry* 25:179–85; N. Porat, J. Yellin, and L. Heller-Kallai, 1991, Correlation between petrography, NAA and ICP analyses: Application to Early Bronze Egyptian pottery from Canaan, *Geoarchaeology* 6:133–49; J. Linderholm and E. Lundberg, 1994, Chemical characterization of various archaeological soil samples using main and trace elements determined by inductively coupled plasma atomic emissions spectroscopy, *Journal of Archaeological Science* 21:303–14.

19. Welz, *Atomic Absorption Spectroscopy,* pp. 260–61.

20. Gratuze et al., Aport de la méthode ICP-MS couplée à l'ablation laser, pp. 89–91.

21. Porat et al., Correlation between petrography, NAA and ICP analyses, pp. 133–34; Harbottle, Provenance studies using neutron activation analysis.

22. K. Ramseyer, J. Fischer, A. Matter, P. Eberhardt, and J. Geiss, 1989, A cathodoluminescence microscope for low intensity luminescence, *Journal of Sedimentological Petrology* 59:619–22; D. J. Marshall, 1988, *Cathodoluminescence of Geological Materials* (Boston: Unwin Hyman).

23. V. Barbin, K. Ramseyer, D. Decrouez, S. J. Burns, J. Chamay, and J. L. Maier, 1992, Cathodoluminescence of white marbles: An overview, *Archaeometry* 34:175–83.

24. Ibid.

25. G. S. Hurst and V. S. Letokhov, 1994, Resonance ionization spectroscopy, *Physics Today* 47:38–45.

26. L. W. Alvarez, W. Alvarez, F. Asaro, and H. V. Michael, 1980, Extraterrestrial cause for Cretaceous Tertiary extinction, *Science* 208:1095–1108.

27. Hurst and Letokhov, Resonance ionization spectroscopy, p. 43.

28. R. Siegel, 1993, Exploring mesoscopia: The bold new world of nanostructures, *Physics Today* 46:64–68; M. S. DiCapua and O. Marti, 1989, *NATO Advanced Study Institute on Scanning Tunneling Microscopy: Physical Concepts, Related Techniques, and Major Application,* ENS Information Bulletin 89–09 (New York: Office of Naval Research, European Office), pp. 48–62.

29. DiCapua and Marti, *Scanning Tunneling Microscopy,* p. 49.

30. G. Binnig, C. F. Quate, C. Gerber, E. Stoll, and T. R. Albrecht, 1986. Atomic force microscope, *Physical Review Letters* 56:930–33.

31. S. Borman, 1993, Near-field scanning optical microscopy, *Chemical and Engineering News* 71(48):6–7; E. Betag and R. J. Chichester, 1993, Single molecules observed by near-field scanning optical microscopy, *Science* 262:1422–25.

32. Borman, Near-field scanning optical microscopy.

33. M. S. Tite, 1991, The impact of electron microscopy on ceramic studies, in *New Developments in Archaeological Science,* ed. A. M. Pollard (Oxford: Oxford University Press), pp. 111–12.

34. M. Rigler and W. Longo, 1994, High voltage scanning electron microscopy theory and application, *Microscopy Today* 94(5):14.

Chapter 12

1. D. Price and G. Feinman, 1992, *Images of the Past* (Mountain View, Calif.: Mayfield Publishing); A. Marshack,1972, *The Roots of Civilization* (New York: McGraw-Hill).

2. S. H. Ball, 1931, *Economic Geology* 26:681–738.

3. D. D. Carr, ed., 1994, *Industrial Minerals and Rocks,* 6th ed. (Littleton, CO: Society for Mining, Metallurgy, and Exploration).

4. D. D. Carr and M. Herz, eds. 1989, *Concise Encyclopedia of Mineral Resources* (Oxford: Pergamon Press).

5. J. W. Barnes, 1988, *Ores and Minerals* (Philadelphia: Open University Press), p. 19.

6. Shephard, 1993, *Ancient Mining* (London: Elsevier), pp. 80–83.

7. C. S. Smith, 1960, *A History of Metallography* (Chicago: University of Chicago Press).

8. Carr and Herz, p. 241.

9. R. Shepherd, 1993, *Ancient Mining* (London: Elsevier); J. Temple, 1972, *Mining: An International History* (New York: Praeger); R. J. Forbes, 1955, *Studies in Ancient Technology* (Leiden: Brill), vols. 7–9.

10. R. S. Solecki, 1971, *Shanidar: The First Flower People* (New York: Alfred Knopf).

11. Barnes, *Ores and Minerals,* p. 1.

12. P. T. Craddock, ed., 1980, *Scientific Studies in Early Mining and Extractive Metallurgy* (London: British Museum); R. S. Forbes, 1964, *Studies in Ancient Technology* (Leiden: Brill), vol. 7; R. F. Tylecote, 1962, *Metallurgy in Archaeology* (London: Edward Arnold Publishers), pp. 1–13.

13. Carr and Herz, p. 155.

14. Carr and Herz, p. 87.

15. G. Rapp, Jr., J. Albert, and E. Henrickson, 1984, Trace-element distribution of discrete sources of native copper, in *Archaeological Chemistry,* ed. J. B. Lambert, American Chemical Society, Advances in Chemistry Series, (Washington D.C.: American Chemical Society), No. 205, pp. 273–94.

16. M. L. Wayman, Neutron activation analysis of metals: A case study, in *History of Technology: The Role of Metals,* ed. S. J. Fleming and H. R. Schenck, (Philadelphia: The University Museum of Archaeology and Anthropology, University of Pennsylvania), MASCA Research Papers in Science and Technology, p. 67.

17. A. P. McCartney and D. J. Mack, 1973, Iron utilization by Thule Eskimos of central Canada, *American Antiquity* 38:328–38; A. P. McCartney, 1991, Canadian Arctic trade metal: Reflections of prehistoric to historic social networks, in *Metals in Society: Theory Beyond Analysis* (Philadelphia: The University Museum of Archaeology and Anthropology, University of Pennsylvania), pp. 27–43.

18. R. J. Braidwood, H. Çambel, and W. Schirmer, 1981, Beginnings of village-farming communities in southeastern Turkey: Cayönü Tepesi, 1978 and 1979, *Journal of Field Archaeology* 8:249–58.

19. McCartney, Canadian Arctic trade metal, pp. 28–29.

20. B. M. Fagan, 1980, *People of the Earth* (New York: Little, Brown), p. 128.

21. J. D. Muhly, 1988, The beginning of metallurgy in the Old World, in *The Beginning of the Use of Metals and Alloys,* ed. R. Maddin (Cambridge: MIT Press), p. 7.

22. J. A. Charles, 1978, The development of the useage of tin and tin-bronze: Some problems, in *The Search for Ancient Tin* (Washington, D.C.: Government Printing Office), p. 28.

23. T. A. Wertime, 1977, The search for ancient tin: The geographic and historic boundaries, in *The Search for Ancient Tin,* ed. A. D. Franklin, J. S. Olin, and T. A. Wertime (Washington, D.C.: Government Printing Office), p. 1.

24. T. Stech and R. Maddin, 1988, Reflections on early metallurgy in Southeast Asia, in *The Beginning of the Use of Metals and Alloys,* ed. R. Maddin (Cambridge: MIT Press), pp. 163–74.

25. N. van de Merwe and D. H. Avery, 1982, Pathways to steel, *American Scientist* 70:146–55.

26. R. F. Tylecote, 1976, *A History of Metallurgy* (London: Metals Society), p. 6.

27. H. Lechtman, 1976, A metallurgical site survey in the Peruvian Andes, *Journal of Field Archaeology* 3:41; Tylecote, *Metallurgy in Archaeology,* p. 34; P. D. Glumac and J. A. Todd, 1991, Early metallurgy in southeast Europe: The evidence for production, in *Recent Trends in*

Archaeometallurgical Research, ed. P. Glumac, MASCA Research Papers in Science and Archaeology, vol. 8, pt. 1 (Philadelphia: University Museum), p. 9.

28. G. Weisberger and A. Hauptmann, 1988, Early copper mining and smelting in Palestine, in *The Beginning of the Use of Metals and Alloys,* ed. R. Maddin (Cambridge: MIT Press), pp. 52, 61.

29. Tylecote, *supra,* p. 36.

30. J. Du Plat Taylor, 1952, A Late Bronze Age settlement at Apliki, Cyprus, *Antiquaries Journal,* 32:133–67.

31. Glumac and Todd, Early metallurgy in southeast Europe, p. 11.

32. Tylecote, *A History of Metallurgy,* p. 7.

33. A. Hauptmann, G. Weisgerber, and H.-G. Bachmann, 1989, Ancient copper production in the area of Feinan, Khirbet En-Nahas, and Wadi El-Jariye, Wadi Arabah, Jordan, in *History of Technology: The Role of Metals,* ed. S. J. Fleming and H. R. Schenck, MASCA Research Papers in Science and Technology, 6, p. 15.

34. A. Hauptmann, 1985, *500 Jahre Kupfer in Oman,* vol. 1: *Die Entwicklung der Kupfer metallurgie vom 3. Jahrtausend bis nur Neuzeit,* Der Anschitt, Beiheft 4.

35. A. Hauptmann, G. Weisgerber, and H.-G. Bachmann, 1988, Early copper metallurgy in Oman, in *The Beginning of the Use of Metals and Alloys,* ed. R. Maddin (Cambridge: MIT Press), p. 37.

36. F. Czedik-Eysenberg, 1958, Beiträge zur Metallurgie des Kupfers in der Urzeit, *Archäologica Austrica* 3:1–18.

37. Tylecote, *Metallurgy in Archaeology,* p. 201.

38. Tylocote, *A History of Metallurgy,* pp. 55–56.

39. Ibid., p. 61.

40. Z. Baoquan, H. Youyan, and L. Benshan, 1988, Ancient copper mining and smelting at Tonglushan, Daye, in *The Beginning of the Use of Metals and Alloys,* ed. R. Maddin (Cambridge: MIT Press), pp. 125–29.

41. G. A. Wagner and G. Weisgerber, 1988, Antike Edel- und Burtmetalgewinnung auf Thasos (Bochum: Deutschen Bergbau-Museums), p. 56.

42. Forbes, R. J. Studies in Ancient Technology, Vol. 7. (Leiden: E. J. Brill) 1956. *supra,* p. 104.

43. J. Needham, 1959, *Science and Civilization in China* (Cambridge: University Press), vol. 3, pp. 673–80.

44. G. I. Quimby, 1960, *Indian Life in the Upper Great Lakes* (Chicago: University of Chicago Press), pp. 43–63.

45. P. E. Martin, 1990, Mining on Minong, *Michigan History* 74(3):19–21.

46. Forbes, *Studies in Ancient Technology,* p. 123.

47. Baoquan, Youyan, and Benshan, Ancient copper mining, p. 126.

48. G. Weisgerber and A. Hauptmann, 1988, Early mining and smelting in Palestine, in *The Beginning of the Use of Metals and Alloys,* ed. R. Maddin (Cambridge: MIT Press), pp. 53–59.

49. D. Hosler, 1988, The metallurgy of ancient west Mexico, in *The Beginning of the Use of Metals and Alloys,* ed. R. Maddin (Cambridge: MIT Press), pp. 340–41; H. Lechtmann, 1980, The central Andes: Metallurgy without iron, in *The Coming of the Age of Iron,* ed. T. Wertime and J. Muhly (New Haven: Yale University Press), pp. 267–334; H. Lechtmann, 1988, Traditions and styles in central Andean metalworking, in *The Beginnings of the Use of Metals and Alloys,* ed. R. Maddin (Cambridge: MIT Press), p. 344.

50. Wertime, Search for ancient tin, p. 1.

51. Ibid., p. 2; Charles, Development of the Usage of Tin and Tin-Bronze, p. 28.

52. G. Kyrle, 1912, Die zeitliche Stellung der prähistorischen Kupfergruben auf dem

Mitterberg bei Bischofshofen. *Mitt. Wiener Anthrop. Ges.;* Forbes, *Studies in Ancient Technology,* p. 138.

53. Shepherd, *Ancient Mining.*

54. C. Conophagus, 1980. Le Laurium antique et la technique grecque de production de l'argent (Athens: Ekdotike Hellados).

55. N. Van der Merwe and Avery, Pathways to steel; T. Wertime and R. Maddin, eds., 1980, *The Coming of Iron* (New Haven: Yale University Press).

56. J. Chaplin, 1961, Notes on traditional smelting in northern Rhodesia, *South Africa Archaeological Bulletin* 62:53–60; P. Schmidt, 1981, *The Origins of Iron Smelting in Africa: A Complex Technology in Tanzania,* Brown University Department of Anthropology Research Papers no. 1 (Providence); van der Merwe and Avery, Pathways to steel; S. T. Childs, 1991, Transformations: Iron and copper production in central Africa, in *Recent Trends in Archaeometallurgical Research,* ed. P. Glumac, MASCA Research Papers in Science and Archaeology, vol. 8, pt. 1 (Philadelphia: University Museum), pp. 33–46; D. Killick, 1991, The relevance of recent African iron-smelting practice to reconstructions of prehistoric smelting technology, in *Recent Trends in Archaeometallurgical Research,* ed. P. Glumac, MASCA Research Papers in Science and Archaeology, vol. 8, pt. 1 (Philadelphia: University Museum), pp. 47–54.

57. D. E. Miller and N. van der Merwe, in press, Early Iron Age metal working at the Tsodilo Hills, northwestern Botswana, *Journal of Archaeological Science.*

58. D. Wenner and N. van der Merwe, 1987, Mining for the lowest grade ore: Traditional iron production in northern Malawi, *Geoarchaeology* 2:199–216.

59. Van der Merwe and Avery, Pathways to steel.

60. R. M. Ehrenreich, ed., 1991, *Metals in Society: Theory beyond Analysis,* MASCA Research Papers in Science and Archaeology, vol. 8, pt. 2 (Philadelphia: University Museum).

Chapter 13

1. P. B. Vandiver, O. Soffer, B. Klima, and J. Svoboda, 1989, The origins of ceramic technology at Dolní Věstonice, Czechoslovakia, *Science* 246:1002–8.

2. K. D. Vitelli, 1989, Were pots first made for foods? Doubts from Franchthi, *World Archaeology* 21:17–29.

3. Z. Goffer, 1980, *Archaeological Chemistry* (New York: John Wiley).

4. J. B. Dixon and S. B. Weed, eds., 1989, *Minerals in Soil Environments* (Madison: Soil Science Society).

5. A. O. Shepard, 1965, *Ceramics for the Archaeologist,* Carnegie Institution, Publication 609 (Washington, D.C).

6. M. Maggetti, 1982, Phase analysis and its significance for technology and origin, in *Archaeological Ceramics,* ed. J. S. Olin and A. D. Franklin (Washington, D.C.: Smithsonian Institution), pp. 121–33.

7. H. Neff, ed., 1992, *Chemical Characterization of Ceramic Pastes in Archaeology* (Madison: Prehistory Press Monograph 7).

8. P. M. Rice, 1987, *Pottery Analysis: A Sourcebook* (Chicago: University of Chicago Press).

9. M. S. Tite, 1992, The impact of electron microscopy on ceramic studies, in *New Developments in Archaeological Science,* ed. A. M. Pollard, Proceedings of the British Academy 77 (London: Oxford University Press), pp. 111–31.

10. S. J. Vaughan, 1991, Material and technical characterization of Base Ring Ware: A new fabric typology, in *Cypriot Ceramics: Reading the Prehistoric Record,* ed. J. A. Barlow, D. L. Bolger, and B. Kling, University Museum Monograph 74 (Philadelphia: University of Pennsylvania), pp. 119–30.

11. D. C. Kamilli and A. Steinberg, 1985, New approaches to mineral analysis of ancient ceramics, in *Archaeological Geology,* ed. G. Rapp, Jr., and J. A. Gifford (New Haven: Yale University Press), pp. 313–30.

12. M. J. Hughes, M. R. Cowell, and D. R. Hook, eds., 1991, Neutron activation and plasma emission spectrometric analysis in archaeology, *British Museum Occasional Paper* 82.

13. W. R. Biers and P. E. McGovern, eds., 1990, *Organic Contents of Ancient Vessels: Materials Analysis and Archaeological Investigation,* MASCA Research Papers in Science and Archaeology, vol. 7 (Philadelphia: University Museum).

14. M. J. Blackman, 1992, The effect of human size sorting on the mineralogy and chemistry of ceramic clays, in *Chemical Characterization of Ceramic Pastes in Archaeology,* ed. H. Neff, Monographs in World Archaeology 7 (Madison: Prehistory Press), pp. 113–24.

15. V. Kilikoglou, Y. Maniatis, and A. P. Grimanis, 1988, The effect of purification and firing of clays on trace element provenance studies, *Archaeometry* 30:37–46.

16. A. Gibson and A. Woods, 1990, *Prehistoric Pottery for the Archaeologist* (New York: John Wiley); J. S. Olin and A. D. Franklin, eds., 1982, *Archaeological Ceramics* (Washington, D.C.: Smithsonian Institution).

Chapter 14

1. G. Faure, 1986, *Principles of Isotope Geology,* 2nd ed. (New York: John Wiley).

2. P. Budd, A. M. Pollard, B. Scaife, and R. G. Thomas, 1995, The possible fractionation of lead isotopes in ancient metallurgical processes, *Archaeometry* 37:143–50.

3. Faure, *Principles of Isotope Geology.*

4. Ibid.

5. R. D. Russell and R. M. Farquhar, 1960, *Lead Isotopes in Geology* (New York: Wiley Interscience); N. H. Gale and Z. Stos-Gale, 1981, Lead and silver in the ancient Aegean, *Scientific American* 6:176–92.

6. N. H. Gale, 1989, Lead Isotope analyses applied to provenance studies—A brief review, in Archaeometry—Proceedings of the 25th International Symposium, ed. Y. Maniatis (Amsterdam: Elsevier), pp. 469–502.

7. Faure, *Principles of Isotope Geology.*

8. R. H. Brill and J. M. Wampler, 1967, Isotope studies of ancient lead, *American Journal of Archaeology* 71:63–77.

9. H. Craig and V. Craig, 1972, Greek marbles: Determination of provenance by isotopic analysis, *Science* 176:401–3.

10. N. Herz, 1987, Carbon and oxygen isotopic ratios: A data base for classical Greek and Roman marble, *Archaeometry* 29:35–43; N. Herz and D. B. Wenner, 1978, Assembly of Greek marble inscriptions by isotopic methods, *Science* 199:1070–72.

11. N. Herz, 1985, Isotopic analysis of marble, in *Archaeological Geology,* ed. G. Rapp, Jr., and J. A. Gifford (New Haven: Yale University Press), pp. 331–51.

12. Ibid.; Gale and Stos-Gale, Lead and silver in the ancient Aegean.

13. A. W. Smith, 1978, Stable carbon and oxygen isotope ratios of malachite from the patinas of ancient bronze objects, *Archaeometry* 20:123–33.

14. Faure, *Principles of Isotope Geology.*

15. N. Herz and N. E. Dean, 1986, Stable isotopes and archaeological geology: The Carrara marble, northern Italy, *Applied Geochemistry* 1:139–51.

16. W. E. Krumbein and C. Urzi, 1992, Biologically induced decay phenomena of antique marbles: Some general considerations, in *The Conservation of Monuments in the Mediterranean*

Basin, ed. D. Decrouez, J. Chamay, and F. Zezza, Proceedings of the 2nd International Symposium, Geneva, 1991 (Geneva: Musée d'Art et d'Histoire), pp. 219–35.

17. Herz, Carbon and oxygen isotopic ratios.

18. N. J. Shackleton and C. Renfrew, 1970, Neolithic trade routes realigned by oxygen isotope analysis, *Nature* 228:1062–65.

19. N. R. Shaffer and K. B. Tankersley, 1989, Oxygen isotopes as a method of determining the provenience of silica-rich artifacts, *Current Research in the Pleistocene* 6:47–49; D. N. Stiles, R. L. Hay, and J. R. O'Neil, 1974, The MNK chert factory site, Olduvai Gorge, Tanzania, *World Archaeology* 5:285–308.

20. J. R. O'Neil and R. L. Hay, 1973, $^{18}O/^{16}O$ ratios in cherts associated with the saline lake deposits of East Africa, *Earth and Planetary Science Letters* 19:257–66.

21. S. C. Carrier, N. Herz, and B. Turi, 1996, Isotopic strontium: its potential for improving the determination of the provenance of ancient marble, International Symposium on Archaeometry, program and abstracts (Urbana: University of Illinois), p. 21. N. H. Gale, 1981, Mediterranean obsidian source characterization by strontium isotope analysis, *Archaeometry* 23:41–51; N. H. Gale, H. H. Einfalt, H. W. Hubberten, and R. E. Jones, 1988, The sources of Mycenean gypsum, *Journal of Archaeological Science* 15: 52–72.

22. W. H. Burke, R. E. Denison, E. A. Hetherington, R. B. Koepnick, H. F. Nelson, and J. B. Otto, 1982, Variation of seawater $^{87}Sr/^{86}Sr$ throughout Phanerozoic time, Geology 10: 516–19.

23. Gale, Mediterranean obsidian source characterization.

24. Gale et al., Sources of Mycenean gypsum.

25. Gale and Stos-Gale, Lead and silver in the ancient Aegean.

26. G. A. Wagner, W. Gentner, H. Gropengiesser, and N. H. Gale, 1980, Early Bronze Age lead-silver mining and metallurgy in the Aegean: The ancient workings on Siphnos, *British Museum Occasional Paper* 20:63–85.

27. Gale and Stos-Gale, Lead and silver in the ancient Aegean.

28. R. M. Farquhar and I. R. Fletcher, 1984, The provenience of galena from Archaic/Woodland sites in northeastern North America: Lead isotope evidence, *American Antiquity* 49:774–84.

29. H. P. Schwarcz and M. J. Schoeninger, 1991, Stable isotope analysis in human nutritional ecology, *Yearbook of Physical Anthropology* 34:283–321.

30. N. Tuross, M. L. Fogel, and P. E. Hare, 1988, Variability in the preservation of the isotopic composition of collagen from fossil bone, *Geochimica et Cosmochimica Acta* 52:929–35.

31. T. D. Price, J. A. Johnson, J. A. Ezzo, J. Ericson, and J. H. Burton, 1994, Residential mobility in the prehistoric southwest United States: A preliminary study using strontium isotope analysis, *Journal of Archaeological Science* 21:315–30.

32. N. J. van der Merwe and J. C. Vogel, 1977, Isotopic evidence for early maize cultivation in New York State, *American Antiquity* 42:238–42.

33. M. J. DeNiro, 1987, Stable isotopy and archaeology, *American Scientist* 75:182–91; P. Aharon, 1982, Stable oxygen and carbon isotope techniques in archaeology and museum studies, in *Archaeometry: An Australasian Perspective,* ed. W. Ambrose and P. Duerden (Canberra: Australian National University Press), pp. 156–72; H. W. Krueger, 1982, Strontium isotopes and Sr/Ca in bone by mass spectroscopy, in *1984 Symposium on Archaeometry* (Washington, D.C.: Smithsonian Institution), p. 85.

34. M. Fizet, A. Mariotti, H. Bocherens, B. Lange-Badré, B. Vandermeersch, J. P. Borel, and G. Bellon, 1995, Effect of diet, physiology, and climate on carbon and nitrogen stable isotopes of collagen in a late Pleistocene anthropic palaeoecosystem: Marillac, Charente, France, *Journal of Archaeological Sciences* 22:67–79.

35. R. Park and S. Epstein, 1960, Carbon isotope fractionation during photosynthesis, *Geochimica et Cosmochica Acta* 21:110–26.

36. N. J. van der Merwe, 1982, Carbon isotopes, photosynthesis, and archaeology, *American Scientist* 70:596–606.

37. M. J. DeNiro and S. Epstein, 1981, Influence of diet on the distribution of nitrogen isotopes in animals, *Geochimica et Cosmochimica Acta* 45:341–51.

38. DeNiro, Stable isotopy and archaeology.

39. Van der Merwe, Carbon isotopes, photosynthesis, and archaeology.

40. J. C. Vogel, 1978, Recycling of carbon in a fast environment, *Oecologia Plantarum* 13:89–94.

41. M. J. DeNiro and S. Epstein, 1978, Influence of diet on the distribution of carbon isotopes in animals, *Geochimica et Cosmochicha Acta* 42:495–506; DeNiro and Epstein, Influence of diet on the distribution of nitrogen isotopes in animals.

42. N. J. van der Merwe, A. C. Roosevelt, and J. C. Vogel, 1981, Isotopic evidence for prehistoric subsistence change at Parmama, Venezuela, *Nature* 292:536–38.

43. C. H. Sullivan and H. W. Krueger, 1981, Carbon isotope analysis of separate chemical phases in modern and fossil bone, *Nature* 292:333–35.

44. M. J. Stenhouse and M. S. Baxter, 1977, Bomb ^{14}C and human radiation burden, *Nature* 267:825–827.

45. Van der Merwe, Carbon isotopes, photosynthesis, and archaeology.

46. Carnegie Institution of Washington, 1987, *Yearbook 86: The President's Report 1986–87*.

47. R. S. MacNeish, 1967, A summary of the subsistence, in *The Prehistory of the Tehuacan Valley*, ed. D. S. Byers (Austin: University of Texas Press), vol. 1, pp. 290–309.

48. DeNiro and Epstein, Influence of diet on the distribution of nitrogen isotopes in animals.

49. B. S. Chisholm, E. E. Nelson, and H. P. Schwarcz, 1982, Stable-carbon isotope ratios as a measure of marine versus terrestrial protein in ancient diets, *Science* 216:1131–32.

50. P. L. Walker and M. J. DeNiro, 1986, Stable nitrogen and carbon isotope ratios in bone collagen as indices of prehistoric dietary dependence on marine and terrestrial resources in southern California, *American Journal of Physical Anthropology* 71:51–61.

51. DeNiro, Stable isotopy and archaeology.

52. M. J. DeNiro and C. A. Hastorf, 1985, Alteration of $^{15}N/^{14}N$ and $^{13}C/^{12}C$ ratios of plant matter during the initial stages of diagenesis: Studies utilizing archaeological specimens from Peru, *Geochimica et Cosmochimica Acta* 49:97–115.

53. DeNiro, Stable isotopy and archaeology.

54. D. Lubell, M. Jackes, H. Schwarcz, M. Knyf, and C. Meiklejohn, 1994, The Mesolithic-Neolithic transition in Portugal: Isotopic and dental evidence of diet, *Journal of Archaeological Science* 21:201–16.

55. H. P. Schwarcz and M. J. Schoeninger, 1991, Stable isotope analysis in human nutritional ecology, *Yearbook of Physical Anthropology* 34:283–301.

56. Chisholm, *supra.*

57. N. J. van der Merwe, J. A. Lee-Thorp, J. F. Thackeray, A. Hall-Martin, F. J. Kruger, H. Coetzee, R. H. V. Bell, and M. Lindeque, 1990, Source-area determination of elephant ivory by isotopic methods, *Nature* 346:744–46.

58. N. J. van der Merve and E. Medina, 1989, Photosynthesis and $^{13}C/^{12}C$ ratios in Amazonian rain forests, *Geochimica et Cosmochimica Acta* 53:1091–94.

59. H. D. Morris, 1970, Walnut Creek Village: A ninth-century Hohokum-Anasazi settlement in the mountains of central Arizona, *American Antiquity* 35:49–61.

60. M. Bell and M. J. C. Walker, 1992, *Late Quaternary Environmental Change: Physical and Human Perspectives* (London: Longman Group), p. 41.

61. I. Friedman, 1983, Paleoclimatic evidence from stable isotopes, in *Late Quaternary Environments of the United States,* vol. 2: *The Holocene,* ed. H. E. Wright, Jr. (Minneapolis: University of Minnesota Press), pp. 385–89.

62. N. J. Shackleton, 1973, Oxygen isotope analysis as a means of determining season of occupation of prehistoric midden sites, *Archaeometry* 15:176–93.

63. Aharon, Stable oxygen and carbon isotope techniques.

64. P. Aharon, 1983, 140,000-yr isotope climatic record from raised coral reefs in New Guinea, *Nature* 304:720–23.

65. R. I. Dorn and M. J. DeNiro, 1985, Stable isotope ratios of rock varnish organic matter: A new paleoenvironmental indicator, *Science* 227:1472–73.

66. Friedman, Paleoclimatic evidence from stable isotopes.

67. S. Epstein, C. J. Yapp, and J. H. Hall, 1976, The determination of the D/H ratio of nonexchangeable hydrogen in cellulose extracted from aquatic and land plants, *Earth and Planetary Science Letters* 30:241–51.

68. C. J. Yapp and S. Epstein, 1977, Climatic implications of D/H ratios of meteoric water over North America (9500–22,000 B.P.) as inferred from ancient wood cellulose C-H hydrogen, *Earth and Planetary Science Letters* 34:333–50.

69. R. S. Harmon, H. P. Schwarcz, and J. R. O'Neil, 1979, D/H ratios in speleothem fluid inclusions: A guide to variations in isotopic composition of meteoritic precipitation, *Earth and Planetary Science Letters* 42:254–66.

Index